The Manila Clam
Ruditapes philippinarum

AF720008

Estuarine and Coastal Sciences Series

The Manila Clam *Ruditapes philippinarum* Science and Significance of a Globalised Species

John Humphreys
Visiting Professor, Institute of Marine Sciences,
University of Portsmouth, Portsmouth, England;
President, Estuarine & Coastal
Sciences Association (ECSA)

ELSEVIER

Series Editors: Steve Mitchell and Michael Elliott

Elsevier
Radarweg 29, PO Box 211, 1000 AE Amsterdam, Netherlands
50 Hampshire Street, 5th Floor, Cambridge, MA 02139, United States

Copyright © 2026 Elsevier Inc. All rights are reserved, including those for text and data mining, AI training, and similar technologies.

Publisher's note: Elsevier takes a neutral position with respect to territorial disputes or jurisdictional claims in its published content, including in maps and institutional affiliations.

Books and Journals published by Elsevier comply with applicable product safety requirements. For any product safety concerns or queries, please contact our authorised representative, Elsevier B.V., at productsafety@elsevier.com.

For accessibility purposes, images in electronic versions of this book are accompanied by alt text descriptions provided by Elsevier. For more information, see https://www.elsevier.com/about/accessibility.

No part of this publication may be reproduced or transmitted in any form or by any means, electronic or mechanical, including photocopying, recording, or any information storage and retrieval system, without permission in writing from the publisher. Details on how to seek permission, further information about the Publisher's permissions policies and our arrangements with organizations such as the Copyright Clearance Center and the Copyright Licensing Agency, can be found at our website: www.elsevier.com/permissions.

This book and the individual contributions contained in it are protected under copyright by the Publisher (other than as may be noted herein).

Notices

Knowledge and best practice in this field are constantly changing. As new research and experience broaden our understanding, changes in research methods, professional practices, or medical treatment may become necessary.

Practitioners and researchers must always rely on their own experience and knowledge in evaluating and using any information, methods, compounds, or experiments described herein. In using such information or methods they should be mindful of their own safety and the safety of others, including parties for whom they have a professional responsibility.

To the fullest extent of the law, neither the Publisher nor the authors, contributors, or editors, assume any liability for any injury and/or damage to persons or property as a matter of products liability, negligence or otherwise, or from any use or operation of any methods, products, instructions, or ideas contained in the material herein.

ISBN: 978-0-443-36704-5

For Information on all Elsevier publications visit our website at https://www.elsevier.com/books-and-journals

Publisher: Peter Llewellyn
Acquisitions Editor: Maria Elekidou
Editorial Project Manager: Bhaviya Koul
Production Project Manager: Fizza Fathima
Cover Designer: Gopalakrishnan Venkatraman

Working together to grow libraries in developing countries
www.elsevier.com • www.bookaid.org

Typeset by Aptara, New Delhi, India

Dedicated to the Fishers

Contents

About the Author

After 10 years as Pro Vice-Chancellor at the University of Greenwich, London, for whom his work in Africa won a Queen's Award, **Professor John Humphreys** moved to the south coast of England, where he has served as non-executive director of a port and harbour authority and as chair of a UK inshore fisheries and conservation authority. He is a visiting professor at the University of Portsmouth's Institute of Marine Sciences and President of the Estuarine and Coastal Sciences Association. His other books for Elsevier include *Marine Protected Areas: Science, Policy, and Management.*

Preface

Some years ago, while conducting a benthic survey of Poole Harbour with Richard Caldow and other colleagues, a clam appeared in our sieves which did not convincingly key out. As usual, the local fishers were the best source of intelligence. It was the Manila clam, and they had been catching it since one of them, Gary Wordsworth, had been encouraged by a government agency to introduce it for aquaculture in the late 1980s. It turned out that some scientists did know it was there, not least Antony Jensen at Southampton, but its presence as a naturalized non-native arrival had not been reported in the scientific literature, and it had not found its way into any identification keys: a contributing factor likely being its close resemblance to the native *Ruditapes decussatus.*

Publishing some early data in 2004 and realizing that in Poole Harbour, at that time, we probably had the most northerly European population, we embarked on a study of its population dynamics and used the data to examine its trophic significance in that estuarine community. Subsequently, we recruited a doctoral student, Matt Harris, and with colleagues from Portsmouth, Bournemouth, and elsewhere, investigated its larval dispersal and discovered it had also naturalized in other estuaries of Southern England. Around this time, as Chair of the regional statutory fisheries authority I also became involved in the management and regulation of the Poole Harbour fishery.

As we progressed in our research, I came to appreciate the significance of this species from many perspectives. A retired fisher found an old press cutting about the introduction by the British government of this "alien monster." Colleagues, including Rob Clark, then Chief Fisheries Officer at our regional fisheries authority, were concerned about how its high value was encouraging illegal activity and were working on a new management regime. And of course, quickly, the global distribution and economic pre-eminence of the species became apparent. So, I embarked on a series of study tours and site visits ranging from its ostensible and actual indigenous locations in the Philippines, China, Japan, and Korea, to Hawaii, the American West Coast, and mainland Europe. Acknowledged below are some of the many people I must thank, both as funders and as academic and industry experts from whom I learnt so much, especially those who guided me to many of the classic Manila clam sites mentioned in this book.

During these travels, I realized that a general survey on the species could be useful: ranging across its biology and fisheries but also addressing what it reveals about the policy and practice of science and the environment. The volume of work on the species is now considerable, and not all of it can be included in a single reasonably sized monograph. Therefore, the book is designed to be comprehensive in scope rather than taking the form of a formal systematic review of each specialist field. Nevertheless, it is hoped that each chapter provides routes into an extensive and growing literature. Those wanting a taste of this literature will find a good range in a special 2025 Manila clam issue of the journal *Estuarine, Coastal and Shelf Science,* which I edited with Kwang-Sik Choi and Satoshi Watanabe of South Korea and Japan, respectively. I have also benefited greatly from special clam symposia in Korea and France, the latter organized by Nathalie Caill-Milly of Ifremer.

I have been motivated in my travels by a strong conviction that for such a widely dispersed species, only a geographically balanced account will suffice. I hope I have provided such a treatment, giving equal weight to sites, literature, and issues pertaining to all continents. Only Chapter 24 is exclusively about the British experience of introduction, but I felt this was useful in elucidating issues of more general relevance to the policy and politics of non-native species.

In this book, the science of the species has been my main preoccupation, but its globalization forces consideration of many contemporary coastal issues which I have not ignored: From socio-economics and food security to estuarine pollution, conservation and non-native species, to illegal and unregulated fishing, and the already irreversible implications of climate change. All of these issues are illuminated by the global spread and exploitation of *Ruditapes philippinarum.*

Note on Terminology

Non-native species: This term is used to signify species that have become established in areas beyond their natural geographical distribution, expedited by human activity. It is synonymous with alternatives such as "non-indigenous" and "alien" and, in some usages, "invasive." Based on their use in everyday life, some of these terms have been criticized as value-laden and implicitly negative, despite evidence that only a small proportion of such arrivals are in fact damaging to indigenous species or ecosystems. An alternative term, "neobiota," avoids any scientifically unjustified connotations. Originating in Germany, it is defined as referring to anthropogenically expedited arrivals since 1492. It is now increasingly used in continental Europe, but not yet widely in English-speaking countries. Since this book is written in English, the term non-native is used throughout, and the term neobiota is only introduced in the final chapter.

Manila clam: In Chapter 2, it is argued that the species *Ruditapes philippinarum* is misnamed, both in the vernacular (Manila) and scientifically (*philippinarum*), due to a series of persistent errors beginning with its first scientific recognition. In light of its verified indigenous distribution, the more appropriate Japanese "Asari" is gaining wider international recognition. However, among the many vernacular names for the species, Manila clam is the most universally recognized worldwide and is therefore used in this book.

Acknowledgments

This book was written over 3 years, coinciding with journeys to many of the centers of Manila clam distribution and exploitation. I have been facilitated in these travels by support from the Estuarine and Coastal Sciences Association (ECSA), the publisher Elsevier, and Jeju National University in South Korea. In each country visited, I have benefited enormously from discussions with scientists and industry participants, many of whom committed considerable time escorting me to locations from Manila clam field sites to hatchery and aquaculture operations, wholesale enterprises, and local fisheries. It has also been gratifying and informative to meet many of these people again at International Manila clam symposia in Kunsan, South Korea, and Arcachon, France. Below, some of these individuals are acknowledged by name. My greatest concern is that despite my extensive notes, there will be too many accidental omissions from this list.

In the Philippines: Joseph Foronda and Annabelle Del Norte Campos of the National Museum of Natural History, Manila; Apple Tordecillas at Panay Island.

In Korea: Kwang-Sik Choi, Jong-Seop Shin, Nobuhisa Kajino, Thatchaneshkanth, and Jeong-Hwa Kim at Jeju National University; Kyung-Il Park and colleagues at Kunsan National University.

In Japan: Satoshi Watanabe, Masashi Kodama, and Yoshitake Takada at the Japan Fisheries and Research Agency, Yokohama.

In China: Yun Tao, Second Institute of Oceanography, Xiu Jun Sun; Weiwei Jiang, Yellow Sea Fisheries Research Institute, Chinese Academy of Fishery Sciences. Tu Kang of the Putian Ocean and Fisheries Bureau. Huang Xi Zhong and colleagues at the Fujian Province Zhong Lin Aquaculture Development Company. Zheng Wen Ping of the Zhang Zhou Biotech Company. Chen Si Qing of the Miao Run Aquaculture Company.

In the United States, including Hawaii: Austin Doctor of Taylor Shellfish, Washington; Bill Dewey, Isaac Nemes, Julie Wilkinson-Rouseau, Rich Strangeland of Chuckanut Shellfish (a great day out on Clamdango). Blair Paul, Skokomish Tribe shellfish manager. Greg Jakob of Taylor Shellfish, Hawaii.

In France: Ifremer: Nathalie Caill-Milly; Florentine Riquet; Florence Sanchez; Christine Paillard; Elvire Antajan; Franck Lagarde. Noelle Bru de l'Université de Pau et des Pays de l'Adour. Fishers at Port de la Hume, Arcachon Bay: Delphine Moulin; Maud Vial; Yannick Vial. Also, fishery manager Anthony Gueguen.

In the United Kingdom: Gary Wordsworth of Othniel Oysters; Rob Clark (AIFCA); Mike Bailey, clam fisher; Kent and Essex IFCA: Will Wright, Phil Haupt; Judith Stoutt, Kat Stuart; Ellie Wyatt, and Thames Fisher Dave Ferris. Simon Cragg from Portsmouth Institute of Marine Science and our doctoral student Matt Harris. John Taylor and Tom White of the Natural History Museum, London.

Thanks also to Jim Carlton, Konstantin Lutaenko, Walid Medhioub, and Youk Greeve. Series editors, Mike Elliott and Steve Mitchell, and the team from Elsevier, including: *Bhaviya Koul, Maria Elekidou, Shilpa Kumar, Subash Balakrishnan, Komal Pal* and *Fizza Fathima*.

The views expressed in this book, along with all errors, are the exclusive responsibility of the author.

PART

Origins and arrivals

I

CHAPTER

1 Origins

1.1 The emergence of the Venus clams

The evolutionary history of the Manila clam *Ruditapes philippinarum* (Adams & Reeve, 1850) (Fig. 1.1) can be traced back to the early Paleozoic era when many marine taxa developed hard calcareous skeletons. About 520 million years ago in the Cambrian period, a group of early mollusks emerged, in which an ancestral single shell in a laterally compressed taxon became separated into two parts by the loss of calcareous components along the dorsal midline. This left the organic outer shell layer (periostracum) to form a flexible hinge connecting distinct right and left valves. Along with earlier established adaptations, such as a planktonic trochophore larva enabling effective dispersal, this shell modification resulted in a highly derived ancestral form, which provided the basis of the first great bivalve diversification. By the Ordovician period (about 500–440 million years ago), most of the major bivalve lineages were established.

The articulated shell is considered well suited to a life within sediments (Yonge, 1953), and the ancestral bivalve is thought to have lived a relatively sedentary adult life occupying benthic infaunal habitats of sand and mud, not dissimilar to *Ruditapes philippinarum*. Although within the constraints of their body plan, bivalves have subsequently evolved to occupy other habitats, such as the epibenthic mussels and oysters, most modern forms, including *R. philippinarum,* continue to occupy the ancestral sediment habitat.

Both genomics and paleontology suggests that the "Venus clams" of the family Veneridae, to which *R. philippinarum* belongs, originated during the early Permian period, about 290 million years ago (Wang et al., 2020). Although not unique to the Venerids, a key feature of their success was the evolution of elaborated "eulamellibranch" ctenidia (gills), highly efficient for both gas exchange and filter feeding. A second major bivalve radiation occurred after the end of the Permian mass extinction, which also marks the end of the Paleozoic era. Before this time, another bivalved, suspension-feeding invertebrate phylum dominated soft benthic sediments, the brachiopods. Afterward, while both brachiopod and bivalve fossil taxonomy was "reset," the relative abundance of brachiopods over bivalves in continental shelf sediments was reversed (Carlston, 2016; Gould & Calloway, 1980), and many new taxa of advanced molluscan bivalves emerged (Stanley, 1968).

During the following Mesozoic era (250–265 million years ago), an emerging range of Veneroid forms came to dominate the soft substrates of shallow seas. It has

The Manila Clam *Ruditapes philippinarum:* Science and Significance of a Globalised Species.
DOI: https://doi.org/10.1016/B978-0-443-36704-5.00005-5

Copyright © 2026 Elsevier Inc. All rights are reserved, including those for text and data mining, AI training, and similar technologies.

FIGURE 1.1

External (above) and internal (below) views of the right valve of the original type specimen of "*Venus philippinarum*" (= *Ruditapes philippinarum*) (NHMUK 1968161). This specimen is almost certainly the holotype collected by Arthur Adams during the voyage of H.M.S. Samarang to the "Eastern Archipelago" described in Adams and Reeve (1850). Chapter 5 provides labeled illustrations of the main shell features of the species.

Courtesy John Taylor and the Natural History Museum, London.

been suggested that as the great continent of Pangea separated and the new continental landmasses drifted toward their current configuration, the emergence of new, more extensive, and widely separated continental margins facilitated speciation through geographic and climatic isolation. Consequent "taxon provincialisation" (Giribet, 2008; Morton, 1996), may have been instrumental in a Cretaceous diversification, about 110 million years ago (Wang et al, 2020). In any event, on the soft sediments of continental margins, a diversification of shallow burrowing, tumid, and strongly ribbed bivalves radiated many of the modern forms (see also Chapter 4).

Today, the Veneridae make up a family of closely related suspension-feeding forms, mostly living close to the surface of coastal sediments. In contrast to deposit feeders, a high proportion of the suspended material entering the mantle cavity through inhalant siphons can be utilized as food. Consequently, the Venerids lack the ctenidial and other adaptations associated with deposit feeding. Similarly, other specializations associated with a deep burrowing or epibenthic habit are also unnecessary. Because of this mode of life, the Venerids have evolved along relatively unspecialized lines, becoming a large and highly successful bivalve group occupying most types of soft marine substrate.

Nevertheless, a largely sedentary infaunal life, while offering protection, creates challenges, one of which is the need to pump water down from the overlying water column for filter feeding. The chaetopterid polychaete worms achieve this by maintaining a U-shaped burrow in the sediment, but in those groups with inhalant and exhalant siphons, the bivalve animal itself effectively becomes the U-shaped tube. The emergence of the siphonate eulamellibranch heterodonts during the Triassic period marks an advanced condition which makes bivalves, such as *R. philippinarum*, "*of all marine animals... the most perfectly adapted for life within the sediment*"(Yonge, 1949).

Dall (1902) referred to the Veneridae as "*the culmination of pelecypod* (bivalve) *evolution so far as this may be represented by any single family*." More recently, the Venerids have been recognized as supremely successful and mechanically superb (Yonge, 1976), and amongst all the bivalves, they are the most species-rich family with over 800 extant species in approximately 170 genera (Mikkelsen et al., 2006).

1.2 Ancient subsistence to global pre-eminence

Early humans began eating marine mollusks at least 160,000 years ago (Jerardino & Marean, 2010), and since the development of coastal human societies, bivalves have contributed a significant food source. The exploitation of shellfish has been a feature of East Asian maritime communities for at least 35,000 years (Fujita et al., 2016). Although early Holocene postglacial sea level rise has submerged many archaeological sites, mid-Holocene shell deposits referred to as middens have confirmed the widespread consumption of bivalves in what is now China, Korea and Japan (He et al., 2023). Fifteen thousand years ago in Futowo, near the entrance to Ariake Bay in the modern Japanese prefecture of Kyushu, the hunter-gatherer people of the Jomon culture fished *R. philippinarum* in large numbers, depositing the empty shells in a "kitchen midden" outside their village (Shimoyame et al, 2015).

At Sakatsuji in Aichi Prefecture, a 4500-year-old midden measuring over 24 m long, 6 m wide, and 1.6 m high, consisting almost entirely of *R. philippinarum* shells, has been found located on the coast of an ancient shallow sea. Adjacent to this midden are over 50 stone "furnaces" believed to have been used for boiling the clams before the removal and possibly drying for preservation of the soft parts. The Sakatsuji midden is one of seven in the region and, in contrast to the much older Futowo site, does not appear to have been located near a settlement. The midden structure and location suggest that people traveled to the site seasonally for the purpose of clam collection and processing (Archaeology Magazine, 2018). Such discoveries are suggestive of a progressively industrialized Manila clam economy in which, from an original subsistence fishery, the species became a commodity, processed and traded to other communities (Habu, 2004).

Much of the progressive sophistication of the clam industry remains obscure, but there can be little doubt that over the millennia local communities began to take steps to optimize their crop through early forms of aquaculture, perhaps like the indigenous "clam gardens" of eastern Pacific coasts. In any event, stratigraphic evidence at Ariake Bay indicates that a Tsunami in 1792 changed the fine muddy character of the inner Bay area to a sandier sediment, making it a more suitable substrate for *R. philippinarum*. By around 1830, the species was also to be found on the innermost shores of the bay, and it has been suggested that it was introduced there by local communities to develop their coastal shellfishery (Shimoyame et al., 2015). Later in the same century, the Manila clam's global dispersal began.

A first wave of globalization was triggered by the opening up of East Asian countries, which enabled migrations eastward to Hawaii and the western coasts of North America. These migrants took their cultural penchant for bivalves with them, bringing in Manila clams purposefully or accidentally, with oyster imports. Chapter 3 details these arrivals along with a second, more recent phase of anthropogenic globalization, focused initially on the maritime states of north-west Europe. This second phase was not culturally driven but rather consisted of attempts to bolster declining native bivalve fisheries with nonindigenous species such as the Pacific oyster *(Magallana gigas)* and *R. philippinarum*, both seen as faster-growing and more robust alternatives. In many such places, the Manila clam showed itself capable of providing a lucrative aquaculture crop, but also of naturalization and natural dispersal.

Subsequent anthropogenic introductions have ensured that *R. philippinarum* is now the only truly cosmopolitan commercial bivalve species (Manzi & Castagna, 1989). Its worldwide distribution ranges from its indigenous centers in China, Japan, Korea, and Russia, across to the temperate coasts of the northeastern Pacific, to the European east Atlantic and Mediterranean coastal seas. An annual worldwide aquaculture production of over four billion tons generates a US dollar equivalent direct value of over USD 7 billion. This makes the species the most globally valuable cultured molluscan species and second only to the whiteleg shrimp amongst all invertebrates. Even among the oysters, no other single molluscan species can match its global pre-eminence, and amongst all cultured marine species only the Atlantic salmon and the whiteleg shrimp make a more valuable (albeit less sustainable, Chapter 18) contribution to worldwide seafood production (see Chapter 16 and FAO, 2025). Today, the global range and ever-increasing value of the species have resulted in an exceptional quantity and quality of research from scientists in institutions across the

Table 1.1 Classification, description, and synonyms of *Ruditapes philippinarum.*

Accepted name. *Ruditapes philippinarum* (Adams & Reeve, 1850) Aphia ID 231750
Original name. *Venus philippinarum* (Adams & Reeve, 1850)
Alternate name: *Venerupis philippinarum* (Adams & Reeve,1850) ITIS preferred: 81,477
Original description:
"*Venus Philippinaeuu.* Pl. XXII. Fig. 10. V. testa oblongo-ovata, antice truncata, fulva, fusco variegata, obscure radiata, radiatim lirata, liris nmnerosis, subundatis, ad latera decussatim nodulosis; intus partini violacea. Hab. Philippine Archipelago" (Adams & Reeve, 1850).
Original illustration: See Fig. 1.2.
Classification:
Phylum: Mollusca; Class: Bivalvia; Subclass: Autobranchia; Infraclass: Heteroconchia; Superorder: Imparidentia; Order: Venerida; Superfamily: Veneroidea; Family: Veneridae; Subfamily: Tapetinae; Genus: *Ruditapes*
Synonyms: *Amygdala philippinarum* (Numura, 1940); *Paphia bifurcata* (Quayle, 1938); *Tapes (Ruditapes) philippinarum* (Adams & Reeve, 1850); *Tapes biradiata* (Deshayes, 1854); *Tapes denticulata* (Sowerby II, 1852); *Tapes ducalis* (Römer, 1870); *Tapes indica* (Sowerby II, 1852); *Tapes japonica* (Deshayes, 1854); *Tapes philippinarum* (Adams & Reeve, 1850); *Tapes philippinarum okupi* (Bryan, 1919); *Tapes semidecussata* (Reeve, 1864); *Tapes violascens* (Deshayes, 1854); *Venerupis philippinarum* (Adams & Reeve, 1850); *Venerupis semidecussata* (Reeve, 1864); *Venerupis japonica* (Deshayes, 1853); *Venus (Amygdala) philippinarum* (Adams & Reeve, 1850); *Venus analis* (Philippi, 1851); *Venus philippinarum* (Adams & Reeve, 1850); *Venus tessellata* (Adams & Reeve, 1850).

Sources *(Goulletquer, 1997).*
https://www.molluscabase.org/aphia.php?p=taxdetails&id=231750#links
https://www.marinespecies.org/aphia.php?p=taxdetails&id=231750
https://www.itis.gov/servlet/SingleRpt/SingleRpt#null
https://www.cabi.org/isc/datasheet/61697#toidentity
Note: *Other synonyms without authorities to be found in some published lists are not included here.*

northern hemisphere (Humphreys et al., 2025), much of whose work has contributed to the chapters in this monograph.

1.3 Scientific recognition and nomenclatural volatility

On April 7, 1844, during its mission to survey the coasts of the "Eastern Archipelago," the British corvette H.M.S. Samarang sailed into the Philippine waters of Mindanao (Belcher, 1848). On board as Assistant Surgeon was the physician and naturalist Arthur Adams, to whom we must credit the collection and establishment of the species within the purview of modern scientific literature. In their 1850 report on the zoology of the voyage of H.M.S. Samarang, Arthur Adams, and Lovell Reeve provided the first formal description of *R. philippinarum* (Table 1.1), which they attributed to the Philippine Archipelago and named with what is now recognized as its original synonym: *Venus philippinarum*.

FIGURE 1.2

Illustration of "*Venus philippinarum*" from the report of the zoology of the voyage of H.M.S. Samarang during the years 1843–1846. (Plate XXII, Fig. 10 from Adams and Reeve, 1850). This illustration closely resembles the specimen shown in Fig. 1.1.

The original Adams and Reeve (1850) specimen can be found in the collection of the Natural History Museum, London (NHM), catalogued as NHMUK 1968161 (Fig. 1.1). A 1968 label attributes its collection to the British expedition of H.M.S. Samarang (1843–1846) and states that it is the "holotype." This shell is a good match with Plate XII, Fig. 10 in the Samarang zoological report (Adams & Reeve, 1850), although its illustration shows a mirror image of the right valve, a common phenomenon resulting from the lithographic printing techniques employed at the time. This original illustration is shown in Fig. 1.2.

Several issues are apparent with the provenance of the supposed London type specimen. Firstly, the 1968 label also refers to the collection of Hugh Cuming, an important 19th-century shell collector. Secondly, there is no direct contemporary evidence linking the shell with either the Samarang expedition or its commander Edward Belcher. While the locality "Philippine Archipelago" was given in the original Adams and Reeve description, the London type specimen is labeled with the additional specific locality "Mindanao."

Since then, the provenance of both the original London specimen and its type locality has been the subject of doubt and debate (Humphreys, 2025). In particular, the tropical Philippine origin of the specimen is now considered erroneous, as is (by implication) the appropriateness of both the widely used vernacular name (Manila) and the scientific species name (*philippinarum*). These questions over the status of the specimen and the implications of the error are examined in detail in Chapter 2. Notwithstanding these questions, *R. philippinarum* (Adams & Reeve, 1850) is established as the widely accepted binomen, with the authority "Adams and Reeve, 1850" sometimes written as A. Adams to distinguish between Arthur and his brother Henry Adams, who was also active in the field.

Since the original binomial *Venus philippinarum* was published, instability surrounding the scientific name has created difficulties, not least for field biologists (Beninger & Backeljau, 2019). As early as 1869, Arthur Adams himself was identifying the clam as *Tapes (Cuneus) philippinarum* while continuing to refer to the original Adams and Reeve (1850) "Plate 2 Fig. 10" (Fig. 1.2) illustration (Adams, 1869). Such variations are generally the consequence of rationalizations informed by progress in understanding the phylogeny of the bivalve mollusks and consequent recognition of synonymic relationships between different published binomens. In the 19th and 20th centuries, many independently attributed names have been recognized as synonymous based largely on conchologically informed taxonomic revisions: Notably, that of Fischer-Piette and Metivier (1971), who in a comprehensive study included a chronological list of reports and synonyms published since 1850, many of which are shown in Table 1.1.

Since the 1990s, mitochondrial and nuclear molecular sequences have informed bivalve systematics, leading to further revisions in the taxonomy of the species. While a subjective element in the interpretation of both morphology and genome remains, some consensus on the binomen *Ruditapes philippinarum* has emerged in recent years based on detailed work on the Family Veneridae (e.g., Mikkelsen et al., 2006).

Nevertheless, as a legacy of taxonomic instability, even across quite recent literature, both the genus and species names vary. This problem is partly but not entirely mitigated by the development of taxonomic databases (Vandepitte et al., 2015), including the European Aphia platform and web-based registers such as MolluscaBase, which also provides source information for other registers, including the World Register of Marine Species (WoRMS) and the Global Register of Introduced and Invasive Species (GRIS). In 2022, one such database with an interactive facility for field observations indicated that, as of August 2022, 84.7% of their presence reports cited the clam as *R. philippinarum*, 13.8% as *Venerupis philippinarum*, and 1.4% as *Tapes philippinarum* (GBIF.org/species/4372686, accessed August 14, 2022).

Despite the proliferation of taxonomic databases now available and the problem of harmonization this creates, especially for biodiversity studies (Grenié, 2022), the availability of web-based synonym lists can serve to reduce the potential for confusion, expedite nomenclatural clarity, and encourage convergence in the literature. At best, such databases promulgate original authorities and sources, rather than cascading data by citing other databases, but when the latter occurs, the provenance and quality of information can prove problematic. This is particularly manifest in geographical presence observations for *R. philippinarum* and is discussed further in Chapter 2.

As must be expected, the vernacular name of *R. philippinarum* varies geographically. In Japan, the species is named Asari, and in Korea, Bajirak. On the extensive coast of China, vernacular names vary with region. With its global spread, new names have emerged, including Japanese littleneck clam and Japanese carpet shell, which are also widely used. The littleneck and carpet references derive from the siphon length and shell surface texture, respectively (see Chapters 5and 6). Geographical variants also include the French "Palourde japonaise," Spanish Almeja japonesa, and German Japanische muschel. Goulletquer (1997) provided a list of 19 common names across many countries.

References

Adams, A. (1869). On the species of Veneridae found in Japan. In *Annals and magazine of natural history Vol III-fourth series.* (pp. 229–335). Taylor and Francis.

Adams, A., & Reeve, L. (1850). Mollusca. In A. Adams (Ed.), *The zoology of the voyage of H.M.S. Samarang; under the command of Captain Sir Edward Belcher, C.B., F.R.A.S., F.G.S. during the years 1843–1846.* Reeve, Benham, and Reeve.

Archaeology Magazine (2018) 4,500 year old clam processing site examined in Japan. 22 January issue. Archaeological Institute of America.

Belcher, E. (1848). *Narrative of the voyage of H.M.S. Samarang, during the years 1843–46; employed surveying the islands of the eastern archipelago; accompanied by a brief vocabulary of the principal languages*: 2 volumes. Reeve, Benham, and Reeve.

Beninger, P. G., & Backeljau, T. (2019). Understanding taxonomic and nomenclatural instability – A case study of the Manila clam. *Aquaculture, 504*, 375–379. https://doi.org/10.1016/j.aquaculture.2019.02.016.

Carlson, S. J. (2016). The evolution of Brachiopoda. *Annual Review of Earth and Planetary Sciences, 44*, 409–438.

Dall, W. H. (1902). Synopsis of the family Veneridae and of the North American recent species. *Proceedings of the United States National Museum, 26*(1312), 335–412. https://doi.org/10.5479/si.00963801.26-1312.335.

FAO. (2025). Fishery and aquaculture statistics – yearbook 2022. *FAO yearbook of fishery and aquaculture statistics*. FAO https://doi.org/10.4060/cd4312en.

Fischer-Piette, E., & Metivier, B. (1971). Revision des tapetinae. *Memoires du Museum National D'Histoire Naturelle. Serie A, Zoologie*. Editions du Museum Tome LXX1.

Fujita, M., Yamasaki, S., Katagiri, C., Oshiro, I., Sano, K., Kurozumi, T., Sugawara, H., Kunikita, D., Matsuzaki, H., Kano, A., Okumura, T., Sone, T., Fujita, H., Kobayashi, S., Naruse, T., Kondo, M., Matsu'ura, S., Suwa, G., & Kaifu, Y. (2016). Advanced maritime adaptation in the western Pacific coastal region extends back to 35,000–30,000 years before present. *Proceedings of the National Academy of Sciences, 113*(40), 11184–11189. https://doi.org/10.1073/pnas.1607857113.

Giribet, G. (2008). Bivalvia. In W. F. Ponder, & D. R. Lindberg (Eds.), *Phylogeny and evolution of the Mollusca*. University of California Press.

Gould, S. J., & Calloway, C. B. (1980). Clams and brachiopods—Ships that pass in the night. *Paleobiology, 6*(4), 383–396. https://doi.org/10.1017/s0094837300003572.

Goulletquer, P. (1997). A bibliography of the Manila clam *Tapes philippinarum. IFREMER, 122.* http://ifremer.fr.docelec/doc/1997/rapport-3221.pdf.

Grenié, M., Berti, E., Carvajal-Quintero, J., Dädlow, G. M. L., Sagouis, A., & Winter, M. (2022). Harmonizing taxon names in biodiversity data: A review of tools, databases and best practices. *Methods in Ecology and Evolution*, 1–14. https://doi.org/10.1111/2041-210X.13802.

Habu, J. (2004). *Ancient Jomon of Japan*. Cambridge University Press.

He, K., Sun, G., Wang, Y., Zheng, Y., Zhang, J., Yu, X., Shen, C., & Lu, H. (2023). Earliest Neolithic occupation and maritime adaptation on the West Pacific coast. *Journal of Archaeological Science, 160*, 105874. https://doi.org/10.1016/j.jas.2023.105874.

Humphreys, J. (2025). Revisiting the indigenous distribution of the "Manila clam" *Ruditapes philippinarum*: Mechanisms of persistent error. *Estuarine, Coastal and Shelf Science, 314*, 109152. https://doi.org/10.1016/j.ecss.2025.109152.

Humphreys, J., Choi, K.-S., & Watanabe, S. (2025). *Ruditapes philippinarum*: The $7 billion mollusc. *Estuarine, Coastal and Shelf Science, 325*, 109481. https://doi.org/10.1016/j.ecss.2025.109481.

Jerardino, A., & Marean, C. W. (2010). Shellfish gathering, marine paleoecology and modern human behavior: Perspectives from cave PP13B, Pinnacle Point, South Africa. *Journal of Human Evolution, 59*(3-4), 412–424. https://doi.org/10.1016/j.jhevol.2010.07.003.

Manzi, J. J., & Castagna, M. (1989). Introduction. In M. Manzi, & M. Castagna (Eds.), *Clam mariculture in North America*. Elsevier.

Mikkelsen, P. M., Bieler, R., Kappner, I., & Rawlings, T. A. (2006). Phylogeny of Veneroidea (Mollusca: Bivalvia) based on morphology and molecules. *Zoological Journal of the Linnean Society, 148*(3), 439–521. https://doi.org/10.1111/j.1096-3642.2006.00262.x.

Morton, B. (1996). The evolutionary history of the Bivalvia. In J. Taylor (Ed.), *Origin and evolutionary radiation of the Mollusca*. The Malacological Society of London.

Shimoyama, S. (2015). Historical occurrence of the short-neck clam, *Ruditapes philippinarum* (Adams & Reeve, 1850), on the sandy flats of Ariake Bay, Kyushu, western Japan. *Plankton Benthos Research, 10*(4), 202–214.

Stanley, S. M. (1968). Post-palaeozoic adaptive radiation of infaunal bivalve molluscs: A consequence of mantle fusion and siphon formation. *Journal of Palaeontology, 42*(1), 214–229.

Vandepitte, L., Vanhoorne, B., Decock, W., Dekeyzer, S., Verbeeck, A. T., Bovit, L., Hernandez, F., & Mees, J. (2015). How aphia—The platform behind several online and taxonomically oriented databases—Can serve both the taxonomic community and the field of biodiversity informatics. *Journal of Marine Science and Engineering, 3*(4), 1448–1473. https://doi.org/10.3390/jmse3041448.

Wang, Y., Yang, Y., Liu, H., Kong, L., Yu, H., Liu, S., & Li, Q. (2020). Phylogeny of Veneridae (Bivalvia) based on mitochondrial genomes. *Zoologica Scripta, 50*(1), 58–70. https://doi.org/10.1111/zsc.12454.

Yonge, C. M. (1949). *The sea shore*. Collins New Naturalist Series No.12.

Yonge, C. M. (1953). The monomyarian condition in the Lamellibranchia. *Transactions of the Royal Society of Edinburgh, 62*(2), 443–478.

Yonge, C. M. (1976). Shallow and deep burrowing bivalves. In C. M. Yonge, & T. E. Thompson (Eds.), *Living marine molluscs* (pp. 185–204). Collins.

CHAPTER

Indigenous distribution and the Philippines revision

2

2.1 Main centers of indigenous distribution

The principal areas of indigenous distribution in Japan, China, and Korea are well established, and the economic significance of the species has ensured that much of what is now known about the species has resulted from research programs in these countries. The northern limit in Russia has also been reliably documented by Russian scientists. However, the southern limits of indigenous distribution, including the ostensible Philippine type locality, have long been the subject of contradictory and often erroneous literature and are considered separately below (Humphreys, 2025).

Both population dynamics and physiological energetics suggest the species finds its optimal conditions on the western Pacific coasts of the Sea of Japan (East Sea) and the Yellow and East China seas. Consequently, the greater proportion of its global value as a fishery commodity comes from the coasts of Japan, Korea, and, most of all, China.

2.1.1 Japan

Whereas much of Japan benefits from the warmer water of the northward-flowing Kuroshio and (on the west coast) Tsushima currents, northern regions, including Hokkaido in particular, are bathed in the south-flowing sub-Arctic waters of the Oyashio current to the east, and the Liman current to the west. Consequently, the coastal areas of Japan can be divided into northern (colder) and southern (warmer) provinces. While the major areas of "asari" clam harvesting are found in the warmer southern provinces, the clam is found on all Japanese coasts, and local distribution is determined by the availability of large bays and shallow coastal areas, rather than oceanography (Toba, 2016). Therefore, *Ruditapes philippinarum* extends across all the main Islands, including the large bays of Akkeshi on Hokkaido and Ariake on the southern island of Kyushu. The species is harvested both recreationally and commercially, with major areas for the latter including Tokyo Bay, Mikawa Bay, Ise Bay, Seto Inland Sea, and Ariake Bay (Toba, 2015).

In recent years declining populations in many parts of Japan have been attributed to various factors including tidal flat reclamation; hypoxia due to eutrophication; oligotrophication of coastal waters; flood events; winter storms; the introduction of alien

The Manila Clam *Ruditapes philippinarum*: Science and Significance of a Globalised Species
DOI: https://doi.org/10.1016/B978-0-443-36704-5.00006-7

Copyright © 2026 Elsevier Inc. All rights are reserved, including those for text and data mining, AI training, and similar technologies.

predators, and food availability (Shindo & Hibino, 2025; Shiba et al., 2021; Toba et al., 2016, 2020, 2025; Tsutsumi, 2006; Watanabe et al., 2009). These population declines and their causes are discussed further in Chapter 14. Stock enhancement measures have included seed clam culture and release (Sugaya, 2006).

2.1.2 China

China's 1800 km coastline extends from the temperate Yellow Sea at a latitude of around 40°N, through the East China Sea and south beyond the Tropic of Cancer, reaching furthest south in the vicinity of the Island of Hainan Dao at 19°N. As such, the Chinese coasts range from temperate to subtropical latitudes. *R. philippinarum* can be found along the entire coast and all 11 coastal provinces, along with Taiwan. In the southern provinces, the species is sympatric with the tropical *Venerupis aspera*, both of which are sold as "Gezi" and are not distinguished.

The Yellow Sea includes at its northwestern extent the Bohai Sea, into which flow the waters of the Yellow, Hai, and Liao rivers. Further to the east, the Yalu River flows into Korea Bay. The nutrients and sediments discharged by these major rivers supply sediment shores and rich phytoplankton communities, both of which are conducive to thriving populations of *R. philippinarum.* Consequently, the Yellow Sea provinces of Liaoning and Shandong are the major producers of *R. philippinarum,* although other provinces make significant contributions (Gou et al., 1999).

In 2021, the Yalu Estuary Manila clam fishery became the first Chinese clam fishery to receive certification with the Marine Stewardship Council's global standard for sustainable fishing (Chapter 19). For this fishery, seed clams are obtained from Putian (southern Fujian Province) on the Taiwan Strait, then transported and laid in the Yalu estuary. Movement from south to north is common, with seed clams from breeding facilities in southern China taken north for fattening (Zheng et al., 2023). This approach means that southern hatcheries can operate effectively without the expense of heating water to expedite larval development, while cooler northern bays provide optimal conditions for seabed growth.

Traditionally, clams have been farmed in the intertidal zone, but due to high demand, they are now also farmed subtidally to a depth of 10 m in the Yellow Sea provinces, where growth rates are higher but predation greater (Fang, 2015). Hatchery operations and pond culture operate around Putian and Zhangzhou (both in the southern Fujian Province), and large-scale offshore seeded fisheries include Hong Dao Harbour near Qingdao in the northerly Shandong Province (Chapter 17).

2.1.3 Korea

The eastern Yellow Sea boundary includes the Korean Peninsula, whose west coast has many protected bays with extensive tidal flats and natural clam populations. Whereas information on the species in North Korea is sparse, the western coast of South Korea is well documented and includes major Manila clam production areas from Ongjin Bay in the north to Gangjin Bay in the south (Park & Choi, 2016). *R. philippinarum* is also found on the east coast, being reported in Gangwon Province from Gamchu Bay south to Deoksan Port (Lutaenko & Noseworthy, 2019). Manila clam production also extends to the South Korean Jeju Island (Lutaenko et al., 2021).

In recent years, South Korea has also suffered declining Manila clam harvests, which have led to the importation of seed clams of length 2–2.5 mm from China, which are grown-on for 2–3 years (Park & Choi, 2016). Further information on aquaculture and fishery practices in these East Asian countries is provided in Part C.

2.2 Russian distribution and the northern limits

The northern limit of native distribution is found on Russian sub-Arctic coasts around the Sea of Okhotsk. The northernmost reported population is in the estuary of the Amur River and the adjacent northern reach of the Tartary Strait, which lies between the mainland and the Island of Sakhalin. Further south on Sakhalin, populations can be found on the west coast from Terpeniya Bay (Gulf of Patience) southwards. On the opposite side of the Sea of Okhotsk, the species is found on Kunashir Island (the southernmost of the Kuril Islands chain, which lies to the NW of Japanese Hokkaido). Other Russian populations are found in Vostok Bay (Ponurovskii & Selin, 1988) and the shallow inlets of Peter the Great Bay, where a fishery has existed since the 1930s, and where research has been conducted into laboratory breeding (Leskova & Rybnikova, 2011; Rakov, 1986).

The Russian northern limit of indigenous distribution falls around latitude 53°N; however, latitudes taken alone can obfuscate the actual temperature means, ranges, and durations to which estuarine and coastal infaunal bivalves are exposed. For example, the Amur estuary and Tartar Strait populations both fall within the influence of the great Amur River, whose main catchment lies further south and whose discharge is characterized by high summer temperatures (up to 20°C–21°C), which are anomalous for that otherwise cold-water (6°C–8°C) northwestern Pacific region (Kamenev & Nekrasov, 2012).

Detailed information on northern distribution, including the Sea of Japan, has been provided by Kolpakov et al. (2024). Fig. 2.1 shows this northern distribution based on both historical and recent field reports. It omits only the Amur River estuary population described by Kamenev and Nekrasov (2012). From its northern limit, the species extends south along the Russian continental coast to Slavyansky Bay near the border with North Korea, this being the southernmost locality available to *R. philippinarum* on the Russian east coast (Ponurovskii & Yakovlev, 1992).

The east coast of Sakhalin is bathed by the Sea of Okhotsk, beyond which is the Kuril Island chain. Manila clams can be found on the southeast coasts of Sakhalin in the large bays of Aniva and Terpeniya, but not further north. Eastwards across the Okhotsk, the Kuril chain islands of Kunashir and Shikotan also support Manila clam populations. In Russia, the species provides for various fisheries such as in the

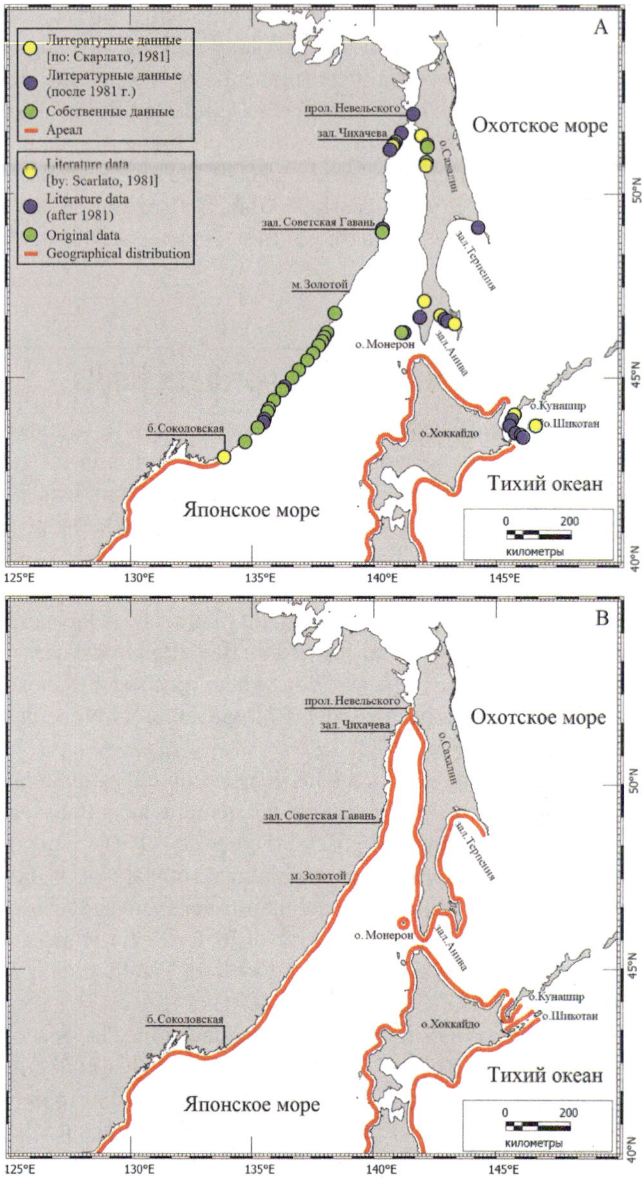

FIGURE 2.1

Northern indigenous distribution of *Ruditapes philippinarum* on Russian and Japanese coasts. (A) Presence data compiled from literature reports and original observations. (B) Inferred extent of geographical distribution in appropriate habitats. Russian language sea areas: Top right: Sea of Okhotsk; bottom left: Sea of Japan; bottom right: Pacific Ocean.

With permission from Kolpakov et al. (2024).

Losos Bight and Busse Lagoon areas at the southern end of Sakhalin in Aniva Bay. Kim (2019) estimated the total stock of *R. philippinarum* in the Aniva Bay fisheries at 240 tons.

2.3 Southern limits: The Philippines question

In their report on the voyage of H.M.S. Samarang, Adams and Reeve (1850) reported the collection of *Venus philippinarum* with the type locality as the Philippine Archipelago. In the years following, many other *de novo* species were reported and then recognized as ostensible synonyms of *R. philippinarum*. As a result of this process, new localities were attributed to the species. Even as he described the distribution as "Andamans, Ceylon, Bombay, Karachi, Philippines, China, Formosa, Korea, Japan, Sachalin, South Sea Islands…" Lynge (1909) commented, "The variability of this species has not yet been sufficiently investigated. It comes near *Tapes variegatus*, Sow…" (*Tapes variegatus* is a synonym of the similar but tropical *Ruditapes aspera*). This wide geographical range, including tropical seas, has been a persistent feature of the literature. In their *Revision Des Tapetinae*, Fischer-Piette & Metivier, 1971 listed the distribution "According to the information given by the bibliography" as including the same tropical localities along with Java (Lynge, 1909).

Consistent with a Philippine distribution, in a full catalogue of over 3100 species of marine mollusks of the Philippines, Hidalgo (1904–1905) included *R. philippinarum* (as its synonym *Tapes philippinarum)*. Silina & Popov, 1989 emphasized that the species is "most common in the Philippines," and many other sources have indicated the native distribution to include these islands (FAO, 2023; Poppe, 2011; Poutiers, 1998; Qi, 1987). Numerous online species registers, including CABI: Bonham et al. (2022), NEMESIS (n.d.), ORBIS (n.d.), WoRMS, Ruditapes, Adams, & Reeve, 1850), GBIF Backbone Taxonomy, 2023, and Mollusca Base (n.d.), have also recorded the species as extant in the Philippines. Arabaca et al. (2022) published the results of a field survey at three sites on the central Philippine Island of Panay. They recorded over 8000 bivalves, including 28 species, nine of which were Venerids. *R. philippinarum* was reported from Ajuy Bay. In this context, it is also of interest that the National Museum of Natural History (NMNH) of the Philippines in Manila has catalogued collections ostensibly of *R. philippinarum*, collected from sites including Panay Island. These specimens, kindly made available to the author, are considered further below.

In contrast and despite these sources, elsewhere the species' presence in the archipelago is rejected, questioned, or more passively unasserted. Kuroda, Habe, & Oyama, 1971 considered the Philippine type locality as erroneous, suggesting South Korea as an alternative type locality. Higo et al. (1999) describe the clam's Philippine status as "possibly erroneous," and Huber (2010) stated categorically that the original type locality is erroneous. Suarez and Raven (2020) state that "Despite its scientific name, it does not occur in the Philippines." Less categorical but consistent with such statements, in the first guide to the mollusks of the Philippines, *R. philippinarum* was not included (Springsteen & Leobrera 1986). Similar inferences from other sources, which describe the species only as "subtropical-low boreal," are consistent with an erroneous Philippine type locality (Lutaenko, 2006; Lutaenko & Noseworthy, 2019). Lutaenko (2016), who has studied and reviewed the biodiversity of the opposite western coast of the South China Sea, believes that reports of *R. philippinarum* are likely to have been misidentifications (Konstantin Lutaenko, personal communication).

2.4 Location error

Lutaenko (2016) has provided a summary of the numerous 19th-century European research expeditions to the South China Sea and surrounding areas. These include the Samarang voyage (1843–1846), during which the London type specimen of *R. philippinarum* was collected, reportedly off the Philippine Island of Mindanao. The original Adams and Reeve (1850) specimen (Chapter 1, Fig. 1.1) matches Plate XII, Fig. 10 in Adams and Reeves' (1850) Samarang report (Chapter 1, Fig. 1.2).

However, several issues are apparent with the provenance of this supposed holotype (NHMUK 1968161). First, a 1968 specimen label also refers to the collection of Hugh Cuming, an important 19th-century shell collector (Fig. 2.2). Secondly, there is no direct contemporary documentary evidence linking the shell with either the *Samarang* expedition or its commander Edward Belcher. While the locality "Philippine Archipelago" was given in the original Adams and Reeve (1850) description, the 1968 label has the additional specific locality "Mindanao." While the London Natural History Museum (NHM) Mollusca section registers record a purchase of specimens from Belcher's collection via Stevens Sale Room in July 1851 (Zoological Accessions 1850–1854.7), this record refers only to material identified to the genus *Venus*, and nothing explicitly identifies the type specimen of *V. philippinarum*.

Hugh Cuming's collection was acquired after his death in 1866 by the NHM (then still the British Museum) and formed a significant core of the Mollusca collections. However, much of the Cuming material was formally accessioned much later, and NHMUK 1968161 apparently in 1968, following examination by Édouard Fischer-Piette. His subsequent publication noted: *"The type of Adams and Reeve was not seen; we came across a specimen appearing entirely consistent with the Fig., but marked as coming from the Cuming collection, Mindanao, which seemed to prevent one from wanting to consider it as being the type. However, Mr Tebble, to whom we submitted it, was of the opinion that it was the type and the labelling "Mus. Cuming" was wrong* (translation from Fischer-Piette & Metivier, 1971, p. 33).

In these circumstances, and based on strict taxonomic convention, it may be more appropriate to consider NHMUK 1968161 as a syntype rather than the holotype (Tom White, personal communication). Nevertheless, the good match between this specimen and the Adams and Reeve illustration suggests a high probability that it is the holotype.

It is possible that Cuming, who purchased and integrated shell collections into his own, acquired material from Belcher or which otherwise originated from the

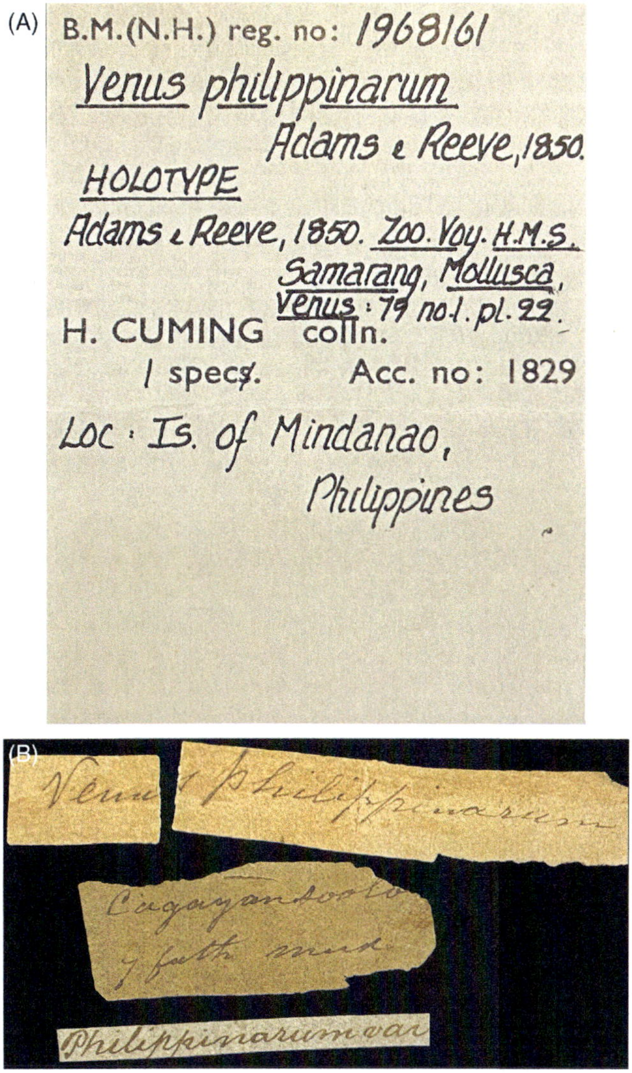

FIGURE 2.2

Labels associated with the London type specimen of *Ruditapes philippinarum*. (A) 1986 label claiming holotype status but referring also to the collection of Hugh Cumming. (B) Fragments of another earlier label which records the localities "Cagayan Soolo". Cagayan is a north-eastern province of the northern Philippine Island of Luzon, and the Sooloo (Sulu) archipelago lies in the south-west Philippines (see text).

Samarang. However, no documentary evidence survives that can demonstrate such a connection for NHMUK 1968161. Fragments of another, possibly 19th-century, label also associated with the specimen record the localities "Cagayan Soolo" (Fig. 2.2B). Hugh Cuming himself traveled extensively during his own Philippine expedition (1836–1840) and visited both Mindanao and Cagayan during his voyage (Merrill, 1926).

Contemporary accounts of Cumings' treatment of his collection criticized his neglect of proper labelling, especially with regard to locality. Such accounts have led to the observation that Cumings must have provided ample opportunity for the introduction of errors, as did some careless treatment of the Cuming collection after it had been purchased in 1866 by the British Museum (Dance, 1966). It is also of interest that Arthur Adams, who served on Belcher's Samarang expedition and with Lovell Reeve prepared the zoological report, also described hundreds of new species from the Cuming collection, many of which were allegedly "wrongly localized," most "unillustrated and insufficiently diagnosed," and with errors perpetuated through "slipshod writings" (Dance, 1966).

While the fact remains that all of the locations associated with the type specimen are consistent with a Philippine locality. Huber (2010), informed by biogeographical considerations, rejects any Philippine presence and generally criticizes the reliability of the molluscan section of the Samarang report (Adams & Reeve, 1850). He suggests that of all new bivalve species collected on the expedition, approximately half were given erroneous type localities. These included locations such as Panama, which was not visited during the voyage but was visited during an earlier voyage of the Samarang and also by H.M.S Sulphur, one of Belcher's earlier commands (Adams, 1848). Based on an "enormous number of erroneous localities." Huber (2010) concluded that "Definitely a Samarang locality is a high danger signal...."

2.5 Identification error

In his comment on *R. philippinarum* in the Philippines, Huber (2010) stated that *"No true philippinarum was found on the many Philippine Islands visited, nor were any seen reliably from there. Many of these philippinarum records represent instead the tropical aspera."* Although genome studies have shown them to be distinct species and can provide for definitive identification, *V. aspera* shares many conchological phenotypic traits with *R. philippinarum* (Liu et al., 2022; Tang et al., 2022). Although it is generally thought of as having a somewhat longer shell shape, these clams are known to vary according to conditions, and this is not a reliable diagnostic feature (Tang et al., 2022). Similarly, while the color pattern of the valve interiors is often included in species descriptions, a range of variation can be found even within a single series of specimens.

While on average, shell features do vary between these two species, they are more reliably distinguished on the basis of different soft tissue anatomy, not least the extent of siphon fusion (see Appendix). Nevertheless, identification errors remain a problem,

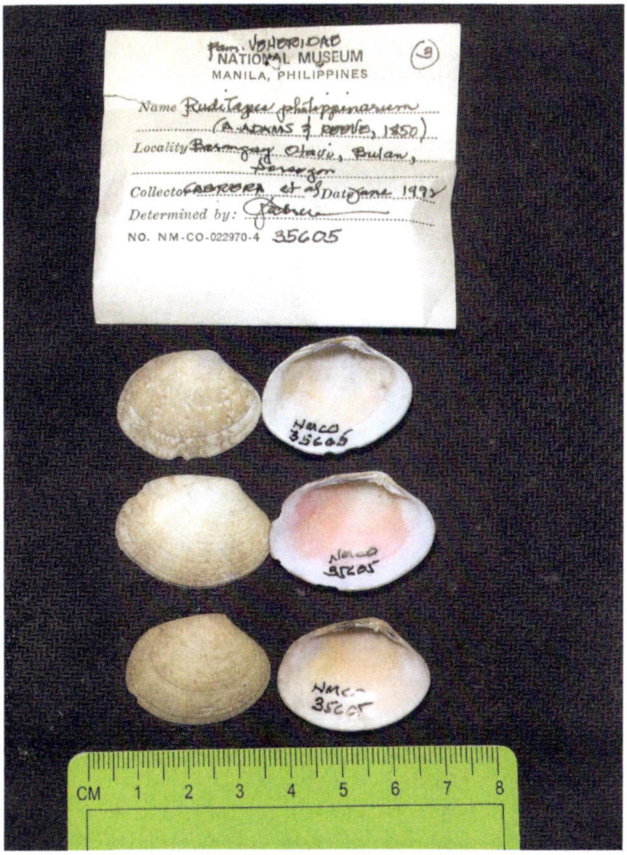

FIGURE 2.3

Shell specimens from Bulan municipality in the Philippine Province of Sorsogon. (Series NMP 35605), consisting of three pairs of matching, although disarticulated, valves. Examination of these shells and others in the collection indicates they are the tropical species *Venerupis aspera.*

Courtesy the National Museum of Natural History, Manila.

especially when based only on shell material. In a case where the original material was still available, Qi (1987) confirmed that a clam originally identified in Fujian Province, China, as *V. aspera* was actually *R. philippinarum*.

The catalogue of the Philippine National Museum of Natural History in Manila (NMNH) includes 13 series of specimens listed as *R. philippinarum* and dated from 1977 to 1997. Three of these have been located and made available to the author for examination. Fig. 2.3 shows one such series (Humphreys, 2025). None of the available specimens included soft tissues. Based on shell features, the Philippine NMNH specimens labeled as *R. philippinarum* have been identified by the author and independently verified as *V. aspera*. Subsequently, in a visit by the author to Panay Island, the source of a cluster of the NMNH specimens, no evidence of *R. philippinarum* was found either on beaches or in local wet markets where many shellfish are available.

In addition to similar shells, misidentification risk factors also include the taxonomic volatility of the species (Vecchione et al., 2000). In the case of the Manila clam, this risk also applies, along with the general "hypertrophied and oversaturated synonymy" of the mollusks (Lutaenko, 2016), which has contributed to considerable nomenclatural instability (Table 1.1). Guidance on the reliable identification of *R. philippinarum* is provided as an Appendix.

2.6 Error cascades

Routinely, species distribution statements in the literature are based on earlier cited works. Lynge (1909) and Faustino (1928), for example, cite Adams and Reeves' Samarang report on *R. philippinarum*. These two cases are scientifically legitimate as the cited authority is an original field observation report which can be (and in this case has been) directly scrutinized and, when necessary, challenged. However, it is not uncommon for distribution citations to be indirect and unsubstantiated in the sense that checking the citation reveals it not to be an original report. This practice increases the risk of "error cascades" as described by Bortolus (2008). For example, *R. philippinarum* distribution statements, including the Philippines, can be found in the introductions of peer-reviewed journal papers (Breber 2002; Silina & Popov, 1989), sometimes with no attempt made to cite any sources (Bae et al., 2021; Kang et al., 2016)

Alternatively, an apparently authoritative source may be cited because it has been cited for the same purpose in another publication, without being properly considered. Scarlato's (1981) comprehensive 480-page volume in Russian on the bivalves of the temperate regions of the western Pacific Ocean provides an example of this phenomenon. While Scarlato references some Philippines records he is in fact skeptical of a Philippine distribution and describes the species as subtropical to low boreal (p. 383), having earlier noted that: *"According to literature the southern border of its range passes into the area of the South China Sea and Philippine Islands ...No other species of bivalve from the studied regions has such a range extending in latitude from 10 to 50°C. It is possible that the named species is a composite and that in fact there are at least two similar species: one is actually tropical, the other is subtropical-low boreal"* (translated from Scarlato (1981), p. 37). Yet Scarlato is a commonly cited source when asserting a Philippine distribution. Raising the possibility that authors have not had direct access to it.

In another example, Poutiers (1998) is often cited to the effect that the species is extant in the Philippines; however, this document is a grey literature secondary source that provides no cited authorities. Such cases are illustrative of the problem that tracking back through the literature can reveal chains of sources for which there is no ultimate traceable safe observation. Poutiers (1998) similarly describes a distribution west to India and Sri Lanka and east to Micronesia. Many such cascades appear to route back to early sources now recognized as erroneous, although often no explicit connection is made (Lynge, 1909).

It is perhaps not surprising that such practices, which superficially appear to follow the established citation conventions of scientific literature, are common. While they can seem like relatively innocuous shortcuts, they can compound errors when somewhere in the chain a source is questionable or misrepresented, but which, through such a cascade, is amplified. Such mechanisms can work both ways. For example, Huber (2010) makes a categorical assertion that the original type locality is erroneous, citing only a source in which the clams' presence in the Philippines is described more tentatively as "possibly erroneous," and whose authors acknowledge that their catalogue is itself based on other literature which in this case they do not cite and is in any event a catalogue of the Mollusca of Japan (Higo et al., 1999).

These examples illustrate that citation chains often do not substantiate distribution statements on a global distribution for *R. philippinarum* that includes the Philippines. A contributory factor to the frequency of this difficulty may be that the name *philippinarum* powerfully encourages an inference that the species distribution must include the (erroneous) Philippine type locality and that an authoritative citation is therefore assumed less necessary.

2.7 Biogeography and energetics evidence bearing on tropical distribution

Biogeographical boundaries, whether informed by climate, species assemblages, or level of endemism, do not attempt to imply the limits of all the species present within them. Similarly energetic optima and the thermal tolerances of individual species may vary in response to selection pressures on different populations. Nevertheless, they provide relevant contextual information bearing on distribution. Further information on physiological energetics is included in Chapter 8.

The South China Sea is bounded in the west by the coasts of southern China and Vietnam, to the south-west by the Malaysian mainland, and in the east by a line that runs from the east coast of Taiwan to the east coasts of the Philippine islands of Palawan and Luzon and down to the north-west coast of the Island of Borneo. The Philippines are biologically diverse, but knowledge of its marine fauna is described as "relatively incomplete" (Teh et al., 2019). The Archipelago includes over 7000 islands, most of which lie eastward of the South China Sea but all of which fall entirely within the tropics, stretching roughly from latitude 4° to 21° North. The islands are divided into three groups: The northern Luzon Group, the southern Mindanao Group, and the central Visayan Group.

Although the East and South China seas each have geographical and political significance, they are not unified biogeographical provinces. Various biogeographical boundaries have been proposed for the seas of the western Pacific. In an early investigation into the biogeography of the mollusks, Woodward (1890) identified a "Japonic Province" within which he listed *R. philippinarum* (as the synonym *Tapes japonica)* based on "Mr. Adams's researches" (Adams, 1869). However, despite the earlier Samarang report, he omitted *R. philippinarum* from his largely tropical Indo-Pacific province, which included the Philippine archipelago. Many of Woodward's provinces have remained approximately familiar, although refined, in later work.

The system of marine coastal biogeographical realms is informed by taxonomy, ecology, evolutionary history, patterns of dispersal, and isolation (Spalding et al., 2007). Based on consensus in the literature along with fisheries data, the Manila clam's indigenous distribution is most clearly centered on Spalding et al.'s (2007). Temperate Northern Pacific, and the clam's northern limit corresponds more or less to the northern boundary of that biogeographic realm. However, the Philippines falls within their Central Indo-Pacific realm, which lies almost entirely within the tropics. These authors regard the area around Taiwan and Hainan Dao as a transition zone reaching into the northern latitudes of the tropics but no further.

Huber (2010) proposed a pattern of 17 biogeographical "core areas" based on a survey of all 7200 bivalve species then recognized, each area with a high degree of endemism. While this scheme differs from some earlier and less bivalve-specific patterns of distribution, Huber, like Woodward (1890) and many others since, distinguishes between Japan and the Indo-Pacific realms, although with an intermediary zone between, which stops short of the tropical Indo-Pacific Philippine archipelago. Huber's (2010) rejection of *R. philippinarum's* presence in the Philippines, while explained by misidentifications, is consistent with what might be predicted from his (and others') biogeographical findings. Moreover, as Skarlato (1981) recognized, an established southern range so deep into the fully tropical Visayas group and the putative Mindanao type locality would indicate a peculiarly wide climatic tolerance for a species found as far north as the sub-Arctic Sea of Okhotsk.

Nevertheless, the southern distribution of *R. philippinarum* is known to extend at least as far south as Hong Kong (Lam, 1980; Lutaenko & Volvenko, 2013) and the southern Chinese island of Hainan Dao (Tang et al., 2022), which correspond to the northern boundary of the South China Sea and whose latitudes overlap with the northern tip of Luzon. Lutaenko (2016), in a review of the biodiversity and biogeography of bivalve mollusks on the western side of the South China Sea, identified two distinct provinces: the subtropical Sino-Japanese to the north and the tropical Indo-Polynesian to the south, with the boundary point around Hong Kong.

Further south, along the western coast of the South China Sea, it appears that *R. philippinarum* is not commonly reported (Tu et al., 2019). A field based checklist of bivalves of Chem Island, Vietnam, reports *V. aspera* but not *R. philippinarum,* as it does for the Gulf of Thailand (Sanpanich, 2011). Likewise, fisheries and aquaculture reports, while emphasizing the economic importance of other Venerids, do not mention *R. philippinarum* (Thang, 1999). Lutaenko (2016), who has studied this coast and found only *V. aspera,* regards reports of *R. philippinarum* in Vietnam as misidentifications (Konstantin Lutaenko, personal communication). Nevertheless, reports of *R. philippinarum* persist. Zvonareva et al. (2019) identified *R. philippinarum* as present, although rare, at Dam Bay on the central Vietnamese Tre Island, at a latitude corresponding to that of the central Philippine Visayas group (latitude around 12°N).

Whether or not *R. philippinarum* reaches much further south than Hainan Dao, such a finding cannot be taken to add plausibility to a possible Philippine distribution. According to Lui (2013), the benthic faunal assemblages and oceanographic data of the South China Sea indicate a difference in biogeography between its western and eastern coasts. While the north Vietnamese coast to latitudes around 13°N falls into a subtropical zone, on the opposite side of the sea, the Philippines (in their entirety) fall within a fully tropical subregion. This biogeographical difference between corresponding latitudes of the west and east coasts shows that here, as in other coastal seas, oceanography rather than latitude is a more reliable consideration in terms of marine climate and distribution.

Other evidence relevant to the question of tropical distribution is provided by studies on the physiological energetics of the species. Kang et al. (2016) quantified a number of physiological responses to experimentally varied suspended particulate (food) concentrations at temperatures up to 23°C. They found *R. philippinarum* to be well adapted to the wide environmental ranges found on the temperate tidal flats of Japan, reporting a high level of physiological plasticity, which they suggested (having asserted the clam's presence in the Philippines) allows the species to be successful over a wide variety of geographical areas. However, experimental temperatures up to 23°C are markedly lower than the tropical sea surface temperatures of the Philippines, whose seasonal means vary from around 26 to 30 and are increasing with climate change (Yu et al., 2019).

Moreover, Kang et al. (2016) reported an optimum physiological energetic response at 18°C, which, consistent with other work, suggests that assimilation efficiency drops at temperatures above 20°C (Han et al., 2008). Also of interest is that Kang et al. (2016) concluded the measures of respiration rate reveal a lack of any acclimatory adjustment of metabolic rate to rising temperature; a feature which, for an ectothermic animal, implies that increasing temperatures require a corresponding increase in food consumption for survival. In a naturalized UK population, it has been hypothesized that mass mortalities of naturalized *R. philippinarum* in temperate waters may be attributed to seasonally warming water temperatures preceding the spring phytoplankton bloom, leaving the species energetically stressed (Franklin et al., 2020; Humphreys et al., 2007).

Regardless of its *philippinarum* name, assertions that an invertebrate species whose northern limit is sub-Arctic is extant in tropical waters require convincing substantiation. Until such time as definitive evidence emerges, skepticism is appropriate. If, based on adequate published and verified identification evidence from direct field surveys, *R. philippinarum* were demonstrably present, the matter would be resolved. However, such evidence remains elusive. A combination of the known unreliability of historic sources, ease of misidentification, and lack in modern published reports of the morphological or molecular basis of identifications, all when considered in wider biogeographical and physiological contexts, indicates that the indigenous range of *R. philippinarum* does not extend into the Philippine archipelago or other fully tropical regions.

2.8 Implications of erroneous distributions

Erroneous distribution statements are not unique to *R. philippinarum.* Zenetos et al. (2010) have defined the status of species in the Mediterranean Sea according to a terminology that distinguishes between secure and unsafe records. Presence reports are regarded as unsafe based on three criteria: Old records of species that have not been subsequently found, new records not verified by experts, or species with unresolved taxonomic status. At various times since the first presence report in 1850, all three criteria have applied to tropical distribution reports for *R. philippinarum.*

The mechanisms of error, including error cascades identified above for *R. philippinarum,* relate to the basic practices of biological identification, recording, and citation. As such, they are potentially relevant to all species, especially those with phenotypically similar relatives. Aside from a general biological importance for biodiversity and community studies, accurately understanding a species' native extent reflects on physiological optima, tolerances and the ecological implications of new arrivals. The correct identification of species is a fundamental requirement in assessments of biological heritage and as a basis for conservation and management (Ng & Tan, 2000). Secure knowledge of the limits of indigenous distribution can also contribute to policy decisions on anthropogenic introductions, which must balance ecological risk with economic opportunity.

Moreover, increasingly, species distributions are used to provide monitoring baselines in investigations into the impacts of climate change, with respect to biodiversity, biogeographical boundary changes, and the arrival of non-native species. Similarly, studies designed to determine rates of non-native arrivals or predict future distributions are commonly based on current reported distributions and inferences on climatic tolerances. Such studies are dependent on reliable data with respect to current indigenous distributions for their validity.

For example, Sebens et al. (2017) using indirect sources such as an online species register for a meta-analysis assessing the global dynamics of alien species accumulations, used a range for *R. philippinarum* with tropical distribution extending from the Philippines to the Anderman Islands and as far as Karachi, including Mumbai and Sri Lanka to the west, and the "South Sea Islands" to the east (Sebens et al., 2017, supplementary materials). In this case, *R. philippinarum* was one of many species considered, so in itself may have had little impact on the overall conclusion. However, if the phenomenon of unsafe records is widespread, it would compromise conclusions, and in any event, the case of *R. philippinarum* raises questions about the wider quality of information on even reputable web-based species registers and their uses in such studies.

In another example, Johnson et al. (2025) have predicted the future worldwide distribution of *R. philippinarum* under climate change. Among the information used in their modeling was presence data records from the Global Biodiversity Information Facility (GBIF), which has also suffered from the error cascade mechanism described above, consequently including erroneous tropical distribution reports. Here, the single species focus of the study makes the validity of some conclusions much more directly susceptible to erroneous information (see Chapter 23).

The mechanisms of persistent error and amplification discussed above suggest that correcting the record will not be straightforward, especially in the West, where these problems with the reported indigenous distribution of *R. philippinarum* are less familiar and recognized. Some improvement would result from authors and species registers excluding the Philippines from summaries and statements on the native distribution of *R. philippinarum*. While some authors indicate an exclusively temperate distribution, sources (and online species registers in particular) showing a Philippine distribution, or referencing the ostensible type-location, could at least add a qualifying statement acknowledging the unsafe status of such reports. Similar care should be taken for all statements that locate *R. philippinarum* in any fully tropical habitat.

Kurzel et al. (2022) and Ramirez et al. (2022) have emphasized the difficulties and importance of robust species identification, advocating "integrative taxonomy" including both morphological and molecular approaches to identification. The increasing ubiquity of molecular sequencing technology in laboratories suggests this has become a more widespread approach (as have environmental DNA techniques), but for many multispecies biodiversity studies, not least in the emerging economies adjacent to the South China Sea, molecular approaches are often not yet a feasible solution. Vecchione et al. (2000), among others, suggest the importance of taxonomic specialist involvement in field survey work, but such expertise is also not universally available (Ng & Tang, 2000).

However, even in the absence of such specialist expertise and facilities, basic good practice can deliver improvements. In particular, identification should include scrutiny of phenotypic features of both shell and soft tissues, especially siphon anatomy, and authors should justify their diagnosis with such evidence. For example, in contrast to work reported forthe Philippines (Arabaca et al., 2022), (Tang et al., 2022) in Japan included good photographic evidence of gross siphon morphology as a basis for convincing identification. Where possible, whole preserved specimens should be retained, rather than just shells.

Over time, improved identification and citation practices would establish a more widespread consensus on the indigenous distribution of *R. philippinarum*. However, definitive resolution of its southern limit would benefit from a program of research focused on tropical and boundary regions. This would involve the systematic re-examination of specimens held in collections, with particular focus, where possible, on siphon anatomy and diagnostic molecular sequencing. Field searches at key tropical sites could, when possible, involve environmental DNA.

Through such measures, should negative evidence continue to accumulate against a tropical distribution for *R. philippinarum*, a case should be made for a change of the type locality. In this respect, it is of interest that the International Code of Zoological Nomenclature includes Recommendation 76A.2.A that "*A statement of a type locality that is found to be erroneous should be corrected*" (Ferraris & Eschmeyer, 2000). In this particular case, such a change would also suggest correction of the *philippinarum* species name. However, while this could further reduce the risk of the positive feedback type of errors described above, in terms of taxonomic convention, a species name change on this basis would appear more difficult.

References

Adams, A. (1848). *Notes from a journal of research into the natural history of the countries visited during the voyage of H M S Samarang, under the command of Captain Sir Edward Belcher*. Reeve, Benham and Reeve Also, facsimile edition by Rediscovery Books Ltd in association with the Royal Geographical Society 2006.

Adams, A. (1869). On the species of Veneridae found in Japan. In *Annals and magazine of natural history: vol. III-Fourth Series* (pp. 229–335). Taylor and Francis.

Adams, L. A. and Reeve, L. 1850. Mollusca. In: A. Adams, (Ed.) The zoology of the voyage of H.M.S. Samarang; under the command of Captain Sir EdwardBelcher, during the years 1843–1846. Reeve, Benham and Reeve.

Arabaca, M. L. C., Ilano, A., & Taunan, V. O. (2022). Diversity of marine macro-Molluscan gastropods and bivalves in the intertidal areas of Ajuy, Iloilo, Western Visayas, Philippines. *11th International and National Seminar on Fisheries and Marine Science. IOP Conference Series: Earth and Environmental Science, 1118*(2022), 012014. doi:10.1088/1755-1315/1118/1/012014.

Bae, H., Im, J., Joo, S., Cho, B., & Kim, T. (2021). The effects of temperature and salinity stressors on the survival, condition and valve closure of the Manila Clam, Venerupis philippinarum in a holding facility. *Journal of Marine Science and Engineering, 9*(7), 754. doi:10.3390/jmse9070754.

Bonham, V., Humphreys, J., Yang, N., Haiying, O., & Caiping, Y. (2022). Ruditapes philippinarum *Japanese carpet shell*. CABI Compendium. doi:10.1079/cabicompendium.61697.

Bortolus, A. (2008). Error cascades in the biological sciences: The unwanted consequences of using bad taxonomy in ecology. *Ambio, 37*(2), 114–118. doi:10.1579/0044-7447(2008)37[114:ECITBS]2.0.CO;2.

Breber, P. (2002). Introduction and acclimatisation of the Pacific carpet clam, *Tapes philippinarum*, to Italian waters. In E. Leppäkoski, S. Gollasch, & S. Olenin (Eds.), *Invasive aquatic species of Europe. Distribution, impacts and management* (pp. 529–547). Kluwer Academic Publishers.

Dance, S. P. (1966). *Shell collecting. An illustrated history*. Faber and Faber.

Fang, J. (2015). Development of Manila clam industry in China. *The third international symposium on Manila (Asari) Clam*. Japan: Ust-Tsu, Tsu, Mie, Japan https://nria.fra.affrc.go.jp/asari/booklet.pdf.

FAO 2023. Veneridae. Cultured aquatic species information programme. Texte de Goulletquer, P. Division des pêches et de l'aquaculture [en ligne]. https://www.fao.org/fishery/fr/culturedspecies/ruditapes_philippinarum?lang=en. (Accessed 13.10 2025)

Faustino, L.A. (1928). Summary of Philippine marine and freshwater mollusks. Monograph 25. Philippine Islands: Bureau of Science. wikia.org/wikipedia/commons/e/e4/Monograph_%28IA_acw7735.0001.025.umich.edu%29.pdf. (Accessed, 13.10.2025).

Ferraris, C. J., & Eschmeyer, W. N. (2000). International Code of Zoological Nomenclature, fourth edition. *Coepia, 2000*, 907–908. doi:10.1643/0045-8511(2000)000[0907:BR]2.0.CO;2.

Fischer-Piette, E., & Metivier, B. (1971). Revision of the Tapetinae. Memoires du Museum national d'histoire naturelle. *Serie A, Zoologie. Editions du Museum. Tome, LXX1.*

Franklin, D. J., Herbert, R. J. H., Chapman, I., Willcocks, A., Humphreys, J., & Purdie, D. A. (2020). Consequences of nitrate enrichment in a temperate estuarine marine protected area: response of the microbial primary producers and consequences for management. In J. Humphreys, & R. W. E. Clark (Eds.), *Marine protected areas. Science, policy and management* (pp. 685–701). Elsevier.

GBIF Backbone Taxonomy (2023) *Ruditapes philippinarum* (A. Adams & Reeve, 1850) in: GBIF Secretariat. Retrieved September 14, 2023, from https://www.gbif.org/species/4372686.

Guo, X., Ford, S. E., & Zhang, F. (1999). Molluscan aquaculture in China. *Journal of Shellfish Research, 18*(1), 19–31.

Han, K. N., Lee, S. W., & Wang, S. Y. (2008). The effect of temperature on the energy budget of the Manila clam, *Ruditapes philippinarum. Aquaculture International, 16*, 143–152.

Hidalgo, J. G. (1904–1905). *Catalogue of the Testaceous Molluscs of the Philippine , Jolo and Marian Islands*. Imprinta de la Gaceta de Madrid.

Higo, S., Callomon, P., & Goto, Y. (1999). *Catalogue and bibliography of the marine shell-bearing Mollusca of Japan. Gastrpoda, Bivalvia, Polyplacophora, Scaphopoda* (p. 749). Elle Scientific Publications. (Catalogue No B556). Photographic plates are available as a separate document (Type Figs. see page 178.

Huber, M. (2010). *Compendium of bivalves*. ConchBooks.

Humphreys, J. (2025). Revisiting the indigenous distribution of the "Manila clam" *Ruditapes philippinarum*: Mechanisms of persistent error. *Estuarine, Coastal and Shelf Science, 314*, 109152. doi:10.1016/j.ecss.2025.109152.

Humphreys, J., Caldow, R. W. G., McGrorty, S., West, A. D., & Jensen, A. C. (2007). Population dynamics of naturalised Manila clams *Ruditapes philippinarum* in British coastal waters. *Marine Biology, 151*(6), 2255–2270. doi:10.1007/s00227-007-0660-x.

Johnson, A. S., Bridges, A. E. H., & Knights, A. M. (2025). Predicting the future distribution of a commercially important clam (*Ruditapes philippinarum*) in a changing climate. *Estuarine, Coastal and Shelf Science, 320.* doi:10.1016/j.ecss.2025.109307.

Kamenev, G. M., & Nekrasov, D. A. (2012). Bivalve fauna and distribution in the Amur River estuary-a warm-water ecosystem in the cold-water Pacific region. *Marine Ecology Progress Series, 455*, 195–210. doi:10.3354/meps09679.

Kang, H. Y., Lee, Y.-J., Choi, K.-S., Park, H. J., Yun, S.-G., Kang, C.-K., & Song, L. (2016). Combined effects of temperature and seston concentration on the physiological energetics of the Manila clam *Ruditapes philippinarum. PLoS One, 11*(3), e0152427. doi:10.1371/journal.pone.0152427.

Kim, A. C. (2019). State of resources and fishery of Manila Clam *Ruditapes Philippinarum* (Adams and Reeve, 1848) in the Losos Bight and Busse Lagoon (Aniva Bay, Okhotsk Sea) in 2010–2016. *Izvestiya TINRO, 197*, 108–117. doi:10.26428/1606-9919-2019-197-108-117.

Kolpakov, E. V., Sokolenko, D. A., & Andryushina, A. D. (2024). Distribution pattern of the subtropical-low boreal bivalve mollusk *Ruditapes philippinarum*. *The Bulletin of the Russian Far East Malacological Society, 28*(1), 124–137.

Kürzel, K., Kaiser, S., Lörz, A.-N., Rossel, S., Paulus, E., Peters, J., Schwentner, M., Martinez Arbizu, P., Coleman, C. O., Svavarsson, J., & Brix, S. (2022). Correct species identification and its implications for conservation using Haploniscidae (Crustacea, Isopoda) in Icelandic waters as a proxy. *Frontiers in Marine Science, 8*, 795196. doi:10.3389/fmars.2021.795196.

Lam, V. W. W. (1980). Shell form and diagnostic differences in the structure of the siphons and ciliary currents of the ctenidia in coastal species of the Tapetinae (Bivalvia: Veneracea) in Hong Kong. In B. Morton (Ed.), *Proceedings, first international workshop on the malacofauna of Hong Kong and Southern China* (pp. 11–31). University Press.

Leskova, S. E., & Rybnikova, I. G. (2011). Juveniles breeding of *Ruditapes philippinarum* in plant conditions. *Nauchno-Trudy Dal'nevostochnogo Gosudarstvennogo Tekhnologicheskogo Rybokhozyaystvennogo Universiteta, 23*, 17–20.

Liu, R.-Y. (2013). Status of marine biodiversity of the China seas. *PLoS One, 8*, e50719. doi:10.1371/journal.pone.0050719.

Liu, Y., Ma, P., Zhang, Z., Li, C., Hu, L., & Wang, H. (2022). Mitogenomic analyses shed new light on phylogenetic relationships within the genus *Ruditapes* (Bivalvia: Veneridae). *Journal of Molluscan Studies, 88*(1). doi:10.1093/mollus/eyac001.

Lutaenko, K. A. (2006). Bivalve mollusks of Ussuriysky Bay (Sea of Japan). *The Bulletin of the Russian Far East Malacological Society, 10*(2), 46–66.

Lutaenko, K. A. (2016). Biodiversity of bivalve mollusks in the western South China Sea: An overview. In A. Adrianov, & K. A. Lutaenko (Eds.), *Biodiversity of the western part of the South China Sea* (pp. 315–384). Dalnauka.

Lutaenko, K. A., & Noseworthy, R. G. (2019). Contribution to the knowledge of the marine bivalve mollusk fauna of Gangwon Province, Korea. *Journal of Asia-Pacific Biodiversity, 12*(1), 14–44. doi:10.1016/j.japb.2018.07.009.

Lutaenko, K. A., Noseworthy, R. G., & Choi, K.-S. (2021). Marine bivalve mollusks of Jeju Island (Korea). Part 2. *The Korean Journal of Malacology., 37*(4), 197–291. doi:10.9710/kjm.2021.37.4.197.

Lutaenko, K. A., & Volvenko, I. E. (2013). On the fauna of bivalve mollusks of Hong Kong (South China Sea). *Bulletin of the Russian Far East Malacological Society, 17*, 79–141 [In Russian with English abstract].

Lynge, H. (1909). The Danish Expedition to Siam, 1899-1900. IV. Marine Lamellibranchiata. Royal Danish Academy of Science and Letters. Kongelige Danske Videnskabernes Selskabs Skrifter, 7. Raekke. *Naturvidenskabelig og Mathematisk Afdeling, Kjøbenhavn, 5*(3), 97–299.

Merrill, E. D. (1926). Hugh Cuming's letters to Sir William J Hooker. *Philippine Journal of Science, 30*, 153–184.

Kuroda, T., Habe, T., Oyama, K., (1971). Book review: The seashells of Sagami Bay, p. 741 (in Japanese) +489 pp. (in English), 121 pls., Maruzen, Tokyo. Collected by his majesty the Emperor of Japan. B L, Imperial Household (Ed.) Maruzan Co., Ltd.

MolluscaBase n.d. Ruditapes philippinarum. Retrieved September 14, 2023, from https://www.molluscabase.org/aphia.php?p=taxdetails&id=231750.

NEMESIS n.d. *Ruditapes philippinarum.* Retrieved September 14, 2023, from https://invasions.si.edu/nemesis/species_summary/81477#:⟨:text=General%3A,The%20sexes%20are%20separate.

Ng, P. K. L., & Tan, K. S. (2000). The state of marine biodiversity in the South China Sea. *Raffles Bulletin of Zoology, 48*(8), 3–7.

ORBIS, n.d. Ruditapes philippinarum. Retrieved September 14, 2023, from https://obis.org/taxon/231750.

Park, K.-I., & Choi, K.-S. (2016). Country report from South Korea. *Bulletin of Japan Fisheries Research and Education Agency, 42*, 119–120.

Ponurovsky, S. K., & Selin, N. I. (1988). Distribution, population structure, and growth of the bivalve mollusk *Ruditapes philippinarum* in Vostok Bay, Sea of Japan. *Soviet Journal of Marine Biology, 14*(1), 11–15.117 1988.

Ponurovsky, S. K., & Yakovlev, Y. M. (1992). The reproductive biology of the Japanese littleneck, *Tapes philippinarum* (A. Adams and Reeve, 1850) (Bivalvia: Veneridae). *Journal of Shellfish Research, 11*(2), 256–277.

Poppe, G. T. (2011). *Philippine marine mollusks. Vol. 4 (Bivalvia Part 2, Scaphopoda, Polyplacophora, Cephalopoda & Addenda)* (p. 676). ConchBooks.

Poutiers, J. M. (1998). Bivalves. In K. E. Carpenter, & V. H. Niem (Eds.), *The living marine resources of the Western Central Pacific. Volume 1. Seaweeds, corals, bivalves, and gastropods. FAO species identification guide for fishery purposes* (pp. 123–362). FAO.

Qi, Q. (1987). The life history of the clam (*Ruditapes philippinarum*). *Journal of Fisheries of China, 11*(2), 111–119.

Rakov, V. A. (1986). Biological bases of cultivation of the Manila clam in Peter the Great Bay. In *Fifth congress of the all-union hydrobiological society: Abstracts of reports* (pp. 114–116). Volzhskaya Kommuna. 1.

Ramírez, F., Sbragaglia, V., Soacha, K., Coll, M., & Piera, J. (2022). Challenges for marine ecological assessments: Completeness of findable, accessible, interoperable, and reusable biodiversity data in European Seas. *Frontiers in Marine Science*, 802235. doi:10.3389/fmars.2021.802235.

Sanpanich, K. (2011). Marine bivalves occurring on the east coast of the Gulf of Thailand. *ScienceAsia, 37*(3), 195–204. doi:10.2306/scienceasia1513-1874.2011.37.195.

Seebens, H., Blackburn, T. M., Dyer, E. E., Genovesi, P., Hulme, P. E., Jeschke, J. M., Pagad, S., Pyšek, P., Winter, M., Arianoutsou, M., Bacher, S., Blasius, B., Brundu, G., Capinha, C., Celesti-Grapow, L., Dawson, W., Dullinger, S., Fuentes, N., Jäger, H., … Kartesz, J. (2017). No saturation in the accumulation of alien species worldwide. *Nature Communications, 8*, 14435 (And supplementary material). doi:10.1038/ncomms14435.

Shuichi, S., Yoshida, T., Tsurushima, D., Kamohara, S., Tanaka, Y., & Suzuki, T. (2021). A comparative examination of individual growth models, as the method to assess the food environment in relation to the resource level of Manila clam Ruditapes philippinarum. *Journal of Advanced Marine Science and Technology Society, 26*(2), 1–12.

Shindo, A., & Hibino, M. (2025). Review of the decline in Asari (Manila) clam (*Ruditapes philippinarum*) stocks and contributing factors in Mikawa Bay, Japan. *Estuarine, Coastal and Shelf Science, 317*, 109207. doi:10.1016/j.ecss.2025.109207.

Skarlato, O. A., 1981. Dvustvorchatye mollyuski umerennykh vod severo-zapadnoi chasti Tikhogo okeana (Bivalve Mollusks in Temperate Waters of the Northwestern Pacific

Ocean), Nauka, [Opredeliteli po faune SSSR (Keys to the USSR fauna), Zool. Inst. Akad Nauk SSSR, vol. 126].

Silina, A. V., & Popov, A. M. (1989). Investigation of linear growth of the bivalve mollusk *Ruditapes philippinarum* from Peter the Great Bay (Sea of Japan) by the shell structure. *Soviet Journal of Marine Biology, 15*(4), 265–271.

Spalding, M. D., Fox, H., Allen, G., Davidson, N. C., Ferdana, Z. A., Finlayson, M., Halpern, B., Jorge, M. A., Lombana, A., Lourie, S. A., Martin, K. D., McManus, E., Molnar, J., Recchia, C., & Robertson, J. (2007). Marine ecoregions of the world: A bioregionalization of coastal and shelf areas. *Bioscience, 57*(7), 574–583. doi:10.1641/B570707.

Springsteen, F. J., & Leobrera, F. M. (1986). *Shells of the Philippines*. Carfel Seashell Museum.

Suarez, A. A., & Raven, H. (2020). First records of naturalised *Ruditapes philippinarum* (Adams & Reeve, 1850) (Bivalvia: Veneridae) in Asturias (NW Spain), keys for identification, and evidence of hybridisation with native *Ruditapes decussatus* (Linnaeus, 1758) Elona. *Revista de Malacologia iderica, 2*, 67–83.

Sugaya, T. (2006). Organization and development of stock enhancement in Japan. In J. H. Primavera, E. T. Quinitio, & M. R. R. Eguia (Eds.), *Proceedings of the regional technical consultation on stock enhancement for threatened species of international concern* (pp. 91–101). Aquaculture Department, Southeast Asian Fisheries Development Center.

Tang, Y., Huo, Z., Liu, Y., Wang, Y., Zuo, L., Fang, L., Zhao, W., Tan, Y., & Yan, X. (2022). Full mitochondrial genomes reveal species differences between the venerid clams *Ruditapes philippinarum* and *R. variegatus*. *Genes, 13*(11), 2157. doi:10.3390/genes13112157.

Teh, L. S. L., Cashion, T., Alava Saltos, J. J., Cheung, W. W. L., & Sumaila, U. R. (2019). Status, trends, and the future of fisheries in the East and South China Seas. In *Fisheries Centre research reports: 27* (p. 101). Institute for the Oceans and Fisheries, The University of British Columbia.

Thang, H. D. (1999). Aquaculture of shellfish in Vietnam. In *On-site training on mangrove-friendly aquaculture* (pp. 199–203). SEAFDEC Aquaculture Department.

Toba, M., Kobayashi, Y., Kakino, J., Yamakawa, H., Ishii, R., & Okamoto, R. (2016). Stocks and fisheries of Asari in Japan. *Bulletin of the Japanese Fishery Research and Education Agency, 42*, 9–21.

Toba, M., Kakino, J., Tada, K., Kobayashi, Y., & Tsuchie, H. (2020). Production of Asari (Manila) clams, *Ruditapes philippinarum*, during the period of harvest decrease in the 2000s in the Banzu tidal area, Tokyo Bay. *Aquatic Living Resources, 33*, 14 2020.

Toba, M. (2015). Stocks and fisheries of Asari in Japan. *The third international symposium on Manila (Asari) Clam* https://nria.fra.affrc.go.jp/asari/booklet.pdf.

Toba, M., Kobayashi, Y., Hayashi, T., & Kagami, Y. (2025). Possible processes responsible for the decline in the stock of Manila clam based on long-term observations in Banzu tidal flat. *Estuarine Coast Shelf Science, 318*, 109232.

Tsutsumi, H. (2006). Critical events in the Ariake Bay ecosystem: Clam population collapse, red tides, and hypoxic bottom water. *Plankton and Benthos Research, 1*(1), 3–25. doi:10.3800/pbr.1.3.

Tu, D., Anh, L., Anh, L., Sasaki, T., & Tuan, T. (2019). Sample checklist of Gastropoda and Bivalvia in Cham Islands, Vietnam. *Biodiversity Data Journal, 7*(7). doi:10.3897/bdj.7.e32930.

Vecchione, M., Mickevich, M. F., Fauchald, K., Collette, B. B., Williams, A. B., Munroe, T. A., & Young, R. E. (2000). Importance of assessing taxonomic adequacy in determining fishing effects on marine biodiversity. *ICES Journal of Marine Science, 3*, 677–681. doi:10.1006/jmsc.2000.0707 10543139.

Watanabe, S., Katayama, S., Kodama, M., Cho, N., Nakata, K., & Fukuda, M. (2009). Small-scale variation in feeding environments for the Manila clam *Ruditapes philippinarum* in a tidal flat in Tokyo Bay. *Fisheries Science, 75*(4), 937–945. doi:10.1007/s12562-009-0113-1.

Woodward, S. P. (1890). *A manual of the Mollusca* (fourth edition). Crosby, Lockwood and Co.

WoRMS, n.d. Ruditapes philippinarum (A. Adams & Reeve, 1850) World Register of Marine Species. Retrieved September 14, 2023, from https://www.marinespecies.org/aphia.php?p=taxdetails&id=231750

Yu, Y., Zhang, H.-R., Jin, J., & Wang, Y. (2019). Trends of sea surface temperature and sea surface temperature fronts in the South China Sea during 2003–2017. *Acta Oceanologica Sinica, 38*(4), 106–115. doi:10.1007/s13131-019-1416-4.

Zenetos, A., Gofas, S., Verlaque, M., Çinar, M. E., García Raso, J. E., Bianchi, C. N., Morri, C., Azzurro, E., Bilecenoglu, M., Foglia, C., Sikou, I., Violanti, D., Sfriso, A., San-Martin, G., Giangrande, A., Katagan, T., Ballesteros, E., Ramos Espla, A. A., Mastrototaro, F., … Ocaña Vicente, O. (2010). Alien species in the Mediterranean MSFD areas by 2010. Part I. Spatial distribution. *Mediterranean Marine Science, 11*(2), 381–493.

Zheng, S., Zhang, T., Tu, K., Li, L., Liu, Z., Wu, B., Zhou, L., & Sun, X. (2023). Population genetics of Manila Clam (*Ruditapes philippinarum*) in China inferred from microsatellite markers. *Biology, 12*(4), 557. doi:10.3390/biology12040557.

Zvonareva, S. S., Mekhova, E. S., Hà, V. T., & Kantor, Y. I. (2019). Checklist of bivalve molluscs in mangroves of Khánh Hòa Province, Vietnam. *Molluscan Research, 39*(4), 296–312. doi:10.1080/13235818.2019.1638567.

CHAPTER 3

Globalization: Process and sequence

3.1 Globalization and the Manila clam

As commonly defined, globalization is the process by which interactions between human societies and the effects of such interactions occur across global distances with increasing regularity and intensity (Jacobs et al., 2003). The concept has now become widely applied from economic, political, cultural, technological, and environmental perspectives. All of these dimensions of globalization are relevant in the case of the Manila clam. The species' global distribution is a direct consequence of human cultural and economic interaction (this chapter), the technology of Manila clam aquaculture has developed and spread with the clam itself (Part C, Aquaculture and Fisheries), and the policy and politics of the species lies at the intersections of economics, socioeconomics, and food production on the one hand, and conservation, especially relating to non-native and invasive species on the other (Part D, Naturalisation and Invasion).

The environmental dimension of globalization includes changes in distribution patterns, facilitated directly or indirectly by human activity. The Manila clam is just one of many examples. By 2004, globalization in marine ecosystems had resulted in over 800 non-native species introductions to European waters alone, the majority being benthic animals (Streftaris et al., 2005). Meanwhile, the science of the Manila clam has also become globalized. Originally developed in East Asia, it is now an international, often collaborative, effort involving many countries across the world (Humphreys et al., 2025). Among these research programs, the question of the ecological and economic consequences of climate change is a matter of increasing focus in both indigenous and nonindigenous regions (Chapter 23).

Since there is evidence of Manila clam fishing and trading at least since the Japanese Jomon culture 15,000 years ago (Chapter 1), early anthropogenic introductions may have been responsible for dispersal into some east Asian areas where the clam is now regarded as indigenous. This chapter provides an account of the wider global introduction of *R. philippinarum*, identifying the mechanisms and dates of introduction. Aspects of the ecology of the naturalization process are discussed in Chapter 20.

The Manila clam can be described as a globalized species, not only through its contribution to the globalization of marine ecosystems, but more particularly due to the unusually wide global extent of its spread as an individual species, along with its current status as an internationally traded commodity. Moreover, the phases

The Manila Clam *Ruditapes philippinarum*: Science and Significance of a Globalised Species.
DOI: https://doi.org/10.1016/B978-0-443-36704-5.00007-9
Copyright © 2026 Elsevier Inc. All rights are reserved, including those for text and data mining, AI training, and similar technologies.

of the spread of the species have coincided with major general movements in the history of social and economic globalization. The spread of the Manila clam has occurred mainly in two anthropogenic phases: The first across the Pacific from around 1880, following the relaxation of the previously isolationist "Sakoku" policy of the Tokugawa shogunate in Japan. This resulted in an eastward movement of Japanese migrants to the Hawaiian Islands and the US West Coast. The second phase, from the 1970s, was to a significant extent driven by problems with European native bivalve stock collapses, often due to the arrival of new diseases and/or overfishing resulting from growing domestic and international markets.

Although many anthropogenic introductions are well documented, caution must be taken when seeking to determine the exact years of first introduction, especially during the first phase. Initial introductions made without government involvement can be virtually impossible to accurately date and may only be documented much later, often in anecdotal accounts. Questions over accurate identification and contemporaneous record-keeping must also be considered. As discussed in Chapter 2, the Manila clam has a history of identification and recording errors, which began with its first formal description in 1850. Even when the date of first anthropogenic introduction is known, the time of naturalization may not be. In fact, it is occasionally impossible to be sure whether "wild" clams are properly naturalized or remain dependent on a seeded aquaculture population in the vicinity.

Nevertheless, despite some conflicting literature and occasional temporal uncertainties, it is possible to track the historic spread of the species with reasonable confidence from its first arrival in Hawaii to its current global distribution. The account below, along with information summarized in Table 3.1 , ranges from secure documented arrivals to estimates based on the best available information.

The UN Convention on Biological Diversity (CBD) has published categories and pathways of introduction. These relate to efforts to prevent or manage the spread of "invasive alien species" that "threaten ecosystems, habitats and species" (Convention on Biological Diversity, 2014). While it is argued in Chapter 22 that the Manila clam has only rarely been claimed as "invasive" in this sense, the CBD definitions provide a useful framework. Based largely on Hulme et al. (2008), six pathway mechanisms are grouped into three broad categories as follows:

1. Relating to transport of a commodity, including (1) *release*, (2) *escape*, and (3) *contaminant*.
2. Relating to a transport vector, including only (4) *stowaway*.
3. Relating to natural spread from a neighboring region, including (5) *corridor* and (6) *unaided* (Convention on Biological Diversity, 2014)

"Unaided" refers to the natural dispersal of a species originally introduced by means of one of the first five pathways. Corridor refers to a natural spread facilitated through infrastructure, such as shipping canals. Of these six mechanisms, "(1) Release" is regarded as intentional introduction, while all others are unintentional.

Pathways 1, 2, 3, and 6 are all known to have facilitated the current global distribution of the Manila clam. Often the pathway is well documented but for such

Table 3.1 Worldwide distribution of *Ruditapes philippinarum* by country and sea area.

Country	Coastal sea area	First arrival	Notes	Selected sources
Indigenous				
East Asia				
Russia	Sea of Okhotsk Sea of Japan	Indigenous	North to Amur Estuary	Kolpakov et al. (2024)
Japan	Sea of Japan & western Pacific	Indigenous	Throughout	Toba et al. (2016)
Korea	West Sea (= Yellow Sea)	Indigenous	Also, the East Coast and Jeju Island	Park and Choi (2016)
China	Yellow Sea; East China Sea	Indigenous	Also, the northern part of the South China Sea	Zheng et al. (2023)
Non-native Phase 1				
USA (Hawaii)	North Pacific	1880 1918	1880 report anecdotal	Bryan (1915, 1919), Brock (1921)
Canada	Salish Sea	1936	Now north to Hecate Strait	Quayle (1938, 1964), Bourne (1997)
USA (West coast)	Samish Bay, Washington	1924	And elsewhere around Puget Sound	Kincaid (1949)
	Netarts Bay, Oregon	Before 1970		Carlton (1971)
	San Francisco Bay, California	1930 or 1946	Also, other California sites, e.g., Elkorn Slough, etc.	Bonnot (1935), Cohen and Carlton (1995)

(continued on next page)

Table 3.1 Worldwide distribution of *Ruditapes philippinarum* by country and sea area—cont'd

Country	Coastal sea area	First arrival	Notes	Selected sources
Non-native Phase 2				
France	Bay of Biscay, La Manche	1972		Flassch and Leborgne (1992)
UK	English Channel	1980	Now, also the southern North Sea coasts	Humphreys et al. (2015)
Ireland	Atlantic coast	1982	Now, also the Irish Sea	Xie and Burnell (1994), Humphreys (2024)
Spain	Atlantic coast	1983	Also, the Mediterranean coast	Pérez et al. (1985)
Italy	Adriatic lagoons	1983	Introduced later in west coast and in Sardinia	Breber (1985), Cannas, Trentadue, & Manca (2003)
Portugal	Bay of Biscay Atlantic coast	1984		Chiesa et al. (2016)
Norway	NE Atlantic	1987	No evidence of naturalization	Mortensen and Strand (2000)

(continued on next page)

Slovenia	Adriatic lagoons	1993		Lipej et al. (2012)
Netherlands	Wadden Sea	1995		Reise et al. (2024), Bos and Verheyen (2024)
Tunisia	Mediterranean Sea	1996	No evidence of naturalization	Souissi et al. (2005)
Turkey	Sea of Marmara	2004	Also, the northern Aegean Sea	Albayrak (2005)
Germany	Wadden Sea	2011		Reise et al. (2024)
Belgium	Wadden Sea	2011		Reise et al. (2024)
Croatia	Adriatic	2013	Zelena Laguna	Nerlovic et al. (2016)
USA (East coast)	Boston Harbour (NW Pacific)	2022	And Cape Cod area	Carlton (personal communication)
Sweden	Skagerrak	2024		Greeve et al. 2025

Dates refer to the earliest report of the introduction of subsequently naturalized populations, or the first discovery of wild naturalized populations (Some reported countries are omitted for reasons explained in 3.5, 3.6 and Chapter 2).Additional insufficiently verified reports include the Philippine type locality, along with Vietnam, Cambodia, Thailand, Indonesia, Malaysia, Bangladesh, India, Senegal, French Polynesia, and Fiji. Many of these locations may represent unsuccessful introduction attempts or misidentifications (see text).

a high value fishery species it is sometimes impossible to be sure whether spread along a coast has been natural (6. Unaided) or a consequence of informal introduction (1. Release) by fishers seeking to initiate a valuable stock in a new location (Humphreys et al., 2015), or though submerged (suspended) pretransportation storage of a spawning catch in an estuary other than that in which it was caught. The possibility of introduction via transport vectors (4. Stowaway), such as in shipping ballast tanks, is conceivable but speculative.

3.2 First phase: Hawaii and North America

In the second half of the eighteenth century, Japan's policy of isolation was superseded by its desire to develop a modern economy. Japanese migrations eastward across the Pacific appear to have commenced in 1877, and from that time and especially during the first decade of the 20th century, many migrants, including people from Japanese fishing communities (Fukawa & Fukawa, 2009), found employment in the growing North American oyster industry (Fry, 2011). A period of mass migration coincided with the development by 1870 of efficient steam-powered screw-propellor ships, which made largescale transocean shipping economically viable. Such ships have been described as a major factor in the first wave of trade globalization (Pascali, 2017).

3.2.1 Hawaiian Islands

The first step in the globalization of *R. philippinarum* occurred across the 6400 km of ocean from Japan to Hawaii. Bryan (1919) reported an account from an apparently reliable local fisher that the species was plentiful for several years around 1880 and 1890 on mudflats at Ewa, near Pearl Harbor, on Oahu, but their subsequent disappearance was attributed to a period of heavy rainfall and flooding. Thaanum (1921) reported two plantings by a Japanese individual on mudflats at Moanalua, Oahu, the second of which survived and by 1921 was abundant enough to be gathered and put on the market. He also reported that Manila clams were "frequently imported from Japan to Honolulu by the barrel for sale among the Japanese."

Carlton and Eldredge (2009) have detailed the circumstances of these and subsequent introductions. Around 1919, live clams were brought from Japan by a Japanese Hawaiian resident for consumption by the Japanese community, and several attempts to plant live clams were made. Yap (1977) describes how, in 1920, ten barrels of Manila clams were planted in various localities, including Pearl Harbor, where they grew well but subsequently disappeared, except in Kaneohe Bay, where an extensive naturalized population developed. Exploitation by a reported 41,000 clam diggers by 1967 led to the closure of harvesting and a detailed study of the population (Yap, 1977).

3.2.2 West Coast USA

Around the same period, trade between East Asia and the West Coast of America boomed. Between 1870 and the 1930s, millions of tons of living oysters, including the

Pacific or Japanese oyster *Magallana gigas*, were introduced from Japan to the Pacific coast of North America. With these oysters, an estimated 27 species of non-native molluscs inadvertently arrived in bays and estuaries of geologically recent postglacial age, with consequently relatively low indigenous diversity. Carlton (1992) has argued that such environments were relatively susceptible to the establishment of the new arrivals when transported alive and mixed among the vast numbers of imported oysters (see Chapter 20).Carlton's (1971) account of the introduction of new species to the Pacific North American coast remains a significant review of the period before the 1970s, including the spread of the Manila clam in both Canada and the USA (Carlton, 1971, pp. 502–514).

It has been supposed that the first US introductions were unintentional (e.g., Quayle, 1938) (CBD pathway [3] "Contaminant"). Moreover, shipments of oysters from Japan could be highly contaminated. Kincaid (1949), for example, reported that when opening several boxes of seed oysters, he found them "swarming with 10 different species of living molluscs." Nevertheless, in light of the apparently intentional introduction of Japanese Manila clams to Hawaii, along with the fact that many Japanese migrants to the North American mainland were also from fishing communities (Fukawa & Fukawa, 2009), it would seem unwise to rule out the possibility of intentional introduction and planting (CBD Pathway [1] "Release"), following that occurring in Hawaii.

In any event, Kincaid (1949) reported that in 1924 he found "*Paphia philippinarum*" in Samish Bay (near Puget Sound, Washington), amongst a spring laying of oyster seed shipped from Japan. By 1949, he was able to note that it had subsequently spread, being first observed just along the coast, south of Puget Sound in Willapa Bay, and had become a "conspicuous element in the catch of local clam diggers." Kincaid's report generated some debate about the identity of the specimens, with the names *Tapes philippinarum* and *Protothaca semidecussata* both proposed as possibilities. All three names can today be regarded as synonyms of *R. philippinarum,* and Kincaid's 1924 finding appears to be the first scientific report of the species on the United States mainland.

In March 1930, a shipment of 150 boxes of Japanese seed oysters was inspected at Elkorn Slough just north of Monterey, California, and found to include *Paphia (Ruditapes) japonica (=Ruditapes philippinarum)* among 22 species of exotic Japanese mollusks. This report of *R. philippinarum's* arrival in California was made in the context of already established oyster businesses further north in San Francisco and nearby Tomales Bay, both of which received supplies of seed from Japan. The expansion of this industry to other Californian bays considered free of undesirable imported species, especially the "dreaded Japanese drill (*Tritonalia japonica*)," provoked the Elkorn inspection (Bonnot, 1935).

In Oregon state, located between California and Washington, Carlton (1971) reported that although a few specimens had been reported elsewhere, successful establishment had only occurred in Netarts Bay. Robinson (1997) noted growing Manila clam landings in the 1970s as local markets shifted to live "steamer" clams and reported that laboratory-produced and imported adult Manila clams had been introduced into several Oregon estuaries from about 1982.

In California, the first record in San Francisco Bay dates from 1946 (Cohen & Carlton, 1995), since when the Manila clam has been increasing its southerly range and is now to be found in Mission Bay, San Diego, California (32°46′N), just north of the Mexican border (Talley et al., 2015). In Mexico, the species was apparently reported in the Bahia de San Quintin, Baja California, in 1987 and in 1998, the introduction of Korean clams into Mexican waters was considered for fattening before shipment to the United States (Cohen, 2005). However, on the basis of an evaluation of the health status of potential seed clams, which found high levels of *Perkinsus* infection (Chapter 13), the idea was dropped, and no importation of clams occurred (Elston et al., 2004).

The much more recent arrival of Manila clams on the US east coast is addressed in Section 3.4.1 below

3.2.3 Canada

In Canada, Quayle (1938) reported finding a few shells of a new species, "*Paphia bifurcata*," at Ladysmith on the north coast of Vancouver Island, British Columbia. This subsequently was recognized as *R. philippinarum.* Ladysmith Harbor lies adjacent to the Strait of Georgia, which separates Vancouver Island from the Canadian mainland. The strait has been described as a classic north-east Pacific estuarine system whose waters, fed significantly by the Fraser River, are relatively warm and somewhat brackish. Ladysmith Harbor has been reported to have an annual temperature range of 5.5°C–20.6°C, a markedly warmer summer maximum than the eastern end of the strait (Levings & Kaise, 2002).

By the 1960s, the species had reached Queen Charlotte Strait on the NE end of Vancouver Island and further north in the 1970s. Gillespie et al. (2012) have provided sources detailing its spread during the 1950s to 1980s, reaching around both east and west sides of Vancouver Island and beyond 50°N (Bourne & Cawdell, 1992).

In 1988, the Heiltsuk First Nation requested that Fisheries and Oceans Canada formally establish a Manila clam fishery in their territory north of Vancouver Island. Subsequent surveys revealed harvestable densities of Manila clams on beaches in the Waglisla area (latitude 52°N) (Bourne & Cawdell, 1992; Bourne et al., 1994).

By the 1990's, various populations could be found extending up British Columbia's "Inside Passage" to the Seaforth Channel at approximately 52°N (Bourne et al., 1994), and by 2006 the species had been found at Haida Gwaii (formerly the Queen Charlotte Islands), separated from the mainland by the Hecate Strait at around 53°N (Gillespie, 2007).

In the context of increases in water temperature and a reduction of sea ice, a commissioned assessment of the risk of non-native species in the Bering Sea put *R. philippinarum* in the top 10 of a list of "species of concern" (Reimer et al., 2017). However, despite various bivalve fisheries and farms on the Alaskan coast of the United States, including both clams and oysters (Foster, 1997), there appear to be no records of the Manila clam there, suggesting the Hecate Strait to be the most northerly eastern Pacific population at that time, a limit corresponding more or less to the border between Canada and the United States Alaskan territory.

3.3 Second phase: Northwest Europe and the Mediterranean basin

The introduction of the Manila clam to Europe constituted a second, distinct, and much later phase of its global dispersal. Although relating to the transport and release of a commodity (CBD Category 1), in this case, the Manila clam itself, the major European introductions differed significantly from the much earlier North American arrivals in terms of both context and motive. By the beginning of the nineteenth century, several northwest European countries had a long history of bivalve exploitation, along with a propensity to import new species in response to declines in native stocks.

For example, in Britain in the 1830s, a boom in native oyster (*Ostrea edulis*) production was followed by a crash, which stimulated efforts to restore native stocks through fishing controls and breeding projects. The failure of these efforts stimulated the procurement of live "Tagus" (Portuguese) oysters, *Crassostrea angulata,* later in the century, but despite these efforts, a decline in consumption has persisted to the present day (Humphreys 2022; Humphreys et al, 2014). A similar narrative played out in France, where the American Hard Shell clam or Quahog (*Mercenaria mercenaria*) was also repeatedly imported from 1861 (Flassch & Leborgne 1992). Continuing demand along with continuing mortality problems led to increasing attention to oyster hatchery techniques, building on work in the United States (Loosanoff & Davis, 1963) and backed up by laboratory studies. In both countries, further species diversification involved the Pacific oyster *(Magallana gigas),* whose introduction and culture required the development of techniques that provided a useful precursor to the later introduction of the Manila clam.

In contrast to the United Kingdom, France maintained a much stronger tradition of bivalve consumption and investment, with state-level organization and strategy and better-established infrastructure, including hatcheries and leased shellfish grounds (Goulletquer, 1997). Consequently, the European introduction of the Manila clam commenced in France.

3.3.1 France

The first European Manila clams were ordered as spat on November 8, 1972, arriving in France from Puget Sound on May 15, 1973. This was followed by two orders of adult clams in 1974. These were part of a strategic diversification of bivalve shellfish production (Flassch & Leborgne, 1992) in direct response to a collapse of Portuguese oyster (*Crassostrea angulata*) production, due to disease-related mass mortalities (Montaudouin, 2016). Aware of the success of the Manila clam and, in particular, the development of clam production in Puget Sound, USA, the French government concluded that the species was also compatible with environmental and climatic conditions in extensive areas of the French coast.

France was quick to invest in the potential of the Manila clam. From 1975, work on Manila clam culture was considerably facilitated by the established oyster hatchery SATMAR (Société Atlantique de Mariculture), on which was centered the controlled production of Manila clam seed. Following this, four strategic stages of development followed, including culture feasibility from 1975 to 1978 and a "national clam programme" from 1981 to 1984 involving the laying of 6.8 million spat at Arcachon and north to Finisterre. These produced 17 tons of clams of value 654,000F, thereby demonstrating the commercial viability of the species in French waters (Flassch, 1988). Growth comparisons with the native clam *Venerupis decussatus* demonstrated better commercial performance.

Between 1980 and 1986, oyster farmers in Arcachon Bay started to culture Manila clams in the intertidal zone, but high stock mortalities led to an early and serious decline (Caill-Milly et al., 2025; Dang et al., 2010). However, difficulties with cultured production, not least disease-related mortality, did not stop the species from naturalizing in shellfish areas. An apparently unaided natural spread (CBD Category C6) soon resulted in commercial fishing operations exceeding aquaculture production by a factor of three, and French production grew from 2 to 550 tons by 1988 (Flassch & Leborgne, 1992). Arcachon Bay remains a center for Manila clam fishing (see Chapter 19), as does the Gulf of Morbihan (and on the Mediterranean coast, the Berre Lagoon in particular)

3.3.2 United Kingdom

The Manila clam was first imported to the UK in 1980 as a consignment from the US Pacific coast. After quarantine procedures and experimental field trials at a government fisheries laboratory in Conwy, north Wales, the species was released to commercial growers. Between 1983 and 2004, official introductions occurred in 19 sites, mainly along the south and southeast coasts of England but ranging also as far north as Morecambe Bay in NW England and Argyll on the west coast of Scotland. By 2010, naturalized populations were known to be established in 11 estuaries on central southern and southeastern coasts, including some with no history of licensed introduction but a strong tradition of inshore shellfisheries (Humphreys et al., 2015).

The first naturalized population was reported at Poole Harbour, a large microtidal lagoon-like feature of the central south coast, which was one of the early government-facilitated introductions for aquaculture (Humphreys, 2005, 2010). Population densities have remained relatively low compared to warmer, more southerly European lagoons; the clam's growth rate is superior to that of the native species, and while the clam is no longer farmed in Poole, it is now the basis of a sustainably managed wild fishery with Marine Stewardship Council accreditation (Chapter 19). Since 2010, the Manila clam has extended its distribution but at a slow rate, probably involving both anthropogenic and natural planktonic dispersal (Herbert et al, 2011; Humphreys et al., 2015). The UK introduction provides a good illustration of the political tensions between conservation and economic interests. It was resisted by a strong conservation movement which campaigned vigorously, although unsuccessfully, against the introduction of this non-native species (Humphreys, 2025; Chapter 24).

Since the French and British introductions, the spread of the species on European coasts has continued.

3.3.3 Ireland and Northern Ireland

In the Republic of Ireland's (Ireland) Atlantic coast, the species was introduced in 1982 for aquaculture (Xie & Burnell, 1994), with the predominant production areas in the large adjacent northwestern bays of Donegal and Sligo. Drumcliffe Bay, an inlet within Sligo Bay, was the site of Ireland's first commercial clam farm (Drummond et al., 2006). Other areas of cultivation in Ireland include Dungloe and Mulroy bays in County Donegal. Two independent studies of Manila clam gonad development and spawning in Irish waters both concluded that temperatures necessary for successful reproduction and recruitment were marginal: being sufficient to induce spawning, but only so late in the summer season that declining water temperatures were incompatible with larval survival and development (See Chapter 10 on temperature and spawning).

While Xie and Burnell (1994) conducted their work on commercially purchased seed clams suspended near Cork in southern Ireland, Drummond et al. (2006) investigated clams collected from the northwestern bays. Both these studies reported an absence of any field evidence of naturalized populations.

To the east, on the Irish Sea coast of the UK constituent country of Northern Ireland, a large number of Manila clams were found in 2020 on a muddy gravel shore at the northern end of Strangford Lough. Of an area around 150 km^2, this elongated coastal inlet is reportedly the largest in the United Kingdom but connects to the sea through southern narrows only 0.8 km across. A site of oyster aquaculture since the 1970s, the northern region includes 50 km^2 of tidal flats where Manila clams can be found along with native clams and cockles, from the town of Greyabbey north to the Comber River estuary. Here, the Manila clam is now reported as naturalized and spreading rapidly (Anderson, 2022; Humphreys, 2024). Fishing is restricted to hand gathering, with clams available in some Belfast restaurants from Spring. Nevertheless, the Manila clam does not form a major component of the Lough's bivalve shellfisheries (Clements & Service, 2015). Since no records of officially sanctioned introductions are extant, it has been assumed that informal introductions by fishers wishing to establish new fisheries may be responsible for its arrival.

3.3.4 Spain and Portugal

The date of the first arrival of *R. philippinarum* in Spain is difficult to determine. Observing a lack of Spanish controls for the importation of species, Camacho and Cuña (1985) reported the species to have already been introduced to many locations along both Atlantic and Mediterranean coasts in Galicia, Cantabria, Andalucia, and Cataluña. Field work conducted in 1983 also demonstrated that hatchery produced seed thrived in Galician rias. Alonso and Raven (2020) have suggested that unauthorized introductions from France were responsible for the arrival in Spanish waters. Habtemariam et al. (2015) suggested that the motives for introductions related to a combination of increased demand, overfishing, and irregular yields of the native *R. decussatus*.

Since then, the clam is reported as naturalized in both northerly and southerly locations including the Villaviciosa, Carreras and Piedras river estuaries in Andalucia (Alonso & Raven, 2020; Navas et al., 1992); the estuaries of the Eo in Asturias (Cigarria et al., 1997) and in the north, the Bay of Santander, Cantabria (Juanes et al., 2012) and the Rias Baixas (Roman et al., 2023). However, the extent to which these populations are entirely independent of adjacent cultured populations has not always been fully clarified.

As with Spain, there appears to be no formal account of the first or early introductions of *R. Philippinarum* in Portugal, suggesting again that informal introductions may have been made in response to insufficient supply of native Venerid species. According to Chiesa et al. (2016), the first report of the Manila clam in Portugal was in 1984 in the Ria Formosa, Algarve. Today, it is an important commercial species in several additional sites, including the Tagus and Sado estuaries and the coastal lagoons of Aveiro, Obidos, and Albufeira (see Chapter 19). The species first appeared in FAO statistics for Portugal in 2009.

3.3.5 Belgium, Netherlands, Germany, and Denmark

The Wadden Sea (which can be translated as Mud Sea) is a large-scale ecosystem consisting of extensive tidal flats occupying the south-eastern coast of the North Sea adjacent to the Netherlands, Germany, and Denmark. It is a UNESCO World Heritage Site and is claimed as the largest unbroken system of intertidal sand and mud flats in the world. Although most attention to non-natives in the Wadden Sea has been focused on the arrival of the Pacific oyster, in fact, the Manila clam now thrives there (Reise et al., 2024). To the south of the Wadden Sea are the large delta waters of the Scheldt and Rhine, including the extensive Oosterschelde lying just north of the Netherlands/Belgium border (Troost et al., 2023).

In the Netherlands, although Wolf (1995) reported dead shells in 1995, the first live specimens were reported by Faasse and Ligthart (2008) and Titselaar (2008) in the Oosterschelde, from where they spread north into the adjacent Rhine Delta and south into Belgium, first to Zeebrugge (Kerckhof, 2014) and then Oostende (Kerckhof, 2016). Subsequently, imports for aquaculture have rapidly increased the number of wild Manila clams (Bruyne et al., 2013). From the area of the Oosterschelde and Rhine Delta, a dispersal northwards reached Texel by 2017, progressing as far east as Schillig by 2024 (Reise et al., 2024).

Further north on the German coast, the earliest formal record of *R. philippinarum* is dated as 2016 in the North Frisian Island of Sylt, just inside the German border with Denmark. Since 2021, many living specimens have been found from Hallig Hooge to the island of Sylt. Possible vectors include oyster transport followed by natural larval dispersal (Lackschewitz et al., 2022). Riese et al. (2024) report that by 2021, Manila clam sites extended along an almost 100 km stretch of coastline with over 13,000 km^2 of intertidal flats from Sylt, south to the Eiderstedt peninsula. The Wadden Sea distribution is shown inFig. 3.1.

Based on high haplotype diversity (see Chapter 4), Reise et al. (2024) conclude that the various Rhine Delta and Wadden Sea populations are the product of multiple introductions or a single already genetically mixed population.

3.3.6 Norway and Sweden

In 1987, Manila clam broodstock of about 500 individuals was imported from Scotland, UK, to two established hatcheries in Norway. Between 1987 and 1991, 200 million seeds were produced for export to Ireland and Spain. In addition, from 1988 to 1991, Manila clams of length 10–15 mm were released in Norway at six sites from latitudes 59°55′N to 65°22′N. Surveys in 1995 and 1996 revealed clams of size 38–51mm in Vetle Godey (60°05′N) and of size 38–51 mm in Seloy (59°55′N). A single clam of 35 mm and two shells were also found well north of any other reported global distribution at Aunvika (64°16′N). The age of the recovered clams corresponded to the time reported for release, indicating that the specimens were survivors of the original releases.

Mature gonad tissue in the clams from Seloy and Vetle Godoy suggested successful spawning. Both these sites were located south of Bergen and near shallow sun-heated (heliothermic) lagoonal basins (oyster pools) in which summer water temperatures ranged up to 10°C higher than the adjacent fjord waters. Nevertheless, no younger age groups were found (Mortensen & Strand, 2000). In 2003 and 2004, a further 24 Manila clams were recovered from Seloy of 46–62 mm length and estimated to be 15 years old, also suggesting that they were part of the original seeded cohort. Again, no juvenile Manila clams were found by Paillard et al. (2008), suggesting that the species was not yet naturalized in Norwegian waters.

While sea temperature may represent the ultimate limiting factor on northerly distribution, the apparent failure of Manila clams seeded in Norway need not suggest that, at 55°N, populations on the north coast of the Island of Sylt will necessarily mark Europe's northernmost limit. Evidence on the northerly extent of both eastern and western Pacific populations indicates that, given optimal habitat characteristics combined with natural or anthropogenic dispersal, some further northward spread seems likely. Indeed, the species has already reached the Swedish Skagerrak coast in the vicinity of the island of Salto near the Tjarno marine laboratory (Fig. 3.1). This is currently the most northerly European naturalized population (Greeve et al., 2025).

3.3.7 Italy, Slovenia, and Croatia

In the Mediterranean basin, the northern Adriatic coast of Italy includes several large lagoons, notably the Venice Lagoon and lagoons associated with the River Po delta. In 1983, 200,000 Manila clam seeds sourced from an English hatchery were introduced into the southern end of the Venice Lagoon with the support of the Province of

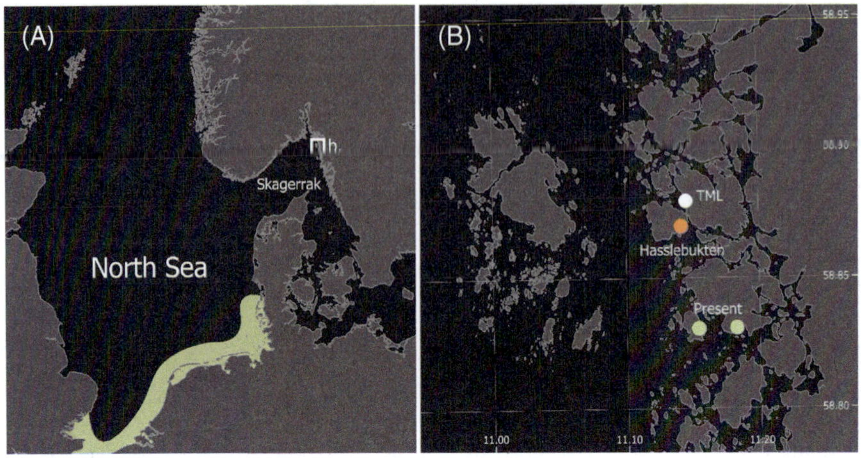

FIGURE 3.1

(A) Overview of the North Sea, including the Wadden Sea coastal area showing the current extent of *R. philippinarum* distribution shaded green. (B) Locations of initial observations of *R. philippinarum* in the Swedish Skagerrak as first reported in 2025 (See also text).

From Greeve et al. (2025).

Venice (Breber, 1985). In 1984, 1.5 million seed clams were distributed between the Venice Lagoon and the Po Delta Caleri Lagoon just to the south. The purpose was to supplement the insufficient production of the native Venerid clam. The success of the early experimental introduction stimulated commercial interest, and by 1986, clam seed from England had been introduced into three additional Po delta lagoons (Vallona, Scardovari, and Goro) and into the Grado-Marano Lagoon in the Gulf of Trieste at the most northerly end of the Adriatic. After 1985, seed was also imported from the US, Spain, and France.

Italian introductions beyond the northern Adriatic lagoons included two sites on the southern and west coasts of the Island of Sardinia in 1985 (Sant'Antioco and Mistras, respectively), and in 1986, the more southerly Adriatic Varano Lagoon, and the Paola lagoon on the Tyrrhenian Sea coast south of Rome (Breber, 2002).

Italian Manila clams naturalized and quickly extended their local distribution to the extent that by 1990, the commercial production from culture beds and wild fisheries had reached 200,000 tons per year, with densities commonly greater than 1000 clams m^{-2} (Breber, 2002).

Since its introduction into the Italian lagoons, the distribution of the Manila clam has further spread "clockwise," reaching further round the Gulf of Trieste to the Bay of Piran in Slovenia, where it was introduced into an abandoned salt basin in 1993 near the Dragonia river estuary, where it is established (Lipej et al., 2012). Further south on the eastern Adriatic, the Manila clam has reached the Zelena Lagoon near Poreč on the Istrian coast of Croatia (Nerlovic et al., 2016).

3.3.8 Turkiye, Tunisia, and Greece

Albayrak et al. (2001) reported three specimens of *R. philippinarum* dredged from a survey of 45 subtidal stations off the Turkish island of Bozcaada in the northern Aegean Sea. Elsewhere, on the Turkish coast, Albayrak (2005) found significant established populations in the adjacent Sea of Marmara at Kucukcekmece and Tuzla, with evidence of ripening and densities up to 11 clams/m^2 at the latter site. Albayrak (2005) has suggested that the Marmara populations may have arrived via ship's ballast water (CBD Pathway 4), from where there has been some natural dispersal into the Aegean on surface water currents flowing from the Sea of Marmara via the Dardanelles. He also notes that these populations may have arrived much earlier and been overlooked due to their similarity with the native *R. decussatus.*

By 2015, *R. philippinarum* had become the focus of intensive commercial harvesting in the Sea of Marmara, although as a wild fishery rather than from cultivated stock. Colakoglu and Palaz (2014) conducted a detailed study of the dynamics of a population in Bandirma Bay on the central south Marmara coast. Here, water temperatures range from around 10°C–25°C, with spawning between May and August after gonad maturation commences in March as water temperatures increase.

Elsewhere in the Mediterranean, the Manila clam has been reported on the African coast of Tunisia. Ostensibly established since 1996 in the southern lagoon of Tunis, it has been reported as rare with restricted distribution and greatly outnumbered by the native *Ruditapes decussatus* (Souissi et al., 2005). Subsequently, it was also reported as established further south in the Gulf of Gabes (Antit et al., 2011). The origin of these Tunisian populations is unknown, but in any event, other authorities have suggested that, due to the very high surface sea temperatures and salinities reaching 40, the Manila clam is not extant in Tunisian waters (Medhioub & Oumaima personal communication).

Finally, informal presence reports include Greece. Although no reliable evidence of this has been found, the species' presence in Turkiye gives some level of plausibility on the basis of possible introductions from other Mediterranean populations. However, as in all such cases, further research is necessary, including genome and population studies.

Zenetos et al. (2010), in a checklist of alien species, reported that *R. philippinarum* has been established in the western, central, Adriatic, and eastern parts of the Mediterranean Sea. Tunisia represents the only report of the species on the African side.

3.4 Additional reported arrivals

3.4.1 East Coast USA

From 2022, reports have been accumulating of *R. philippinarum* in various locations in the New England region of the Eastern USA. These represent the first reports of the species on the coasts of the northwest Atlantic (iNaturalist, n.d.). Initially reported around Boston Harbour, the species now extends from the Salem-Beverley area of

Massachusetts south to Cape Cod in the state of Rhode Island. Backed by external shell photographs, collectively, these reports appear reliable. The species may have been present in the Cape Cod area since the 1980s (Putnam et al., 2026). These reports mark a significant further expansion of the clam's global distribution, with the potential for further dispersal along the eastern US seaboard to New Hampshire and likely beyond, perhaps as far as Chesapeake Bay, where oyster bottom culture is well developed.

3.4.2 Territories associated with France

There are indications of commercially motivated introductions in non-native areas, including Senegal and French Polynesia (see, for example, Poutiers [1998] and various websites including SeaLifeBase and Nemesis). It seems plausible that, following the first European introduction in France, places that maintained close links with France may have received socio-economically motivated introductions, such as the now-independent state of Senegal or French Polynesia, which still falls under French sovereignty. However, no documentary evidence has been found of any naturalized extant populations.

Reports of a population in the former British colony of Fiji might be accounted for similarly, but again, no evidence exists. Moreover, during the author's recent visit to the island, no sign of the species was found.

3.4.3 Other Indo-Pacific reports

Beyond all those discussed above, there have also been assertions of (presumably indigenous) distribution in countries including Brunei, Cambodia, Indonesia, Malaysia, Myanmar, Pakistan, India, Singapore, Sri Lanka, Thailand, Vietnam; Guam; Micronesia; North Mariana Islands and Palau. Chapter 2 has discussed in detail the mechanisms of error, as a consequence of which, presence reports should be considered unsafe. In light of that analysis, it seems likely that many of these other Indo-Pacific reports, especially in warmer seas, represent misidentifications or unsuccessful commercially motivated introduction attempts (Humphreys, 2025).

References

Albayrak, S. (2005). First record of *Tapes philippinarum* (Adams & Reeve, 1850) (Bivalvia: Veneridae) from the Sea of Marmara. *Zoology in the Middle East.*, *35*(1), 108–109. https://doi.org/10.1080/09397140.2005.10638113.

Albayrak, S., Aslan, H., & Balkis, H. (2001). A contribution to the Aegean Sea fauna: *Ruditapes philippinarum* (Adams and Reeve, 1850) (Bivalvia: Veneridae). *Israel Journal of Zoology*, *47*, 299–300.

Alonso, A. A., & Raven, J. G. M. (2020). First records of naturalised *Ruditapes philippinarum* (Adams & Reeve, 1850) (Bivalvia: Veneridae) in Asturias (NW Spain), keys for identification, and evidence of hybridisation with native *Ruditapes decussatus* (Linnaeus, 1758). *Elona, Revista de Malacología Ibérica, 2*, 67–83.

Anderson, R. (2022). Establishment and spread of Manila clam (*Ruditapes philippinarum* (A. Adams & Reeve)) (Mollusca: Veneridae) in Strangford Lough, Co. Down. *Irish Naturalists' Journal, 38*, 48–49.

Antit, M., Gofas, S., Salas, C., & Azzouna, A. (2011). One hundred years after Pinctada: An update on alien mollusca in Tunisia. *Mediterranean Marine Science, 12*(1), 53–73. https://doi.org/10.12681/mms.53.

Bonnot, P. (1935). A recent introduction of exotic species of molluscs into California waters from Japan. *Nautilus, 49*, 1–2.

Bos, O., & Verheyen, S. (2024). Marine biodiversity of the Netherlands: Philippine carpet shell (Ruditapes philippinarum). *Nederlands Soortenregister*. https://www.nederlandsesoorten.nl/linnaeus_ng/app/views/species/nsr_taxon.php?id=174600.

Bourne, N., & Cawdell, G. (1992). Intertidal clam survey of the north coast district of British Columbia - 1990. Canadian technical report of fisheries and aquatic sciences. Biological Sciences Branch, Department of Fisheries and Oceans, Pacific Biological station. 1864; 0706–6457.

Bourne, N. F., Heritage, G. D., & Cawdell, G. 1994. Intertidal Clam surveys of British Columbia, 1991. Canadian technical report of fisheries and aquaculture sciences, No 1972. Biological Sciences Branch, Department of Fisheries and Oceans, Pacific Biological Station. Cat. No. Fs 97-6/1972E ISSN 0706-6457.

Bourne, N. F. (1997). Molluscan fisheries of British Columbia. In C. L. MacKenzie Jr., V. G. Burrell Jr., A. Rosenfield, & W. L. Hobart (Eds.), The history, present condition, and future of the molluscan fisheries of North and Central America and Europe. Vol. 2: Pacific Coast and supplemental topics (pp. 115–130). U.S. Department of Commerce, NOAA Technical Report NMFS 128.

Breber, P. (1985). On-growing of the carpet-shell clam (*Tapes decussatus* (L.)): Two years' experience in Venice Lagoon. *Aquaculture, 44*(1), 51–56. https://doi.org/10.1016/0044-8486(85)90041-9.

Breber, P. (2002). Introduction and acclimatisation of the Pacific Carpet Clam, *Tapes philippinarum*, to Italian waters. In E. Leppäkoski, S. Gollasch, & S. Olenin (Eds.), *Invasive aquatic species of Europe. Distribution, impacts and management* (pp. 120–126). Springer. https://doi.org/10.1007/978-94-015-9956-6_13.

Brock, V. E. (1921). The introduction of aquatic animals into Hawaiian waters. *International Review of Hydrobiology, 45*, 463–480.

Bryan, W. A. (1919). A Hawaiian form of *Tapes philippinarum*. *Nautilus, 32*, 124–125.

Bruyne, R. H., van Leeuwen, S. J., Gmelig Meyling, A. W., & Daan, R. (2013). *Shellfish of the Dutch North Sea region. Ecological atlas of the marine molluscs (Mollusca). (Schelpdieren van het Nederlandse Noordzeegebied. Ecologische atlas van de mariene weekdieren (Mollusca)* (pp. 1–414). Tirion Uitgevers, Utrecht en Stichting Anemoon.

Bryan, W. A. (1915). *Natural history of Hawaii* (p. 596). Honolulu: The Hawaiian Gazette, Ltd.

Caill-Milly, N., Sanchez, F., Lissardy, M., Bru, N. oë, Kermorvant, C., Montaudouin, X., Lapègue, S., Riquet, F., Bouché, L., D'Hardivillé, C. é, Lagarde, F., Chambouvet, A. é, Mayot, N., Dauvin, J.-C., Pezy, J.-P., Basuyaux, O., Guéguen, A., Weiller, Y., Arzul, I., … Carter, C. (2025). Manila clam (*Ruditapes philippinarum*) in France: Fishing activity, governance and present knowledge challenges regarding its biology and ecology. *Estuarine, Coastal and Shelf Science, 317*, 109206. https://doi.org/10.1016/j.ecss.2025.109206.

Camacho, A. P., & Cuña, M. A. (1985). First data on raft culture of Manila clam *Ruditapes philippinarum* in the Ria de Arosa (NW Spain). ICES, C.M.I.F: 43.

Cannas, A., Trentadue, M, & Manca, S (2009). Settlement of Ruditapes philippinarum (Adams and Reeve) in the Olbia Gulf (Sardinia, Italy) in relation with Ruditapes decussatus (L.) (Insediamento di *Ruditapes philippinarum* (Adams & Reeve) nel golfo di Olbia (Sardegna, Italia) e distribuzione in rapporto a *Ruditapes decussatus* (L.)). *Biologia Marina Mediterranea, Bio/07 Ecologia*, 109–129.

Carlton, J. (1992). Introduced marine and estuarine mollusks of North America: An end-of-the-20th-century perspective. *Journal of Shellfish Research, 11*, 489–505.

Carlton, J. T. (1971). *History, biogeography and ecology of the introduced marine and estuarine invertebrates of the Pacific coast of North America. Doctoral Dissertation*. University of California.

Carlton, J. T., & Eldridge, L. G. (2009). Marine bioinvasions of Hawai'i. The introduced and cryptogenic marine and estuarine animals and plants of the Hawaiian archipelago. *Bishop museum bulletins in cultural and environmental studies*: 4. Bishop Museum Press.

Chiesa, S., Chaino, P., & Ruano, F. (2016). Country report from Portugal. *B, 42*, 121–123 *ulletin of the Japanese Fisheries Research and Education Agency*.

Cigarria, J., Rodriguez, C., & Fernandez, J. M. (1997). Impact of Perkinsus sp. on Manila clam *Ruditapes philippinarum* beds. *Diseases of Aquatic Organisms, 29*, 117–120.

Clements, A. J., & Service, M., (2015). Bathymetric and habitat map for Strangford Lough (Special area of conservation and marine conservation zone), Northern Ireland. Report to the Department of the Environment https://doi.org/10.13140/RG.2.1.4377.8008.

Cohen, A. N. 2005–2024 Exotics guide- non-native species of the North American Pacific Coast. https://www.exoticsguide.org/. (Accessed 27.04.2026)

Colakoglu, S., & Palaz, M. (2014). Some population parameters of *Ruditapes philippinarum* (Bivalvia, Veneridae) on the southern coast of the Marmara Sea, Turkey. *Helgoland Marine Research, 68*, 539–548. https://doi.org/10.1007/s10152-014-0410-7.

Cohen, A. N., & Carlton, J. T. (1995). *Nonindigenous aquatic species in a United States estuary: A case study of the biological invasions of the San Francisco Bay and Delta (Technical Report CONN-T-95-002)*. U.S. Fish and Wildlife Service; Connecticut Sea Grant, National Sea Grant College Program https://scispace.com/pdf/nonindigenous-aquatic-species-in-a-united-states-estuary-a-41ldqi144d.pdf.

Convention on Biological Diversity 2014).. *Pathways of introduction of invasive species, their prioritization and management: Note by the Executive Secretary (UNEP/CBD/SBSTTA/18/9/Add.1). Secretariat of the Convention on Biological Diversity*. https://www.cbd.int/doc/meetings/sbstta/sbstta-18/official/sbstta-18-09-add1-en.pdf.

Dang, C., de Montaudouin, X., Gam, M. é, Paroissin, C., Bru, N., & Caill-Milly, N. (2010). The Manila clam population in Arcachon Bay (SW France): Can it be kept sustainable? *Journal of Sea Research, 63*(2), 108–118. https://doi.org/10.1016/j.seares.2009.11.003.

Drummond, L., Mulcahy, M., & Culloty, S. (2006). The reproductive biology of the Manila clam, *Ruditapes philippinarum*, from the North-West of Ireland. *Aquaculture, 254*(1-4), 326–340. https://doi.org/10.1016/j.aquaculture.2005.10.052.

Elston, R. A., Dungan, C. F., Meyers, T. R., & Reece, K. S. (2004). Perkinsus sp. infection risk for Manila clams, *Venerupis philippinarum* (A. Adams and Reeve, 1850) on the Pacific coast of North and Central America. *Journal of Shellfish Research, 23*(1), 101–105.

Faasse, M., & Ligthart, M. (2008). Exotic carpet shells Ruditapes philippinarum in the Netherlands (De exotische tapijtschelp *Ruditapes philippinarum* (Adams & Reeve, 1850) vestigt zich in Nederland). *Het Zeepaard, 68*, 175–179.

Flassch, J. P., & Leborgne, Y. (1992). Introduction in Europe, from 1972 to 1980, of the Japanese Manila clam (*Tapes philippinarum*) and the effects on aquaculture production and natural settlement. *ICES Marine Science Symposium., 194*, 92–96.

Flassch, J.-P. (1988). *The clam breeding file. (La Palourde: Dossier d'elevage)*. Ifremer Institut Francais de Recherche pour l'Exploitation de la Mer.

Foster, N. R. (1997). The Molluscan fisheries of Alaska. In C. L. MacKenzie, V. G. Burrell, A. Rosenfeld, & W. L. Hobart (Eds.), *The history, present condition, and future of the molluscan fisheries of North and Central America and Europe. Vol. 2. Pacific Coast and supplemental topics*. In .: *128* (pp. 131–144). NOAA. NOAA Technical Report NMFS.

Fry, K. W. (2011). Transforming the Tidelands: Japanese labor in Washington's oystering communities before 1942. *Pacific Northwest Quarterly, 102*(3), 132–143.

Fukawa, M., & Fukawa, S. (2009). *Spirit of the Nikkei fleet: BC's Japanese Canadian fishermen*. Harbor Publishing.

Gillespie, D., Bower, S. M., Marcus, K. L., & Kieser, D. 2012 Biological synopses for three exotic molluscs, Manila Clam (Venerupis philippinarum), Pacific Oyster (Crassostrea gigas) and Japanese Scallop (*Mizuhopecten yessoensis*) licensed for aquaculture in British Columbia. Canadian Science Advisory Secretariat, Fisheries and Oceans Canada, Ottawa, Ontario.

Gillespie, G. E. (2007). Distribution of non-indigenous intertidal species on the Pacific Coast of Canada. *Nippon Suisan Gakkaishi, 73*(6), 1133–1137. https://doi.org/10.2331/suisan.73.1133.

Greeve, Y., Lindegarth, M., Panova, M., & Strand, Å. (2025). First report of the Manila clam *Ruditapes philippinarum* on the Swedish Skagerrak coast. *Estuarine Coast Shelf Science, 325*, 109487.

Habtemariam, B. T., Arias, A., García-Vázquez, E., & Borrell, Y. J. (2015). Impacts of supplementation aquaculture on the genetic diversity of wild *Ruditapes decussatus* from northern Spain. *Aquaculture Environment Interactions, 6*(3), 241–254. https://doi.org/10.3354/aei00128.

Herbert, R. J. H., Willis, J., Jones, E., Ross, K., Hübner, R., Humphreys, J., Jensen, A., & Baugh, J. (2011). Invasion in tidal zones on complex coastlines: Modelling larvae of the non-native Manila clam, *Ruditapes philippinarum*, in the UK. *Journal of Biogeography, 39*(3), 585–599.

Hulme, P. E., Bacher, S., Kenis, M., Klotz, S., Kühn, I., Minchin, D., Nentwig, W., Olenin, S., Panov, V., Pergl, J., Pyšek, P., Roques, A., Sol, D., Solarz, W., & Vilà, M. (2008). Grasping at the routes of biological invasions: A framework for integrating pathways into policy. *Journal of Applied Ecology, 45*(2), 403–414. https://doi.org/10.1111/j.1365-2664.2007.01442.x.

Humphreys, J. (2010). The introduction of the Manila clam to British coastal waters. *The Biologist, 57*(3), 134–138.

Humphreys, J. (2024). Continued spread of the Manila clam *Ruditapes philippinarum* in and around Britain: From "alien monster" to "brilliant Brexit bonus". *Bulletin of the Porcupine Marine Natural History Society, 22*, 32–35.

Humphreys, J. (2025). The politics of an Alien Monster: Retrospective analysis of the use of evidence at the science-policy interface. *Ocean & Coastal Management, 266*, 107693. https://doi.org/10.1016/j.ocecoaman.2025.107693.

Humphreys, J., Choi, K.-S., & Watanabe, S. (Eds.). (2025). The Manila clam, *Ruditapes philippinarum.* Special issue of estuarine, coastal, and shelf science. Elsevier.

Humphreys, J., Harris, M. R. C., Herbert, R. J. H., Farrell, P., Jensen, A., & Cragg, S. M. (2015). Introduction, dispersal and naturalization of the Manila clam *Ruditapes philippinarum* in British estuaries, 1980-2010. *Journal of the Marine Biological Association of the United Kingdom, 95*(6), 1163–1172. https://doi.org/10.1017/S0025315415000132.

Humphreys, J., Herbert, R. J. H., Roberts, C., & Fletcher, S. (2014). A reappraisal of the history and economics of the Pacific oyster in Britain. *Aquaculture, 428–429*, 117–124.

Humphreys, J. (2022). Ecology and exploitation of Poole Harbour oysters. In J. Humphreys, & A. E. Hall (Eds.), *Harbour ecology: Environment and development in Poole Harbour*. Pelagic Publishing.

Humphreys, J. (2005). Salinity and tides in Poole Harbour: Estuary or lagoon? In J. Humphreys, & V May (Eds.), *The ecology of Poole Harbour*. Elsevier.

Goulletquer, P., and Heral., M. (1997). Marine Molluscan production trends in France: From fisheries to aquaculture. US Department of Commerce, NOAA Technical Report NMFS 129. p 137–164.

iNaturalist (n.d.) Retrieved August 26, 2025, from https://www.inaturalist.org/observations?place_id=2&taxon_id=343269.

Jacobs, M., Lent, A., & Watkins, K. (2003). *Progressive globalisation*. Fabian Society.

Juanes, J. éA., Bidegain, G., Echavarri-Erasun, B., Puente, A., García, A., García, A. és, Bárcena, J. F., Álvarez, C. é, & García-Castillo, G. (2012). Differential distribution pattern of native *Ruditapes decussatus* and introduced *Ruditapes phillippinarum* clam populations in the Bay of Santander (Gulf of Biscay): Considerations for fisheries management. *Ocean & Coastal Management, 69*, 316–326. https://doi.org/10.1016/j.ocecoaman.2012.08.007.

Kerckhof, F. (2014). A population of the Manila clam Ruditapes philippinarum (Adams & Reeve, 1850) in the harbor of Zeebrugge. De Strandvlo." (Een populatie van de Azlatlsche tapijtschelp Ruditapes philippinarum (Adams & Reeve, 1850) in de Zeebrugse haven). *De Strandvlo, 34*(2), 57–61.

Kerckhof, F. (2016). New Nature: The Bivalve Fauna of the Klein Strand in Ostende and a Second Population of the Manila Clam Ruditapes philippinarum" (Nieuwe natuur: de bivalven fauna van het Klein Strand in Oostende en een tweede populatie van de Filipijnse tapijtschelp Ruditapes philippinarum). *Strandvlo, 36*(1), 6–11.

Kincaid, T. B. (1949). Japanese mollusks in seed oyster boxes from Japan shipped to Willapa Bay. In *Minutes of the conchological club of Southern California: 88* (p. 1). The Club.

Kolpakov, E. V., Sokolenko, D. A., & Andryushina, A. D. (2024). Distribution pattern of the subtropical-lowboreal bivalve mollusk Ruditapes philippinarum (A. Adams & Reeve, 1850) in Russian waters. *The Bulletin of the Russian Far East Malacological Society, 28*(1), 125–137.

Levings, C. D., & Kaise, D (2002). Marine and estuarine alien species in the Strait of Georgia, BC. In R. Claudia (Ed.), *Alien invaders in Canada's waters, wetlands, and forests* (pp. 111–131). Natural Resources Canada, Canadian Forest Service.

Lipej, L., Mavrič, B., Orlando-Bonaca, M., & Malej, A. (2012). State of the art of the marine non-indigenous flora and fauna in Slovenia. *Mediterranean Marine Science, 13*(2), 243–249. https://doi.org/10.12681/mms.304.

Loosanoff, V. L., & Davis, H. C. (1963). Rearing of Bivalve Mollusks. *Advances in Marine Biology, 1*(C), 1–136. https://doi.org/10.1016/S0065-2881(08)60257-6.

Lackschewitz, D., Reise, K., Buschbaum, C., & Karez, R. (2022). *Neobiota der deutschen Nord- und Ostseeküste: Eingeschleppte Arten in deutschen Küstengewässern* (pp. 394–pp). Flintbek, Germany: Landesamt für Landwirtschaft, Umwelt und ländliche Räume Schleswig-Holstein (LLUR SH.

Montadaudoin, X. (2016). Asari clam (*Ruditapes philippinarum*) in France: History of an exotic species 1972–2015. *The Bulletin of Japan Fisheries Research and Education Agency, 42*, 35–42.

Mortensen, S. H., & Strand, O. (2000). Releases and recaptures of Manila clams (*Ruditapes philippinarum*) introduced to Norway. *Sarsia, 85*(1), 87–91. https://doi.org/10.1080/00364827.2000.10414557.

Navas, J. I., Castillo, M. C., Vera, P., & Ruiz-Rico, M. (1992). Principal parasites observed in clams, *Ruditapes decussatus* (L.), *Ruditapes philippinarum* (Adams et Reeve), *Venerupis pullastra* (Montagu) and *Venerupis aureus* (Gmelin), from the Huelva coast (S.W. Spain). *Aquaculture, 107*(2-3), 193–199. https://doi.org/10.1016/0044-8486(92)90067-u.

Nerlovic, V., Korlevic, M., & Mravinac, B. (2016). Morphological and molecular differences between the invasive bivalve *Ruditapes philippinarum* (Adams & Reeve, 1850) and the native species *Ruditapes decussatus* (Linnaeus, 1758) from the northeastern Adriatic Sea. *Journal of Shellfish Research, 35*(1), 1–9. https://doi.org/10.2983/035.035.0100.

Paillard, C., Korsnes, K., Le Chevalier, P., Le Boulay, C., Harkestad, L., Eriksen, A. G., Willassen, E., Bergh, Ø, Bovo, C., Skår, C., & Mortensen, S. (2008). Vibrio tapetis-like strain isolated from introduced Manila clams *Ruditapes philippinarum* showing symptoms of brown ring disease in Norway. *Diseases of Aquatic Organisms, 81*(2), 153–161. https://doi.org/10.3354/dao01950.

Park, K.-I., & Choi, K.-S. (2016). In *Country report from South Korea: 42* (pp. 119–120). Bulletin of the Japan Fisheries Research and Education Agency.

Pascali, L. (2017). The wind of change: Maritime technology, trade, and economic development. *American Economic Review, 107*(9), 2821–2854. https://doi.org/10.1257/aer.20140832.

Poutiers, J. M. (1998). Bivalves. Acephala, Lamellibranchia, Pelecypoda. In K. E. Carpenter, & V. H. Niem (Eds.), *FAO species identification guide for fishery purposes. The living marine resources of the Western Central Pacific. Volume 1. Seaweeds, corals, bivalves, and gastropods* (pp. 123–362). FAO. http://www.fao.org/docrep/009/w7191e/w7191e00.htm.

Putnam, A. B., Nichols, O. C., Clements, J. C. et al. (2026). Completing the journey in the global north: the Manila clam (*Ruditapes philippinarum*) (Bivalvia: Veneridae) arrives in the Northwest Atlantic Ocean. *Biol Invasions 28: 163.*

Quayle, D. B. (1938). *Paphia bifurcate*, a new molluscan species from Ladysmith Harbour, BC. *Journal of the Fisheries Research Board, Canada, 4*(1), 53–54.

Quayle, D. B. (1964). Distribution of introduced marine Mollusca in British Columbia waters. *Journal of the Fisheries Research Board, Canada, 21*(5), 1155–1181.

Reimer, J. P., Droghini, A., Fischbach, A., Watson, J. T., Bernard, B., and Poe, A. 2017. Assessing the risk of non-native marine species in the Bering sea. NPRB Project 1523. Alaska Center for Conservation Science, University of Alaska Anchorage. AK 39 pp.

Reise, K., Wegner, K. M., Borcherding, R., Brand, S., Buschbaum, C., & Waser, A. M. (2024). Manila clams *Ruditapes philippinarum* spreading north and establishing in the European Wadden Sea. *Estuarine, Coastal and Shelf Science, 309*, 108940. https://doi.org/10.1016/j.ecss.2024.108940.

Robinson, A. M. (1997). Molluscan fisheries of Oregon: Past, present and future. In C. L. MacKenzie, V. G. Burrell, A. Rosenfield, and W. L. Hobart (Eds.), The history, present condition and future of the molluscan fisheries of North and Central America and Europe. Volume 2: Pacific Coast and supplemental topics. NOAA Technical Report NMFS 128. Seattle: National Marine Fisheries Service.

Román, S., Olabarria, C., Weidberg, N., Román, M., & Vázquez, E. (2023). Population structure and habitat assessment for two commercial clam species exploited in small-scale fisheries. *Reviews in Fish Biology Fisheries*. https://doi.org/10.1007/s11160-023-09791-6.

Souissi, J. B., Trigui, M., El Menif, N. T., Mahjoub, M. S., Mejri, H., Capapé, C., & Zaouali, J. (2005). On the recent occurrence of marine exotic species in the Tunisian waters. In E. Özhan (Ed.), *Proceedings of the seventh international conference on the Mediterranean coastal environment, MEDCOAST 05, Kusadasi, Turkey*.

Streftaris, N., Zenetos, A., & Papathanassiou, E. (2005). Globalisation of marine ecosystems: The story of non-indigenous marine species across European seas. In B Russel, & P A Todd (Eds.), *Oceanography and Marine Biology: An Annual Review* (pp. 43419–43453).

Talley, D. M., Talley, T. S., & Blanco, A. (2015). Insights into the Establishment of the Manila Clam on a Tidal Flat at the Southern End of an Introduced Range in Southern California, USA. *PLoS One, 10*(3), e0118891. https://doi.org/10.1371/journal.pone.0118891.

Thaanum, D. (1921). *Tapes philippinarum* in the Hawaiian Islands. *Nautilus, 34*, 107.

Titselaar, F. F. L. M. (2009). The carpet shells of Zeeland and the South Holland islands (De tapijtschelpen van Zeeland en de Zuid-Hollandse eilanden). *Spirula, 369*, 103–106.

Toba, M., Kobayashi, Y., Kakino, J., Yamakawa, H., Ishii, R., & Okamoto, R. (2016). In *Stocks and fisheries of asari in Japan: 42* (pp. 9–21). Bulletin of the Japan Fisheries Research and Education Agency.

Troost, K., van Asch, M., Cornelisse, S., Glorius, S., van den Ende, D., van Es, Y., Keur, M., Perdon, K. J., van der Pool, J., Suykerbuyk, W., van Zweden, C., & van Zwol, J. (2023). *Schelpdierbestanden in de Nederlandse kustzone, Waddenzee en zoute deltawateren in 2022. (CVO rapport; No. 23.009)*. Centrum voor Visserijonderzoek (CVO) https://doi.org/10.18174/588755.

Wolf, P. (1995). Venerupis philippinarum op de Dr. Lelykade; een nieuwe immigrant? *Het Zeepaard, 55*(4), 85–88.

Xie, Q., & Burnell, G. M. (1994). A comparative study of the gametogenic cycles of the clams *Tapes philippinarum* (Adams & Reeve 1850) and *Tapes decussatus* (Linnaeus, 1758) on the south coast of Ireland. *Journal of Shellfish Research, 13*(2), 467–472.

Yap, G. (1977). Population biology of the Japanese little-neck clam, *Tapes philippinarum*, in Kaneohe Bay, Oahu, Hawaiian Islands. *Pacific Science, 31*(3), 223–244.

Zenetos, A., Gofas, S., Verlaque, M., Çinar, M. E., García Raso, J. E., Bianchi, C. N., Morri, C., Azzurro, E., Bilecenoglu, M., Froglia, C., Siokou-Frangou, I., Violanti, D., Sfriso, A., San Martin, G., Giangrande, A., Katagan, T., Ballesteros, E., Ramos Espla, A. A., Mastrototaro, F., … Streftaris, N. (2010). Alien species in the Mediterranean MSFD areas by 2010. Part I. Spatial distribution. *Mediterranean Marine Science, 11*(2), 381–493.

Zheng, S., Zhang, T., Tu, K., Li, L., Liu, Z., Wu, B., … Sun, X. (2023). Population Genetics of Manila Clam (*Ruditapes philippinarum*) in China Inferred from Microsatellite Markers. *Biology, 2023*(12), 557. https://doi.org/10.3390/biology12040557.

PART

Biology and ecology

II

CHAPTER

Phylogeny and the genome

4

4.1 The genome and its utility

The genome provides the fundamental basis for the success of the Manila clam, and the techniques of molecular genetics have provided evidence on many aspects of the species' biology and evolutionary history. Significant contributions have been achieved from the analysis of both nuclear and mitochondrial DNA, including insights into the genetic bases of different phenotypes such as shell color; the nature of adaptation to infaunal life; selection pressures for low temperature tolerance, phylogenetic histories and evolutionary divergence times; and the modern processes of dispersal and diversification between and within geographically discrete populations. In addition, gene activation and RNA studies have elucidated the control of metabolic responses to environmental stressors. Various chapters in this volume reflect further on aspects of genomic evidence and gene expression summarized below.

The nuclei of somatic Manila clam cells contain 38 chromosomes in 19 chromosome pairs (Perez-Garcia et al., 2014). Reported whole genome size measured as DNA base pairs (bp) varies according to authority between 2.56 Gbp (where $1 \text{Gbp} = 1 \times 10^9$ base pairs) (Mun et al., 2017) and 1.32 Gbp (Yan et al., 2019), the latter of which is within a range more typical of bivalve genomes (Saavedra & Bachere, 2006). Mun et al. (2017) estimated the nuclear genome to contain 108,034 protein-coding genes.

The metazoan mitochondrial genome typically contains 72 genes, of which 13 are protein-coding, and the complete mitochondrial genome of *Ruditapes philippinarum* is reported to consist of 22,706 base pairs (Tang et al., 2022). Mitochondrial DNA (mtDNA) has particular utility, normally being directly inherited from a single parent, never undergoing recombination during meiosis (therefore changing only through mutation), and having a relatively high mutation rate.

DNA techniques include the examination of groups of alleles that are inherited together (haplotypes) and the use of molecular markers to determine the genetic diversity and differentiation between populations, from which can be inferred not only the routes of dispersal, but also aspects of importance to the breeding and distribution practices of commercial culture. Generally, genes, or portions of genes (molecular markers), are used as genotype proxies, and from the extent to which they are similar or different, relationships can be inferred. Such sequences, once elucidated, are made available on nucleotide databases such as GenBank for use by other researchers.

The Manila Clam *Ruditapes philippinarum*: Science and Significance of a Globalised Species.
DOI: https://doi.org/10.1016/B978-0-443-36704-5.00003-1
Copyright © 2026 Elsevier Inc. All rights are reserved, including those for text and data mining, AI training, and similar technologies.

Many favored gene sequences are now in common use. The mitochondrial gene coding for Cytochrome oxidase I (COI) has elucidated dispersal routes, and sometimes repeated introductions of the Manila clam have been concluded (Chiesa et al., 2017). Molecular markers have been identified and used, including nuclear genes such as ITS, a nonfunctional "Internal Transcribed Spacer," and 16S rRNA, which encodes a structural component of the ribosome (Wei et al., 2023). "Microsatellite" markers consisting of repetitive sequences of DNA are also valuable because they have a sufficiently high mutation rate to make them useful in determining relatedness between populations (Zheng et al., 2023).

Moreover, the power of polymerase chain reaction technology is now such as to allow early notice of the arrival of the species as a non-native through the detection of environmental DNA (eDNA), which can reveal traces of the species in the water column.

4.2 Phylogeny and divergence times

Modern approaches to taxonomy attempt to reflect evolutionary history (phylogeny). Since Darwin, phylogenies have been described in terms of branching lineages such as cladograms indicating patterns of descent from common ancestors (Kitching, 1998). Such patterns are regarded as hypotheses based on the interpretation of evidence. As new evidence accumulates, new and different phylogenies may be proposed and accepted (Ponder & Lindberg, 2008).

Technically, the phylogeny of a species such as *R. philippinarum* extends back through geological history to the earliest life forms. Metazoan phylogenies start with the relationships of the major taxa (Laumer et al., 2019). At the other end of the scale, much phylogenetic work focuses on the relationships between lower taxa such as families, genera, and species.

While traditionally phylogenies have been established based on morphological characters, the development of molecular genetic sequencing since the 1980s has provided many new lines of evidence. This has led to numerous revisions and partly explains changes in the accepted Manila clam binomen, in particular the preferred genus to which it belongs (Chapter 1). However, like morphological analyses, molecular sequences are also subject to homoplasy (similarity which is the result of independent evolutionary change); moreover, in the absence of entire genome elucidations, comparative analysis of different gene sets can suggest conflicting phylogenies. Even at phylum levels of analysis, different gene sets can give support for conflicting phylogenies (Laumer et al., 2019).

From a currently accepted taxonomy, such as is shown in Table 1.1, we can infer an evolutionary history for *R. philippinarum* in which a range of advanced features were achieved over geological time. Such advances, many of which are described in the next chapters, include the fusion of an original set of previously independent gill filaments to form elaborated W-shaped gills (Subclass Autobranchia) and inhalant and exhalent siphons (Infraclass Heterochonchia).

At the family level, the Veneridae exhibit a number of morphological traits that do not necessarily correspond to phylogenetic relationships as revealed by genome studies. Similar traits shared by distantly related species and dissimilar traits shared by closely related species have resulted in a still somewhat unsettled taxonomy. Mikkelson et al. (2006), in a systematic examination of a range of morphological venerid characters across 45 species, found difficulties in inferring phylogenetic relationships. Morphological characters that are securely the result of common evolutionary descent (synapomorphies) among the Bivalvia are sometimes elusive.

In early molecular studies, limits on the extent and nature of genetic material sequenced and the range of species sampled both influenced the reliability of phylogenetic conclusions (Liu et al., 2022). Nevertheless, several published phylogenies that include molecular analyses have concluded the Veneridae to be a largely natural monophyletic group (Mikkelsen et al., 2006), which includes two distinct subgroups, the Venerinae (including *R. philippinarum*) and the Meretricinae (Carter et al., 2001). Lui et al. (2022) have corroborated this division, identifying two distinct clades based on a wide range of species and multiple gene sequences. These authors also suggest that Carter et al.'s (2001) subfamilies should both be raised to family level, with only the first retaining the Venerid name. Either way, *R. philippinarum* remains within the Family Veneridae.

Within the Veneridae, *R. philippinarum* is located in the subfamily *Tapetinae*, which is also (after some refinement) thought to be a valid monophyletic group (Wang et al., 2020). Based on a less conservative gene arrangement and higher transversion rate in gene 16S rRNA, Canapa et al. (1996) have suggested that the *Tapetinae* species evolve more rapidly than other Venerid subfamilies.

Based on the fossil record, the Venerids were once thought to have appeared in the Late Jurassic period; however, molecular genetic techniques now suggest an earlier origin. Wang et al. (2020) have calculated the divergence times for the Veneridae, estimating that the family emerged from a common ancestor in the Permian period around 290 million years ago (mya). The two major Venerid clades separated around 276 mya, also during the early Permian. From that point, the various Venerid subfamilies emerged between 230 and 96 mya, with the *Tapetinae* near the Triassic/Jurassic boundary about 200 mya. Speciation of *R. philippinarum* is estimated to have occurred in the subsequent mid-Cretaceous period, approximately 110 mya.

The subfamily *Tapetinae* is commonly considered to contain 16 or 17 genera, depending on whether *Ruditapes* is given generic status (Huber 2010). Coen et al.'s (2000) catalogue of the bivalves of North America, asserted that the argument for more than subgenus status for *Ruditapes* was overstated considering the variability of shape and sculpture of the genus *Venerupis*. However, while the binomen *Venerupis philippinarum* has been more persistently used in the United States, molecular studies have generally supported *Ruditapes* as a distinct genus, and *R. philippinarum* is now the widely accepted binomen (WoRMS accessed 15.04.2025).

There is a general, although not exclusive, relationship between the *Tapetinid* genera and their biogeographical distributions. For example, *Polititapes spp.* are eastern Atlantic, whereas *Tapes, Paphia, Protapes, Gomphinella, Irus, Katelysia, and*

Marcia are characteristically Indo-Pacific. Moreover, at the species level, based on chromosomal analysis, Perez-Garcia et al. (2014) have observed that the European *R. decussatus* is closer to the *Tapetinid* species coming from the same geographic area (*Polititapes. aureus, Venerupis corrugata,* and *Polititapes rhomboides*) than to its congeneric east Asian *R. philippinarum.* This geographical aspect of phylogeny is consistent with the concept of a post-Permian "provincialisation" of Venerid species during the Mesozoic breakup of continental Pangaea. It also supports the idea that we should not always expect morphologically similar species originating in different biogeographical regions to be closely related.

However, neither should it be assumed that morphologically similar *Tapetinid species* with overlapping biogeography are necessarily closely related. Based on mitogenomic sequencing, Lui et al. (2022) have suggested that the two western Pacific species *Ruditapes aspera* and *Ruditapes philippinarum* do not exhibit a high degree of similarity or a genetic sister group relationship and therefore might better be considered as belonging to different genera. In any event, they are at least clearly distinct species. Notwithstanding much recent progress, the final achievement of definitive phylogenies and corresponding taxonomies of the *Tapetinid* species remains a work in progress.

4.3 The genome and benthic adaptation

As a largely estuarine and temperate benthic bivalve, the Manila clam must survive in habitats that present many challenges, from variations in temperature, salinity, and oxygen to the pathogen and parasite-rich nature of coastal sediments (An et al., 2012). The molecular basis of adaptations that enhance resilience is therefore of interest, not least in the context of commercial breeding programs.

Yan et al. (2019) conducted comparative genome analysis of the Manila clam, the Pacific oyster, and other bivalves to assess genetic differences with a bearing on adaptation to benthic lifestyles. Gene duplication is a type of genetic mutation that is believed to facilitate adaptation by providing available raw material from which new genes and functions can arise, without compromising the original gene function. For example, duplicates may evolve new similar but nonidentical functions, which, if beneficial, are retained by natural selection. Duplication results in "gene families," i.e., similar genes formed by the duplication of a single ancestral original (Zhang, 2013). Compared with other bivalves, *R. philippinarum* gene families related to immunity are relatively expanded. The expansion of stress-related genes has also been observed.

For example, C1q is a protein molecule that contributes to the innate (nonspecific) defense against infections by complementing the ability of phagocytic cells to clear pathogens. It is therefore an important part of the immune system. In *R. philippinarum,* the gene family that includes the C1q gene is greatly expanded, three times more than the epibenthic oyster *Magallana gigas* (and 12 times that of humans). Yan et al. (2019) also reported that, compared with *M. gigas* and three other molluscan species, several other immune-related genes are expanded, including those involved in coagulation, phagocytosis, and heat shock. The relatively large size of several other gene families also suggested species-specific gene family extensions in *R. philippinarum*. The expansion of immune-related genes was taken as evidence of an association between gene duplication and the species' adaptation to a pathogen-rich and challenging environment (Yan et al., 2019).

Analysis of messenger RNA (the transcriptome), through which genes are expressed, provides an indication of gene activity and cell function. Yan et al. (2019) also studied the Manila clam's transcriptome responses by conducting transcriptome analyses of clams subjected to a variety of conditions across different ontogenetic life stages. These challenges also required both immune defenses and stress resistance in conditions of parasite infection, hypoxia, and acidification exposure.

After 2 days of exposure to hypoxia, for example, Yan et al. (2019) found adjustments in gene expression in 102 genes corresponding to the physiological functions of the gills, energy metabolism, and antioxidant and immune defense. Comparing these responses to those of four epibenthic species, including a gastropod (limpet), two species of oyster, and a scallop, these authors concluded that the Manila clam is well adapted specifically to an infaunal benthic lifestyle through relatively high resistance to adverse environmental conditions characteristic of that habitat.

Comparisons with other *Tapetinae* species were not conducted in Yan et al.'s (2019) study, and it seems plausible that such adaptations to infaunal life may also be exhibited by related clam species. But in any event, *R. philippinarum* appears robust and well adapted to infaunal life. Nevertheless, the severity of infaunal challenges in anthropogenically influenced estuaries and intertidal areas is, in some places, sufficient to cause episodes of mass mortality, for example, when temperature or salinity extremes, pollution events, or pathogen epidemics, especially in combination, go beyond sublethal levels (see Chapter 14).

4.4 Genetic diversity and founder effects

In the natural world, genetically diverse populations with a good level of phenotypic variation are considered resilient in the context of environmental challenges and changes. In theory, Darwinian variation will increase the probability that at least some individuals will survive environmental change and volatility. Conversely, restricted genetic diversity in a population carries risks. In the first place, the number of individuals that are homozygous for deleterious recessive alleles will likely be greater, resulting in a reduced ability to survive and reproduce (inbreeding depression). Secondly, lack of Darwinian variation is likely to result in greater susceptibility of a population to challenges such as new disease variants, habitat changes, or predator arrivals. The question of genetically restricted populations is relevant in both wild and cultivated Manila clam populations.

The arrival of a species to a new area by natural dispersal carries risks if a rare settlement event results in the initial establishment of only a relatively small number of mature breeding adults. Population theory predicts that, if an introduction occurs as a single event with only a limited number of individuals, alleles will be fixed and lost at an accelerated rate relative to the source population (Hartl & Clark, 1997; Mayr, 1963).

For the Manila clam, predation intensity can be such that even successful spatfalls may be quickly annihilated or reduced to a small cohort. Moreover, the stochasticity of environmental variables in estuarine tidal flats means that successful spatfalls in new areas may indeed be rare rather than regular events. In any event, relatively small numbers of individuals in an original colonization will represent only a subset of the original parent population, with a correspondingly smaller gene pool and a more restricted range of alleles. Such limited diversity will persist in such a way as to be detectable, even in subsequent generations and as the population grows. This "founder effect" has been detected in genetic studies of non-native Manila clam populations in northern Adriatic lagoons, the Ria de Aveiro in Portugal, and on the Galician coast in Spain (Chiesa et al., 2011, 2014). A similar effect can also occur if an isolated established population is temporarily reduced, such as might also be caused by combinations of challenging stochastic variables (the bottleneck effect).

A version of such phenomena may be manifest during the cultivation of Manila clams in hatcheries (Chapter 17), especially when this involves a relatively small and repeatedly spawned "broodstock," which, even if replenished, may originate from a single source population. This is functionally equivalent to an isolated natural population with the potential inbreeding depression and restricted diversity this entails, even when many millions of larvae or juveniles are produced.

Small "effective population size" (signifying restricted genetic diversity), whether naturally or anthropogenically caused, may also result in genetic drift. This occurs when the frequency of uncommon alleles shifts due to chance operating over generations. The effect is usually manifest in newly introduced or declining populations with consequently limited gene flow. Propensity to genetic drift is therefore a function of low effective population size. In the Manila clam, such circumstances may occur in a small, recently established population, resulting in a chance loss of variation, which limits the efficacy of natural selection. In their study of Chinese populations, Zhang et al. (2023) found most Manila clam populations to have low effective population size, with two populations below their critical value. They postulated the cause as human selection of the best progeny from a small number of parents, such as when selecting hatchery broodstock.

In hatchery conditions, the probability of breeding with genetically similar individuals (inbreeding) may also be a risk during breeding programs in which the extent of phenotypic variation is deliberately attenuated for the purpose of the selection of characters such as shell color deemed advantageous for sales (Chapter 5). Industrial culture techniques can also lead to other unnatural distortions in the genetic diversity of populations. Sequencing genes from a range of coasts from north to south China. Wei et al. (2023) and Zhang (2023) have discovered that, although the overall genetic diversity of the species was high, the genetic differentiation between individual populations did not show patterns that would be expected based on their distance apart and consequent geographical isolation. To explain this, they noted that large numbers of clam seeds produced in southern China had been distributed to many sites across China, resulting in considerable genetic homogeneity between widely distant northern and southern populations (Chapter 17). A similar phenomenon has also been reported for Korean Manila clams (An et al., 2012).

In China, the same authors, Wei et al. (2023) and Zhang et al. (2023), reported a revealing exception to the above phenomenon in two populations around Laizhou Bay in the north, which showed significantly greater genetic differentiation. These populations were divergent from all other northern and southern Chinese populations and are thought to have maintained their naturally higher levels of genetic variability. This has been taken to indicate the persistent natural status of the wild populations in this area.

On Chinese coasts where the Manila clam is indigenous, industrial levels of southern hatchery-reared seed distribution have therefore had a major influence, to the extent that the genetics of wild populations is altered, and in some places wild clam genotypes may actually have been displaced. In any event, the frequent translocation of clam seed risks a decline in the genetic variation, which, in a large and truly wild species, provides the selective potential for long-term survival of environmental and other challenges. These findings suggest the introduction of conservation measures. Firstly, to protect the genetic diversity of remaining wild clam populations, including the regulation of genuinely wild clam fisheries and the protection of natural wild clam habitats. Secondly, to develop aquaculture protocols and breeding strategies with particular regard to broodstocks ("bloodstocks"), which will better maintain genetic diversity and viability in cultured Manila clams (Zhang et al., 2023).

4.5 Environmental DNA and metabarcoding

Genetic differences between *R. philippinarum* and the often sympatric European native *R. decussatus* have been used to corroborate morphologically based identifications of individual clams, for example, Nerlovic et al. (2016) in Croatian waters, where the non-native made up about 12% of the combined populations.

In recent years, the detection of individual species from the complex mixtures of environmental DNA (eDNA) released by animals into marine waters has been demonstrated as effective for monitoring macro-organisms, not least non-native species. Increasing numbers of individual marine species sequences are now available as barcodes in nucleotide databases (Lavrador et al., 2023). Moreover, through the technique of metabarcoding, it is possible to achieve the simultaneous DNA-based identification of many different taxa from a single environmental water sample (Taberlet et al., 2018). While visual discrimination between different closely related bivalve planktonic larvae can be laborious, difficult, and resource intensive, the metabarcoding approach has been used to reliably detect the presence of the planktonic larvae of *R. philippinarum* and the timing of spawning activity.

Using a 400 bp subunit of the rRNA coding gene (18S) and a 313 bp sequence of the mitochondrial DNA Cytochrome Oxidase I coding gene (COI), Couton et al. (2019) successfully demonstrated the utility of metabarcoding for the detection of Manila clam larvae in the Bay of Morlaix, Brittany, France, where the species had not previously been reported, thus demonstrating the value of the approach in biological non-native species surveillance and biodiversity monitoring generally.

4.6 Transcriptome analysis

Gene expression is also susceptible to elucidation by transcriptome analysis. The transcriptome represents those parts of the genetic code which are being actively expressed through the production of RNA. Nie et al. (2020) investigated transcriptome responses to the stresses of alternating aerial exposure and immersion as experienced in intertidal habitats. They found that metabolic and immune-related genes responded to such conditions, including genes involved in water reabsorption and energy metabolism, which were "up-regulated."

In China, Jahan et al. (2022) have investigated gene expression and adaptation in *R. philippinarum* in response to acute temperature challenges, comparing populations from northern and southern latitudes. They reported that activation of genes involved in protein processing in response to temperature stress, with phagosome-related genes activated under high temperature stress. Some gene activation patterns varied by population. For example, genes encoding peroxisomes (which detoxify reactive oxygen species among other compounds) were upregulated in clams from colder waters under heat stress, but not in those from warmer waters.

Transcriptomics is also used for elucidating clam microbiomes from RNA sequences. Such studies include the detection of pathogens, including new pathogens consequent on climate change (Ray-Campos et al., 2025). Conversely, some microbiome species suppress pathogens. "Indigenous" bacteria in the clam microbiome are known to promote survival through competitive exclusion of pathogens and the production of inhibitory compounds. Le Noc et al. (2025) have investigated relationships between specific microbes and the clam's immune system to develop larval treatments involving probiotics to enhance aquaculture production.

References

An, H. S., Park, K. J., Cho, K. C., Han, H. S., & Myeong, J. I. (2012). Genetic structure of Korean populations of the clam *Ruditapes philippinarum* inferred from microsatellite marker analysis. *Biochemical Systematics and Ecology, 44*, 186–195. https://doi.org/10.1016/j.bse.2012.05.007.

Canapa, A., Marota, I., Rollo, F., & Olmol, E. (1996). Phylogenetic analysis of Veneridae (Bivalvia): Comparison of molecular and palaeontological data. *Journal of Molecular Evolution, 43*(5), 517–522. https://doi.org/10.1007/bf02337522.

Carter, J. G., Altaba, C. R., Anderson, L. C., Araujo, R., Biakov, A. S., Bogan, A. E., Campbell, D. C., Campbell, M., Jin-hua, C., Cope, J. C. W., Delvene, G., Dijkstra, H. H., Zong-jie, F., Gardner, R. N., Gavrilova, V. A., Goncharova, I. A., Harries, P. J., Hartman, J. H., Hautmann, M., … Yancey, T. (2011). A synoptical classification of the Bivalvia (Mollusca). *Paleontological Contributions, 4*, 1–47.

Chiesa, S., Lucentini, L., Freitas, R., Nonnis Marzano, F., Minello, F., Ferrari, C., Filonzi, L., Figueira, E., Breda, S., Baccarani, G., & Argese, E. (2014). Genetic diversity of introduced Manila clam *Ruditapes philippinarum* populations inferred by 16S rDNA. *Biochemical Systematics and Ecology, 57*, 52–59.

Chiesa, S., Lucentini, L., Freitas, R., Marzano, F. N., Breda, S., Figueira, E., Caill-Milly, N., Herbert, R. J. H., Soares, A. M. V. M., & Argese, E. (2017). A history of invasion: COI phylogeny of Manila clam *Ruditapes philippinarum* in Europe. *Fisheries Research, 186*, 25–35. https://doi.org/10.1016/j.fishres.2016.07.024.

Chiesa, S., Marzano, F. N., Minervini, G., De Lucrezia, D., Baccarani, G., Bordignon, G., Poli, I., Ravagnan, G., & Argese, E. (2011). The invasive Manila clam *Ruditapes philippinarum* (Adams and Reeve, 1850) in Northern Adriatic Sea: Population genetics assessed by an integrated molecular approach. *Fisheries Research, 110*(2), 259–267. https://doi.org/10.1016/j.fishres.2011.04.013.

Coan, E. V., Scott, P. V., & Bernard, F. R. (2000). Monographs number 2 studies in biodiversity number 2. *Bivalve seashells of Western North America. Marine bivalve mollusks from Arctic Alaska to Baja California*. Santa Barbara Museum of Natural History.

Couton, M., Comtet, T., Cam, S. L., Corre, E., & Viard, F. (2019). Metabarcoding on planktonic larval stages: An efficient approach for detecting and investigating life cycle dynamics of benthic aliens. *Management of Biological Invasions, 10*(4), 657–689. https://doi.org/10.3391/mbi.2019.10.4.06.

Hartl, D. L., & Clark, A. G. (1997). *Principles of population genetics*. Sinauer Associates.

Huber, M. (2010). *Compendium of bivalves volume 1*. Conch Books.

Jahan, K., Nie, H., Yin, Z., Zhang, Y., Li, N., & Yan, X. (2022). Comparative transcriptome analysis to reveal the genes and pathways associated with adaptation strategies in two different populations of Manila clam (*Ruditapes philippinarum*) under acute temperature challenge. *Aquaculture, 552*, 737999. https://doi.org/10.1016/j.aquaculture.2022.737999.

Kitching, I. J. (1998). *Cladistics: The theory and practice of parsimony analysis*. Oxford University Press.

Laumer, C. E., Fernández, R., Lemer, S., Combosch, D., Kocot, K. M., Riesgo, A., Andrade, S. C. S., Sterrer, W., Sørensen, M. V., & Giribet, G. (2019). Revisiting metazoan phylogeny with genomic sampling of all phyla. *Proceedings of the Royal Society B: Biological Sciences, 286*(1906), 20190831. https://doi.org/10.1098/rspb.2019.0831.

Lavrador, A. S., Fontes, J. T., Vieira, P. E., Costa, F. O., & Duarte, S. (2023). Revision and annotation of DNA barcodes of marine invertebrate non-indigenous species (NIS) Occurring in European coastal regions. *Diversity, 15*, 174.

Liu, Y., Ma, P., Zhang, Z., Li, C., Hu, L., & Wang, H. (2022). Mitogenomic analyses shed new light on phylogenetic relationships within the genus *Ruditapes* (Bivalvia: Veneridae). *Journal of Molluscan Studies, 88*(1). https://doi.org/10.1093/mollus/eyac001.

Mayr, E. (1963). *Animal species and evolution*. Harvard University Press.

Mikkelsen, P. M., Bieler, R., Kappner, I., & Rawlings, T. A. (2006). Phylogeny of Veneroidea (Mollusca: Bivalvia) based on morphology and molecules. *Zoological Journal of the Linnean Society, 148*(3), 439–521. https://doi.org/10.1111/j.1096-3642.2006.00262.x.

Mun, S., Kim, Y.-J., Markkandan, K., Shin, W., Oh, S., Woo, J., Yoo, J., An, H., & Han, K. (2017). The whole-genome and transcriptome of the Manila clam (*Ruditapes philippinarum*). *Genome Biology and Evolution, 9*(6), 1487–1498. https://doi.org/10.1093/gbe/evx096.

Nerlovic, V., Korlevic, M., & Mravinac, B. (2016). Morphological and molecular differences between the invasive bivalve *Ruditapes philippinarum* (Adams & Reeve, 1850) and the Native Species *Ruditapes decussatus* (Linnaeus, 1758) from the Northeastern Adriatic sea. *Journal of Shellfish Research, 35*(1), 1–9. https://doi.org/10.2983/035.035.0105.

Nie, H., Jiang, K., Li, N., Li, D., & Yan, X. (2020). Transcriptomic analysis of *Ruditapes philippinarum* under aerial exposure and reimmersion reveals genes involved in stress response and recovery capacity of the Manila clam. *Aquaculture, 524*, 735271. https://doi.org/10.1016/j.aquaculture.2020.735271.

Pérez-García, C., Hurtado, N. S., Morán, P., & Pasantes, J. J. (2014). Evolutionary dynamics of rDNA clusters in chromosomes of five clam species belonging to the family Veneridae (Mollusca, Bivalvia). *BioMed Research International, 2014*, 754012. http://dx.doi.org/10.1155/2014/754012.

Le Noc, M., Even, Y., Petek, S., Hignard, M., Matozzo, V., et al. (2025). Study of the cross-talk between the immune system and the microbiota of the Japanese clam with a view to selecting probiotics for aquaculture. In International Manila and European Clams Conference (SYMPA), Book of Abstracts. Ifremer.

Ponder, W., & Lindberg (2008). *Phylogeny and evolution of the Mollusca*. University of California Press.

Rey-Campos, M., Novoa, B., & Figueras, A. (2025). Metatranscriptomic tools for the detection of pathogens in clams. In *Ifremer: International Manila and European Clams Conference -SYMPA September 8-10* Book of Abstracts.

Saavedra, C., & Bachere, E. (2006). Bivalve genomics. *Aquaculture, 256*, 1–14.

Taberlet, P., Bonin, A., Zinger, L., & Coissac, E. (2018). *Environmental DNA: For biodiversity research and monitoring*. Oxford University Press.

Wang, Y., Yang, Y., Liu, H., Kong, L., Yu, H., Liu, S., & Li, Q. (2020). Phylogeny of Veneridae (Bivalvia) based on mitochondrial genomes. *Zoologica Scripta, 00*, 1–13.

Tang, Y., Huo, Z., Liu, Y., Wang, Y., Zuo, L., Fang, L., … Yan, X. (2022). Full Mitochondrial Genomes Reveal Species Differences between the Venerid Clams Ruditapes philippinarum and R. variegatus. *Genes, 13*, 2157. doi:10.3390/genes13112157.

Wei, D., Zheng, S., Wang, S., Yan, J., Liu, Z., Zhou, L., Wu, B., & Sun, X. (2023). Genetic and haplotype diversity of Manila clam *Ruditapes philippinarum* in different regions of China based on three molecular markers. *Animals, 13*(18), 2886. https://doi.org/10.3390/ani13182886.

Yan, X., Nie, H., Huo, Z., Ding, J., Li, Z., Yan, L., Jiang, L., Mu, Z., Wang, H., Meng, X., Chen, P., Zhou, M., Rbbani, M. d. G., Liu, G., & Li, D. (2019). Clam genome sequence clarifies the molecular basis of its benthic adaptation and extraordinary shell color diversity. *Iscience, 19*, 1225–1237. https://doi.org/10.1016/j.isci.2019.08.049.

Zhang, J. (2013). Gene duplication. In J. Losos (Ed.), *The Princeton guide to evolution* (pp. 397–405). Princeton University Press [Google Scholar].

Zheng, S., Zhang, T., Tu, K., Li, L., Liu, Z., Wu, B., Zhou, L., & Sun, X. (2023). Population genetics of Manila clam (*Ruditapes philippinarum*) in China inferred from microsatellite markers. *Biology, 12*(4). https://doi.org/10.3390/biology12040557.

CHAPTER

The shell: Form, function and color

5

5.1 External form and function

Venerid shells are commonly thick, strong, and markedly sculptured. In these respects, *Ruditapes philippinarum* is no exception, being generally described as having a "solid," "heavy," or "strong" shell, with a combination of comarginal "ribs" (roughly concentric with the shell margin) and radial ridges (Figs. 1.1, 5.5 and 5.1). Stanley (1970) identified an inverse relationship between bivalve shell thickness and depth of burrowing, which is thought to be an adaptation related to greater vulnerability to predation in shallow burrowing species. A heavy shell and consequent whole-animal density are also thought to be an adaptation for a fixed position within a stable sediment, at the expense of speed of burrowing. The shell is divided into distinct right and left "valves" connected by a flexible ligament.

The dorsal tip of each valve is the oldest part of the shell and is termed the beak. The areas immediately behind each beak, deriving from the larval shell, are the umbo (Fig. 5.1). In *R. philippinarum,* each umbo points anteriorly (prosogyrate), and a hypothetical line dropped vertically from them to the ventral edge separates anterior from posterior. Since the umbones are located toward the anterior end, *R. philippinarum* is described as inequilateral, and as the two valves are of similar shape and size, it is described as equivalve. As well as protecting from mechanical challenges, including predation, the valves also play an important role in burrowing (Chapter 6) and provide a skeletal structure that supports the soft tissues.

The *R. philippinarum* shell includes a concave anteroposterior region in front of the forward-facing umbones, which is occupied by the lunule. In *R. philippinarum,* the lunule is a pronounced structural feature corresponding to a somewhat flat depression relative to the surrounding raised shell form and sculpture (Fig. 5.1).

Posterior to the lunule on the dorsal side are various internal and external structures associated with the hinge apparatus. The ligament is an elastic fibrous structure, which in evolutionary terms is seen as a decalcified part of a single ancestral molluscan shell. Although in some deep-burrowing bivalves the ligament is internal, in the Venerids it is external and a clearly visible dorsal feature which helps resist shear stress and acts as a spring to open the valves antagonistically to the action of the adductor (shell closure) muscles. Surrounding and reaching somewhat posterior to the ligament in Venerids is the escutcheon, consisting of a narrow depressed area of shell. The escutcheon is not always bilaterally symmetrical, and in *R. philippinarum* is typically more distinct on

The Manila Clam *Ruditapes philippinarum:* Science and Significance of a Globalised Species.
DOI: https://doi.org/10.1016/B978-0-443-36704-5.00009-2
Copyright © 2026 Elsevier Inc. All rights are reserved, including those for text and data mining, AI training, and similar technologies.

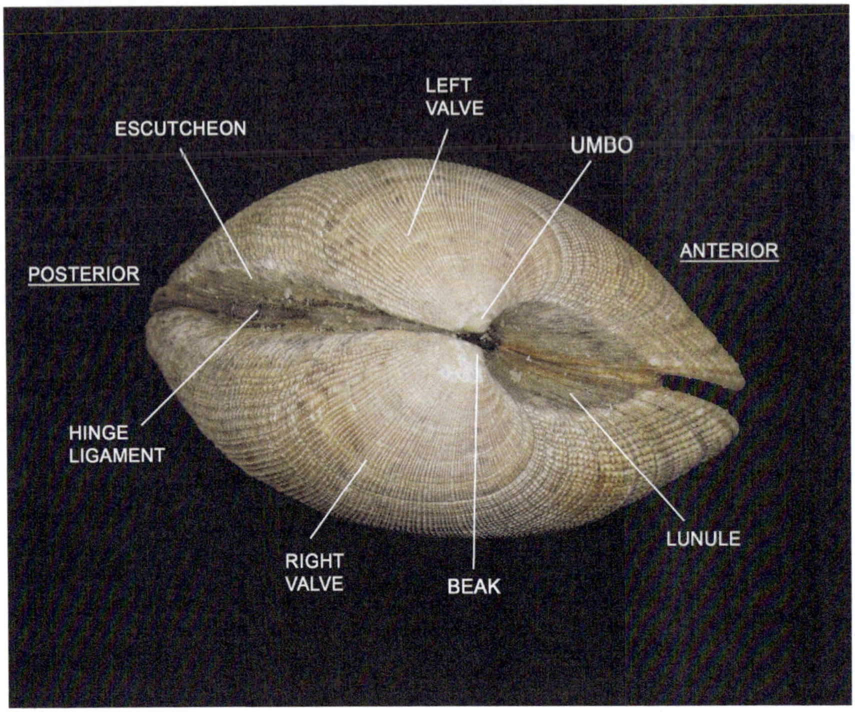

FIGURE 5.1

Dorsal aspect of *Ruditapes philippinarum* showing external shell features.

the left valve than on the right, a feature that in Europe can be used to help distinguish it from the native *R. decussatus,* in which the escutcheon is symmetrical.

The combined existence of both concentric ribs and radial ridges in *R. philippinarum* must be of functional significance, not least as many Venerids possess only the former. It is possible that the degree of anterior rugosity resulting from the interaction of these two patterns further enhances stability during burrowing, but this does not explain the even more accentuated posterior shell rugosity. Moreover, although in some Venerids the radial shell ridges terminate to form pronounced serrations on the external rim of the shell, this is not the case in *R. philippinarum.* Since such serrations have also been interpreted as an adaptation for penetrating the sediment during burrowing, the smooth shell rim of *R. philippinarum* may also signify lower selective priority for burrowing capability.

In any event, since the functional significance of the burrowing of a shell sculpture based on a lattice of ribs and ridges is not clear, alternative explanations have been considered. As strong shells are generally associated with a shallow life position in the substrate, ridges and rugosity have been considered to be primarily an adaptation enhancing general stability. Such rugosity provides resistance both to upward drag from a predator and to turbulent physical instability at the sediment/water interface.

The degree of "inflation" of a bivalve shell (otherwise referred to as tumidity or compression) is also considered ecologically significant. Descriptions such as "moderately strong and inflated" are commonly observed in the literature (Poutiers, 1998). Suarez and Raven (2020) describe the species as "moderate to inflated" compared with the "moderate" inflation of the European sympatric *R. decussata*. More precisely, Mikkelsen (2005) found a ratio of shell inflation to height, which they categorized as "not compressed."

Trueman et al. (1966) reported positive correlations between the degree of tumidity and force required for sediment penetration, with consequences for speed of burrowing. Therefore, conversely, greater tumidity must also imply greater resistance to extraction by a predator. All of which suggests that *R. philippinarum* is adapted for a sedentary existence in which the maintenance of infaunal position is not achieved primarily by fast burrowing after disturbance, but rather by avoiding positional disturbance through enhanced stability within sediments largely free from significant wave energy.

From all this, we can infer that movement, including burrowing in adult *R. philippinarum,* must be energetically demanding. Nevertheless, the conspicuous lunule and resulting anterior shape are of functional significance in relation to burrowing behavior, suggesting that burrowing for reburial or sediment depth adjustment is not without selective utility. Chapter 6 considers the burrowing behavior and capability of *R. philippinarum* and examines its significance in the life of the adult animal.

In summary, it can be inferred that *Ruditapes philippinarum* is well adapted for a sedentary life buried near the surface of largely stable sediments in which fast or frequent reburrowing, as might be required in shifting sands, is not a priority.

Notably, adaptation for stable sediment conditions does not necessarily equate to a narrow range of habitat granularity. Although largely excluding coarse-grained high-energy shifting sands, many sedimentary regimes provide suitably stable substrates, from fine silt-clay mud through sheltered sands to gravel (and including unsorted mixed sediments). In theory, at least, the potential habitat range of *R. philippinarum* is considerable. In Chapter 11 and 10, these inferences will be examined in relation to reported habitats and larval and juvenile behavior.

Although, like the Venerids generally, *R. philippinarum* can be considered an advanced form, its burrowing habit does not demand the more extreme adaptive responses found in both deep burrowing and epibenthic bivalves. This advanced but not highly specialized morphology provides an effective basis for success in *R. philippinarum.*

5.2 Internal shell features

Fig. 5.2 provides a diagram of the interior of a Manila clam valve. Ventral to the hinge apparatus is the internal shell surface, which has a nacreous sheen. Various marks ("scars") are caused by the attachment of soft tissues in life. Of particular significance are the anterior and posterior adductor muscle attachment marks. Connecting these

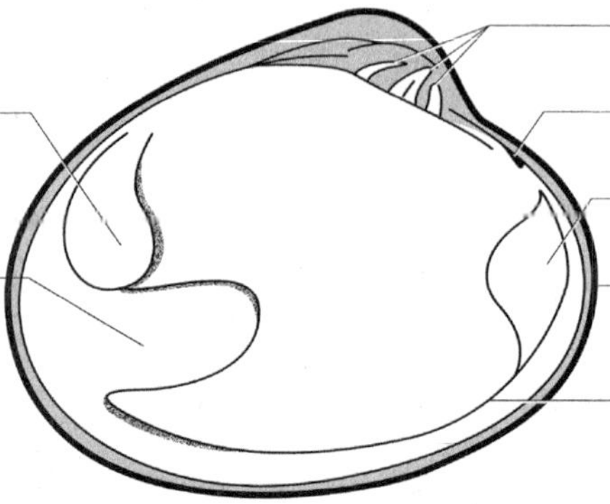

FIGURE 5.2

Major internal shell features of *Ruditapes philippinarum* (left valve).

muscle scars is the pallial line, which runs parallel to the ventral shell margin, except posteriorly, where it is markedly indented to form the pallial sinus, which opens toward the posterior. In addition, a small scar near but separate from the anterior adductor muscle is found tucked under the dorsal shell margin, marking the attachment of the pedal retractor muscle (Fig. 5.2).

The extent to which the pallial sinus reaches forward toward the shell center is significant in identifying the species. Mikkelsen et al. (2005) described the pallial sinus as rounded and of medium size in relation to the Veneriodea superfamily, with a ratio of anterior extent to shell length of between 0.3 and 0.5 (see Appendix).

Some deep burrowing bivalves, such as *Mya* and *Ensis* species, are incapable of fully closing their valves, leaving a marginal gape. In contrast, to be an effective protection against predation, the shells of shallow burrowing bivalves such as *R. philippinarum* are capable of complete closure and resistant to lateral shear stresses, which might cause misalignment of the closed valves. In this context, the hinge apparatus of external ligament and internal "dentition" is vital. In *R. philippinarum,* a "heterodont" dentition characteristic of the Euheterodonta taxon consists of three dorsally positioned interlocking "cardinal" teeth on each valve, immediately below the umbones (Fig. 5.2). The middle cardinal tooth on the left valve has a cleft and is therefore described as bifid. On the right valve, both middle and posterior cardinal teeth are bifid, although the latter less deeply.

As is characteristic of the subfamily Tapetinae, *R. philippinarum* lacks any of the "lateral" teeth of other Venerid subfamilies. It has been suggested that the heterodont hinge is an adaptive feature, especially of shallow burrowers, where, as a relatively low-friction mechanism, it may facilitate rapid valve closure (Stanley, 1970). If so, this might represent a selection-mediated trade-off, in which the absence of lateral teeth, along with the smooth shell rim, leaves only the cardinal teeth with the ligament and adductor muscles in structural opposition to lateral shearing stress.

Alternatively, the absence of lateral teeth in *R. philippinarum* may just reflect the developmental restrictions of the Tapetinae shell form: An adaptive response to which would include the strengthening of the cardinal teeth or their movement to a more central position. Whereas the blunt anterior of *R philippinarum* leaves the cardinal teeth not centrally positioned, they are quite robust, at least in comparison to the European sympatric *Ruditapes decussata* (Suárez & Raven, 2020).

5.3 Microstructure and strength

The bivalve shell consists primarily of calcium carbonate, which in the Venerid Manila clam takes the form of aragonite crystals. However, the first part of the shell to be secreted is a thin proteinaceous layer, the periostracum, which, although often abraded, provides a template at the growing edge of the shell onto which the calcareous layers are deposited (Paillard & Le Pennec, 1993). The average thickness of the periostracum of *R. philippinarum* is 5 μm (Harper, 1997).

The shell of the Manila clam is composed of two phases. In addition to crystalline aragonite, there is an organic matrix largely composed of fibrous protein. A general account of the composition and physical properties of bivalve shells has been provided by Taylor & Laymen (1972).

The calcareous part of the shell consists of three layers, the inner layer being homogeneous, the middle layer having a cross-lamellar structure, and the outer layer a composite prismatic structure (Fig. 5.3). The inner layer is thickest, being around the same thickness as the middle and outer layers combined, and the outside margin of the outer layer follows the characteristic shell sculpture patterns.

The strength of the Manila clam shell is impressive and plays a vital role in protecting the soft tissues from predation and wave hydrodynamic challenges. The compression force necessary to fracture the shell ranges from 100 to 400 N, depending primarily on age (shell thickness) (400 N equates to 40 kg weight). Three-year-old clams can withstand a fracture force twice that of 1-year-old clams. Other factors affecting fracture force include loading orientation and loading speed (Mu et al., 2018).

5.4 Periodicity of shell secretion

The shell is produced by epithelial mantle cells underlying the shell edge. Growth involves the periodic accretion of new layers visible as microgrowth bands. In constant conditions of salinity, temperature, food supply, and water flow, *R. philippinarum* tends to lay down these bands at approximately semidiurnal intervals, i.e., in tidal rather than diurnal daylight rhythms. However, in tidal conditions, the growth of wide bands corresponds to the number of emersions, and Richardson (1987) reported these bands to be laid down during periods of aerial exposure rather than immersion. In contrast, Kanazawa and Sato (2008) described each band as corresponding to a period of tidal immersion. In any event, it appears that in this species there is no discernible influence of diurnal cycles of illumination on band formation, which led

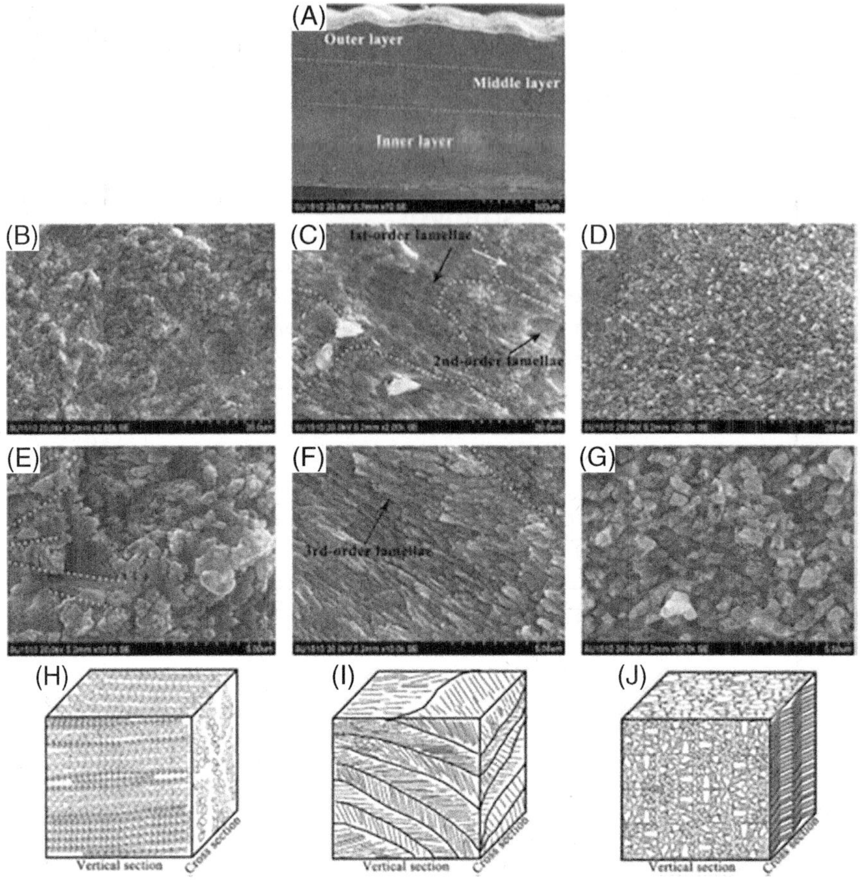

FIGURE 5.3

Scanning electron micrographs of Manila clam shell structure. (A) Overall view of vertical section; (B) microstructure of outer layer on vertical section; (C) microstructure of the middle layer in the vertical section; (D) microstructure of the inner layer in the vertical section; (E) magnified microstructure of outer layer in vertical section; (F) magnified microstructure of the middle layer in vertical section; (G) magnified microstructure of inner layer in vertical section; (H) schematic illustration of outer layer; (I) schematic illustration of the middle layer; (J) schematic illustration of the inner layer.

From Mu et al. (2018).

Richardson (1987) to conclude that growth is subject to a semidiurnal endogenous rhythm entrained by the tidal cycle.

Despite such general conclusions, individual clams vary, and the number of microgrowth bands may significantly exceed (Ricardson, 1987) or fall short of that expected based on tides alone. In fact, both environmental variables and physiological conditions affect patterns of microgrowth such that these bands effectively record both physiological and environmental stresses over time. For example, in individuals in spawning condition. Kanazawa and Sato (2008) report a "spawning break" in which a sequence of wide (over 100 μm) microgrowth rings may be abruptly interrupted and followed by a postspawning period of reduced (less than 60 μm) increments, presumably as resources are committed to recovery.

In nonspawning individuals, growth is positively correlated with water temperature up to 20°C, after which this relationship deteriorates. Essentially, the diurnal and seasonal patterns of microgrowth rings reflect a combination of the effects of, inter alia, temperature, food availability, and the physiological demands of gonad maturation and spawning (Kanazawa & Sato, 2008).

Superimposed on the concentric groove pattern are more pronounced, darker concentric rings, which are often assumed to reflect the cessation of growth in the winter season (Ponurovskii, 2000; Silina & Popov, 1989). However, determination of age by counting these rings is not always reliable in *R. philippinarum*, as growth cessation may not always correspond to an annual cycle, and the relationship is therefore not always clear-cut (Ohba, 1959). Consequently, age determinations informed also by "size-at age" graphs determined for multiple clam specimens from a location provide a more secure approach (Chapter 14), especially when combined with ring pattern estimates

5.5 Shell shape and allometry

Watanabe and Katayama (2010) studied four Manila clam populations from their indigenous range on the coast of Japan. They determined that faster-growing individuals in better nutritional condition have "flatter" (less inflated/tumid) shells than slower-growing individuals. Ohba (1959) also suggested that tumidity relates inversely to food availability. Therefore, the general shape of Manila clam shells can vary with locality. This is at least in part also a function of environmental conditions such as sediment, salinity, current velocity, and elevation. The shells' "roundness" (a function of the relationship between height and width relative to length) appears greater in less optimal habitats. Conversely, more favorable habitats produce more elongated (or "sharp") shells.

Meisterhans et al. (2015) found that individuals from a fully marine habitat had significantly more elongated shells and better condition than those from a more stressful brackish mudflat. When moved between habitats of different suitability, Manila clams correspondingly changed their shape over time (Tomiyama, 2021). Moreover, this author suggests that, while shell height appears to be environmentally affected by habitat suitability, shell width is affected largely by individual growth rate, which is described as an internal factor.

Working on the French Atlantic coasts, Caill-Milly et al. (2014) found morphometric trends along a north-south latitudinal range for *R. philippinarum*. Examining the relation between shell shape and environmental variables, including salinity, temperature, and chlorophyll *a* (food) concentration, they concluded that shell shape may vary in response to a complex interaction between such variables and others, including sediment character. In this latter context, phenotypic plasticity might be seen as an adaptive capability rather than simply a selectively neutral response to environmental conditions. A further complicating factor with shell shape is allometric changes in metric ratios as individuals mature.

Isometric growth occurs if the relationship between two parts or dimensions remains constant as the animal grows. If the ratio is other than 1, as is commonly the case, then growth is allometric. Nerlovi et al. (2016) used the external morphometric measures of length (l), height (h), and width (w) in clams of a range of sizes to infer growth patterns in *Ruditapes philippinarum* and native *Ruditapes decussatus* from the Adriatic. For *R. philippinarum,* all three ratios (h/l, w/l, and w/h) were isometric, whereas for *R. decussatus,* only h/l was isometric, the other ratios being negatively allometric. This suggests that *R. philippinarum* retain their overall shape as they grow, at least on this Croatian Adriatic coast.

However, we cannot assume that this allometric relation will apply across all growth stages. Limitations of Nerlovi et al. (2016) relate to the use of different animals to infer allometric patterns and the restriction of specimen size range (l = 26.10–38.71 mm). Evidence from actual growth measurements on individual wild Japanese specimens by Ohba (1959) suggests three distinct allometric growth stages as revealed by the values of b/l, b/h, and h/l when plotted against l. (where l = length, h = height, and b = breadth in the commoner sense of width). Determinations by Wilbur and Owen (1964) indicate allometry to change at roughly length = 10 mm and again at length = 22 mm.

The thickness of the shell is also a function of growth rate, with slower-growing individuals having thicker shells. Watanabe and Katayama (2010) studied four Manila clam populations from their indigenous range on the coast of Japan and determined that faster-growing individuals in better nutritional condition have lower shell thickness than slower-growing individuals.

5.6 Shell color and pattern

A translation of the Adams & Reeve (1850) original Latin description of the London-type specimen has the external shell color as variegated tawny, the latter being a color adjective describing a mix of yellow, orange, and brown. Usually, superimposed on this general variegation are alternating lighter and darker stripes following the curved dorsoventral orientation of the radial ribs. These colors and patterns are a consequence of varying activity in pigment-producing mantle border cells during the growth phases of the clam shell.

However, external shell color and pattern are highly variable, with shell background colorations from tawny to largely white. The striped pattern is also not universal, being replaced by a range from uniform background coloration to irregular blotches on lighter backgrounds. Such a range can be found within single breeding populations (Wolfling et al, 2015).

FIGURE 5.4

Manila clam color variants collected from a single-seeded clam bed in Samish Bay, Washington State, United States.

Courtesy Bill Dewey, Chuckanut Shellfish (photo by the author).

Considering its year of collection, we can be confident that the London type specimen, at least, is a wild type, unaffected by the sorts of intensive selective breeding programs that are sometimes part of aquaculture operations. As part of the process of industrial production, many clams are spawned, hatched, and grown as juveniles from broodstock in commercial hatcheries and nurseries, sometimes with specific color variants as goals of selection. Such breeding programs may seek to genetically improve Manila clam stock in terms, for example, of improved growth rates or reduced mortality in response to environmental challenges, but such benefits may also correlate with shell color. In China, physiological studies have demonstrated that "zebra" and orange strains showed greater resistance to temperature, salinity, and infection stress compared to the wild type (Nie et al., 2017).

Manila clams cultured in Hawaiian breeding programs for the US company Taylor Shellfish have several color types ranging from blue to white (Chapter 17). Fig. 5.4 shows a not dissimilar diversity of color variants taken from a seeded culture ground in Samish Bay, Washington, United States. In China, a major selective breeding program was initiated in 2005, in which natural color variations have been selected and inbred over many generations, resulting in color and pattern strains such as "zebra" "white and orange," in contrast to the wild type, whose coloration and pattern are more like the Adams and Reeve (1850) type specimen. Within their indigenous range, bright orange and red patterning is not uncommon (see Hong et al.'s 2006 Korean catalogue). However, in areas where hatchery stock is routinely released, it is difficult to estimate the extent of hatchery selection and release as a cause of color variation, although this must be a significant factor in many cases. Although colors and color patterns are

FIGURE 5.5

Shell color change over three age cohorts UK wild Manila clams from Poole Harbour, UK.

sometimes selected for aesthetic and potential value reasons, color variants fade as they grow, exemplifying an apparently ontogenetic progression that is also manifest in wild populations. Fig. 5.5 shows color variation by age in a wild UK population. Overall, *R. philippinarum* is capable of great variations in external shell color and pattern, with genetics, environment, and ontogeny all contributing to this variety.

Wolfling et al. (2015) reported the various color variants not to be genetically distinguishable based on DNA sequencing and suggested them to be a consequence of environmental factors such as pollution. However, among the genes responsible for shell color are those for Tyrosinase (Tyr) and associated transcription factors (Mitf). These are involved in the production of the pigment melanin from tyrosine. Yan et al. (2019) found 21 Tyr genes and 12 homologues of the Mitf gene in the *R. philippinarum* genome and reported variations in the expression of these genes associated with shell color. In the shell color variant described as orange, most of the Mitf mRNAs were upregulated, with three uniquely strongly expressed in comparison to other shell color variants (including "zebra" and "white"). Moreover, upregulation of four Mitf genes in juvenile clams was detected, and the Tyr gene family also showed specific expression patterns at different developmental stages. Findings that appear to corroborate an ontogenetic aspect to shell color.

Since Tyrosine and the Tyr gene are also involved in the Manila clam's immune responses, we cannot assume adaptive benefits of color per se, and indeed, in the wild, any selective advantages of different shell color variants, such as crypsis, are difficult to discern, not least in the context of an infaunal life where camouflage cannot serve as protection from predation.

However, ontogenetic changes in gene expression and shell color may provide some insights. It is possible to speculate that the more pronounced colors and patterns characteristic of settled spat and juvenile clams may provide some protective advantage when the clams are so small as to be effectively mixed in with, and initially attached to, surface sediments of granular appearance. As the clam grows and penetrates deeper into the sediment, the shell color pattern would seem to become redundant as a protection from predators. However, the brighter colors of the small specimens make this seem less plausible, and in light of the tyrosine link, the more likely explanation is that shell color has evolved as an incidental by-product linked to the metabolism of immunity (see also Saenko and Schilthuizen, 2012).

Internal shell color was described by Adams and Reeve (1850) as partly violet and, consistent with their type specimen, *R. philippinarum* commonly has internal violet shading posteriorly (Fig. 1.1). While this is often regarded as symptomatic of the species in wild-type specimens, it is also a variable phenotypic character. Consequently, neither internal nor external shell color is a universal and reliable guide to identification, even though they can provide useful corroborating indications.

Overall, with regard to identification, it should be recognized that taxonomic species descriptions based on shell features alone do not always easily accommodate the true range of phenotypic variation. Species descriptions must necessarily represent typical forms and colors, and, not least in the presence of sympatric phenotypically similar species such as *R. decussatus* and *R. aspera,* identification based on shell characters alone can be misleading (see the Appendix).

References

Adams, A., & Reeve, L. (1850). Mollusca. In A. Adams, The zoology of the voyage of H.M.S. Samarang: Under the command of Captain Sir Edward Belcher, C.B., F.R.A.S., F.G.S. during the years 1843–1846. London: Reeve, Benham, and Reeve.

Caill-Milly, N., Bru, N., Barranger, M. é, Gallon, L., & D'amico, F. (2014). Morphological trends of four Manila Clam populations (Venerupis philippinarum) on the French Atlantic Coast: Identified spatial patterns and their relationship to environmental variability. *Journal of Shellfish Research, 33*(2), 355–372. https://doi.org/10.2983/035.033.0205.

Harper, E. M. (1997). The molluscan periostracum: An important constraint in bivalve evolution. *Palaeontology, 40*(1), 71–97.

Hong, S. Y. (2006). *Marine invertebrates in Korean coasts*. Academy Publishing.

Kanazawa, T., & Sato, S. (2008). Environmental and physiological controls on shell microgrowth pattern of *Ruditapes philippinarum* (Bivalvia: Veneridae) from Japan. *Journal of Molluscan Studies, 74*(1), 89–95. https://doi.org/10.1093/mollus/eym049.

Meisterhans, G., Raymond, N., Girault, E., Lambert, C., Bourrasseau, L., de Montaudouin, X., Garabetian, F., & Jude-Lemeilleur, F. (2015). Structure of Manila Clam (*Ruditapes philippinarum*) microbiota at the organ scale in contrasting sets of individuals. *Microbiological Ecology, 71*(1), 194–206. https://doi.org/10.1007/s00248-015-0662-z.

Mikkelsen, P. M., Bieler, R., Kappner, I., & Rawlings, T. A. (2005). Phylogeny of Veneroidea (Mollusca: Bivalvia) based on morphology and molecules. *Zoological Journal of the Linnean Society, 148*, 439–521.

Mu, G., Duan, F., Zhang, G., Li, X., Ding, X., & Zhang, L. (2018). Microstructure and mechanical property of *Ruditapes philippinarum* shell. *Journal of the Mechanical Behavior of Biomedical Materials, 85*, 209–217. https://doi.org/10.1016/j.jmbbm.2018.06.012.

Nerlovi, V., Korlevi, M., & Mravinac, B. (2016). Morphological and molecular differences between the invasive bivalve *Ruditapes philippinarum* (Adams and Reeve, 1850) and the native species *Ruditapes decussatus* (Linnaeus, 1758) from the northeastern Adriatic Sea. *Journal of Shellfish Research, 35*(1), 1–9.

Nie, H., Liu, L., Huo, Z., Chen, P., Ding, J., Yang, F., & Yan, X. (2017). The Hsp70 gene expression responses to thermal and salinity stress in wild and cultivated Manila clam. *Aquaculture, 470*, 149–156.

Ohba, S. (1959). Ecological studies in the natural population of a clam, Tapes japonica, with special reference to seasonal variations in the size and structure of the population and to individual growth. *Biological Journal of Okayama University, 5*, 13–42.

Paillard, C., & Pennec, M. Le (1993). Ultrastructural studies of the mantle and the periostracal lamina in the Manila clam, *Ruditapes philippinarum. Tissue and Cell, 25*(2), 183–194. https://doi.org/10.1016/0040-8166(93)90018-g.

Ponurovskii, S. K. (2000). Size and age structures of the bivalve mollusk *Ruditapes philippinarum* population in the shallow waters of South Primor'e. *Oceanology, 40*(5), 693–699.

Poutiers, J. M. (1998). Bivalves. In K. E. Carpenter, & V. H. Niem (Eds.), *The living marine resources of the Western Central Pacific. Volume 1. Seaweeds, corals, bivalves and gastropods. FAO species identification guide for fishery purposes* (pp. 123–362). FAO.

Richardson, C. A. (1987). Tidal bands in the shell of the clam Tapes philippinarum (Adams & Reeve, 1850). Proceedings of the Royal Society of London. Series B. *Biological Sciences, 230*(1260), 367–387. https://doi.org/10.1098/rspb.1987.0025.

Saenko, S. V., & Schilthuizen, M. (2012). Evo-devo of shell colour in gastropods and bivalves. *Current Opinion in Genetics & Development, 69*, 1–5.

Silina, A. V., & Popov, A. M. (1989). Investigation of linear growth of the bivalve mollusk *Ruditapes philippinarum* from Peter the Great Bay (Sea of Japan) by the shell structure. *Russian Journal of Marine Biology, 4*, 265–271.

Stanley, S. M. (1970). *Relation of shell form to life habits of the Bivalvia (Mollusca)*. The Geological Society of America.

Suarez, A. A., & Raven, H. (2020). First records of naturalised *Ruditapes philippinarum* (Adams & Reeve, 1850) (Bivalvia: Veneridae) in Asturias (NW Spain), keys for identification, and evidence of hybridisation with native *Ruditapes decussatus* (Linnaeus, 1758) Elona. *Revista de Malacologia Iberica, 2*, 67–83.

Taylor, J D, & Laymen, M (1972). The mechanical properties of bivalve (Mollusca) shell structures. *Palaeontology, 15*, 73–87.

Tomiyama, T. (2021). Ecophenotypic plasticity in shell growth direction of the asari clam *Ruditapes philippinarum. Journal of the Marine Biological Association of the United Kingdom, 101*, 555–560. https://doi.org/10.1017/S0025315421000412.

Trueman, E. R., Brand, A. R., & Davis, P. (1966). The effect of substrata and shell shape on the burrowing of some common bivalves. *Proceedings of the Malacological Society of London, 37*, 97–109.

Watanabe, S., & Katayama, S. (2010). Relationships among shell shape, shell growth rate and nutritional condition in the Manila Clam *Ruditapes philippinarum* in Japan. *Journal of Shellfish Research, 29*(2), 353–359.

Wilbur, K. M., & Owen, G. (1964). Chapter 7. Growth. In K. M. Wilbur, & C. M. Yonge (Eds.), *Physiology of the Mollusca. Volume 1* (pp. 211–242). Academic Press. https://doi.org/10.1016/B978-1-4832-3241-6.50013-5.

Wolfling, M., Rohrhirsch, M., Uhl, B., & Keller, A. (2015). DNA-analyses of Venerupis philippinarum (Adams & Reeve 1850) with a new and constant shell-pattern from Italy, Emilia Romagna. *Malacologia, 86*, 30–32 Mostra mondiale - Cupra Marittima - I/2015, Gennaio, n.

Yan, X., Nie, H., Huo, Z., Ding, J., Li, Z., Yan, L., Jiang, L., Mu, Z., Wang, H., Meng, X., Chen, P., Zhou, M., Rbbani, M. G., Liu, G., & Li, D. (2019). Clam genome sequence clarifies the molecular basis of its benthic adaptation and extraordinary shell color diversity. *Iscience, 19*, 1225–1237. https://doi.org/10.1016/j.isci.2019.08.049.

CHAPTER

Soft tissue anatomy and burrowing

6

6.1 Soft tissues

Within the shell, the body of the clam is enclosed within the two-lateral sheet-like lobes of the mantle, which create an internal mantle cavity. Within this cavity two lateral gills (ctenidia) are suspended on either side of a visceral mass and ventral muscular foot which can be extended as the primary organ of locomotion (Fig. 6.1). Characteristic of the subclass Autobranchia, the ctenidia are of the "eulamellibranch" form: highly evolved W-shaped organs, which function in both feeding and gas exchange and effectively separate the mantle cavity into two compartments. Ctenidial structure and function are considered further in Chapter 8. The absence of a head is indicative of the highly derived molluscan bivalve condition and the phylogenetic ancestral commitment to a sedentary benthic life. Fig. 6.2 shows the general disposition of soft tissues viwed laterally.

The posterior parts of the inner folds of the mantle lobes are fused to form two tubular siphons with the inhalant positioned ventral to the exhalant. These extend to contact the water column above the normally completely buried clam body. The apertures of the siphons of *Ruditapes philippinarum* are surrounded by crowns of numerous tentacles: approximately 34 on the exhalant and 29 on the inhalant (Wang et al., 2021). These function as a filter to stop the influx of large particles and are visible in Fig. 6.3.

During inhalation of water, the innermost row of tentacles on the inhalant siphon is usually held bent inwards over the aperture, with outer tentacles held parallel to the water flow. Since in life they are normally extended to reach the surface of the sediment, the siphons are vulnerable to cropping by predators but can regenerate over a few days. Siphon cropping is considered further in Chapter 12.

In the Veneridae, the mantle margin consists of one outer and three inner folds (Sartori et al., 2008). Fig. 6.3 also shows part of the mantle margin running ventrally and anterior to the siphons. In addition to the secretion of the valves, these folds include short tentacles which serve sensory functions, providing an important interface with external sediment and interstitial water. Although little work has been reported for the Manila clam, it is likely that, consistent with other bivalves, the mantle margin includes mechanoreceptors that can, for example, respond to vibration from approaching predators, chemoreceptors sensitive to water quality changes, and

The Manila Clam *Ruditapes philippinarum*. Science and Significance of a Globalized Species
DOI: https://doi.org/10.1016/B978-0-443-36704-5.00015-8
Copyright © 2026 Elsevier Inc. All rights are reserved, including those for text and data mining, AI training, and similar technologies.

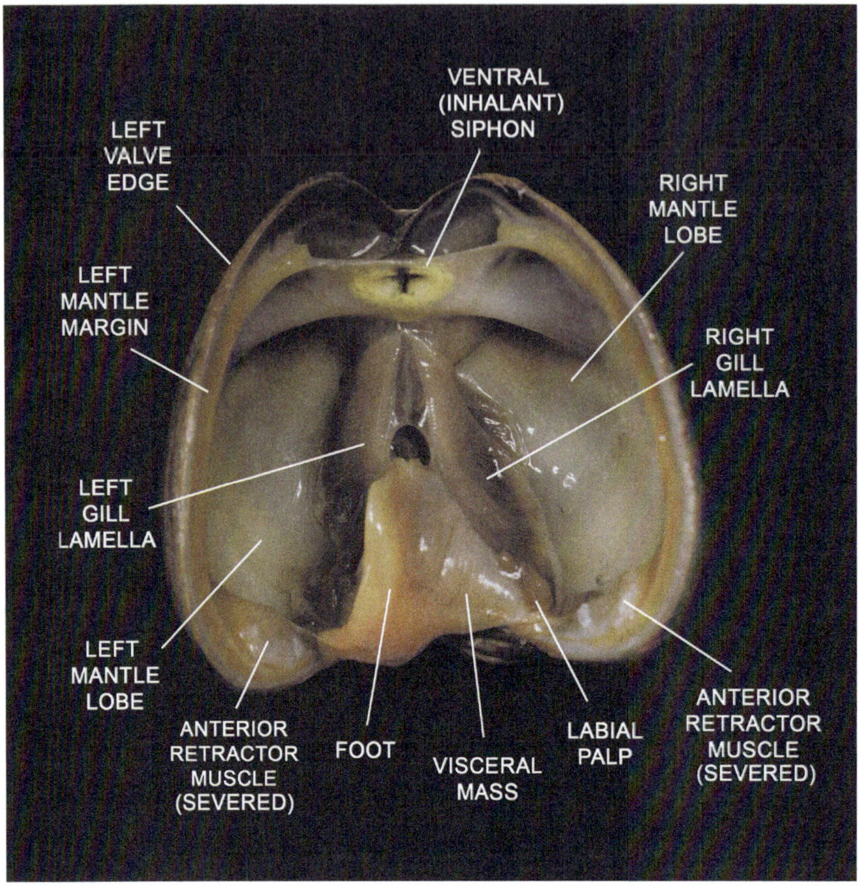

FIGURE 6.1

Ventral view of a freshly opened Manila clam showing the main anatomical features.

The inhalant siphon can be seen to originate from the posterior fusion of the right and left mantle lobes. Elsewhere, the mantle lobes are separate. The dorsal exhalant siphon is concealed below the inhalant siphon. The yellow muscular foot can be seen extending ventrally from the visceral mass to which it is attached. The anterior retractor muscle connects across the two valves but has been severed on opening (photograph by Ester Brown).

simple photosensitive epithelial cells. Sensory information from the mantle fringe is processed in ganglia, which coordinate behavioral responses.

When all soft tissues are removed, the valve interiors show scars that indicate the places of attachment of various organs (Fig. 6.2C). Additionally, the anterior and posterior retractor muscle scars show the attachment positions of powerful muscles that close the valves and strongly resist attempts to open them. A smaller retractor scar

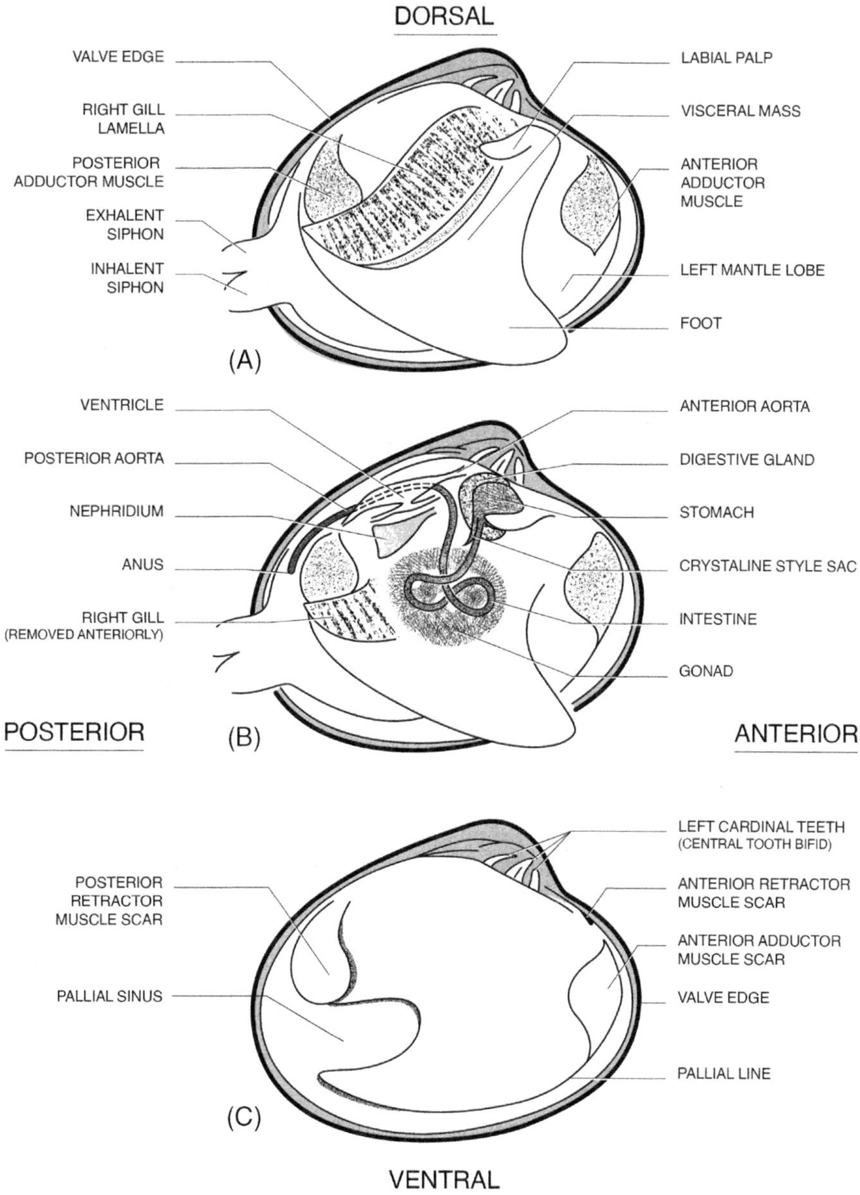

FIGURE 6.2

(A) General disposition of soft tissues after removal of the right valve and mantle lobe. Protruding below the outer gill lamella, the inner lamella is visible (see Chapter 8). (B) Schematic illustration showing the locations of some major organs. (C) View of left valve interior after removal of all soft tissues, showing corresponding tissue attachment scars.

FIGURE 6.3

Right lateral view of live *R. philippinarum* with extended posterior siphons and ventral mantle margin (see text). Scale bar represents 1cm.

Courtesy Nobuhisa Kajino and Kwang-Sik Choi of Jeju National University, South Korea.

marks the point of attachment of a foot retractor muscle. The pallial line corresponds to the mantle attachment, and the pallial sinus to the location of the retracted siphons. The size of the pallial sinus is significant as an indication of the size of the siphons and, therefore, of burrowing depth. In preserved specimens, the siphons are typically short and contracted, but in life, they are highly extendable. Lam (1980) has reported that in life, the siphons may exceed the shell length, a statement corroborated by Fig. 6.4. The maximal extent of the siphons can be revealed by subjecting submerged clams to heat stress and oxygen deprivation. In the laboratory, this can be achieved during shallow submersion under a strong lamp.

The eulamellibranch ctenidia and well-developed siphons constitute an efficient mechanism by which water is drawn into the mantle cavity via the ventral inhalant siphon, passed through and filtered by the ctenidia, and expelled via the dorsal exhalant siphon.

6.2 Burrowing technique

The foot is extensible and can help penetrate the sediment at the beginning of a burrowing movement. By muscular control of internal hydrostatic adjustments, it can

FIGURE 6.4

Manila clams for sale on the quay in Qingdao, China, showing siphon extension capacity greater than the length of the shell under hypoxic stress conditions (photograph by the author).

then be configured to serve as an anchor while also contracted in length to pull the main body and shell after it. Anterior downward movement can be facilitated by valve closure, which ejects water from the mantle cavity, thereby liquefying the surrounding sediment, reducing drag. A downward rotation of the posterior completes the cycle. In summary, each cycle includes the following components:

- Foot probes down into sediment
- Siphons close
- Adductor muscles cause rapid valve closure, ejecting mantle cavity water ventrally
- Simultaneous foot dilation within the sediment
- Foot retractor muscles contract. Anchored foot therefore, pulls the clam into the sediment
- Muscles relax and valves gape
- Cycle repeated

As a consequence of these cycles, burrowing involves a repeated rocking behavior, in which the anterior and then posterior shell is effectively pulled into the sediment in stages and stabilized using the extended foot. The adaptive significance of the relatively flat dorsoanterior lunule (see Chapter 5) relates to its service as a fulcrum for the posterior downward rotation after anterior penetration of the sediment (Germann et al., 2011; Stanley, 1975, 1988). There is also evidence that concentric shell ribs may enhance the stability of the anterior during burrowing (Stanley, 1981).

6.3 Burrowing depth and rate

Burrowing depth was defined by Stanley (1970) as the distance from the sediment surface to the shallowest point on the buried shell. Manila clams normally adopt a posterior-up position, though slightly oblique to the vertical. In a study of the burrowing depths of Japanese infaunal bivalves, Kondo (1987) reported that the depth to which *R. philippinarum* buries varies with shell length and sediment grade. In fine sand, small clams of shell length (L) up to about 1.0 cm have a maximum burrowing depth (D) of 1.5 cm (D = 1.5 × L). However, in larger clams, Kondo found this relationship to break down, such that shell lengths of 2–4 cm were found at depths of roughly 1–3.5 cm. Stanley (1972) has suggested that this "allometric shallowing" by larger individuals may be due to their decreased susceptibility to both hydrodynamic disturbance and predation.

In muddy sand, *R. philippinarum* appears to burrow to shallower depths (D = 0.7 L), while in gravelly sand to markedly greater depths of up to 5 or 6 cm (D = 2.6 L). Other factors, such as hypoxia and temperature, also affect burial depth, but in any event, the species is generally considered a shallow burrower, commonly found less than 5 cm from the sediment surface.

For active shallow burrowers, there is an interspecific relationship between maximum burrowing depth, the length of the siphons, and the depth of the pallial sinus. The pallial sinus index (PSI) is the ratio between the distance from the innermost part of the sinus to the posterior extreme of the shell and the total shell length. In 16 Japanese Manila clams examined by Kondo (1987), the PSI ranged from 0.431 to 0.552. According to Kondo in *R. philippinarum*, this ratio is quite variable, and the species is one of the few in which individual PSI increases with age.

The downward burrowing velocity of Manila clams increases with finer sediments: Kido et al. (1996) reported that in mudflat sediments of median particle diameter (Md_{50}) of 0.079 mm, specimens of mean shell length 32 mm at water temperature between 19.1°C and 20.1°C exhibited a mean downward burrowing velocity of 0.056 mm s^{-1}. In coarser sand sediments (Md_{50} = 0.78 mm), downward mean burrowing velocity was slower at 0.025 mm s^{-1}. In both sediment grades upward, burrowing was reported as markedly faster, although the experimental technique used was different.

6.4 Active control of sediment depth

In Chapter 5, features of shell morphology—such as tumidity and rugosity—are interpreted as adaptations that enhance stability within the sediment, serving to resist predation rather than facilitating rapid burrowing. This suggests that optimal circumstances for mature, established Manila clams involve remaining in a secure and static position in a stable sediment. Certainly, their capacity for repeated reburial if removed from sediments in the laboratory is limited, indicating the activity requires a significant and limiting energetic commitment. Nevertheless, there are various circumstances in the lives of clams that require effective burrowing behavior. These include adjustments of depth in response to hydrodynamically triggered chronic changes in sediment height, acute disturbance of sediments during storms, and partial or complete extraction from sediments during failed predation attempts.

Therefore, although Manila clams favor relatively stable sediments, it should not be supposed that they do not actively control their burial depth. On Banzu tidal flats, Tokyo Bay, Kakino (2000) used field experiments to compare the effect of wave action and sediment erosion on retention of living and dead (formalin-fixed with shells sealed) clams of mean shell length 26–28 mm. Over a period of 7–12 days, 1 cm (by height) of sediment erosion had little effect on the retention of both live and dead clams. At 2 cm of erosion, a marked difference between live and dead clam loss occurred, which continued to 4 cm erosion, at which point most, but not all, live clams were lost. It appears that adult Manila clams can exhibit vertical movement to maintain their position at constant depth in circumstances of gradual sediment erosion over a few days, provided other circumstances are sufficient for the maintenance of condition and therefore burrowing ability. Smaller clams are assumed to be more vulnerable.

Manila clams are also capable of "active escape" from the sediment, which they achieve with an anterior-up vertical orientation, suggesting the foot is used similarly to that during burial (Kondo, 1987). In the field, it is not uncommon to find moribund or recently dead (with flesh) Manila clams on the sediment surface. This could be due to extraction during failed predation attempts or possibly weakened clams being washed out of unstable sediments by wave action. However, the phenomenon can be observed on stable sediments with adjacent clams still buried. Also, it might be expected that healthy clams could at least once regain an infaunal position by reburial. On balance, it seems likely that many such finds result from active upward migration. In the laboratory, recently dead clams are reported both on the surface and buried, with the greater number remaining buried (Roman et al., 2023, 2024, and personal communication).

Liu et al. (2025) illustrate both the active control of sediment depth by Manila clams and two important depth-determining factors. These authors exposed clams to various combinations of temperature from 24°C to 32°C, simulating heat wave conditions, and dissolved oxygen concentrations from 1 mg/L (hypoxic) to 6 mg/L. They reported the species to be capable of rapid behavioral responses to changed conditions. These included increased burial depth with increased temperature and decreased burial depth in response to hypoxia. A combined progression of heatwave and hypoxia conditions led to a transition from an initial downward movement followed by an upward movement corresponding to a shift from aerobic to anaerobic respiration. In response to the lowest dissolved oxygen conditions, upward movement resulted in the clams emerging onto the surface of the sediment, which, under higher temperature conditions, resulted in death. This work provides an alternative explanation for the occasional finding of fully surfaced clams on apparently stable sediments.

Takeuchi et al. (2015) compared the burrowing abilities of *R. Philippinarum* with two other sympatric infaunal Venerid species in Ariake Bay, Japan: *Meretrix lusoria* and *Mactra veneriformis*. Based on laboratory studies, they reported that forced repetitive reburrowing reduced the burrowing performance of *R. philippinarum,* while enhancing or maintaining it in the other two species. They postulated that differences in the resistivity of juvenile clams to unstable sediments explained their overwintering success/failure in the high to mid-tide zone, with *R. philippinarum* least able to survive sediment instability during the north-westerly winter monsoon. At lower shore elevations with more stable sediments, *R. philippinarum* was dominant.

References

Germann, D. P., Schatz, W., & Hotz, P. E. (2011). Artificial bivalves - The biomimetics of underwater burrowing. *Procedia Computer Science, 7*, 169–172. https://doi.org/10.1016/j.procs.2011.09.012.

Kakino, J. (2000). Dispersal of Japanese littleneck clam Ruditapes philippinarum (Adams and Reeve) in relation to changes of bottom level due to wave action on Banzu tidal flats, Tokyo Bay. *Fisheries Engineering, 37*(2), 115–128. https://doi.org/10.18903/fisheng.37.2_115.

Kido, K., Kinoshita, H., Umemori, T., Honda, M., & Isono, R. (1996). Biological studies on coastal artificial structures-2 Biological conditions for possible enlargement of bivalve habitat in the neighbouring coastal area of power plant. *Bulletin on Coastal Oceanography, 33*(2), 133–146.

Kondo, Y. (1987). Burrowing depth of infaunal bivalves- observations of living species and its relation to shell morphologies. *Transactions of the Palaeontological Society of Japan New Series, 148*, 306–323. https://doi.org/10.14825/prpsj1951.1987.148_306.

Lam, V. W. W. (1980). Shell form and diagnostic differences in the structure of the siphons and ciliary currents of the ctenidia in coastal species of the Tapetinae (Bivalvia: Veneracea) in Hong Kong. In B. Morton (Ed.), *Proceedings of the first international workshop on the malacofauna of Hong Kong and southern China* (pp. 11–31). Hong Kong University Press.

Liu, Y. i, Ren, J. S., Wang, X., Chen, J., Wu, W., & Zhang, J. (2025). The impact of marine heatwaves and hypoxia on behavior and physiology of Manila clam *Ruditapes philippinarum*. *Marine Pollution Bulletin, 219*, 118327. https://doi.org/10.1016/j.marpolbul.2025.118327.

Roman, M., Vázquez, E., Viejo, R. M., Woodin, S. A., Wethey, D. S., Román, S., Weidberg, N. F., Troncoso, J. S., Mendez, M. M., & Olabarria, C. (2024). Context-dependent resilience of intertidal seagrass and venerid clams after hyposalinity stress. *Marine Ecology Progress Series*. https://doi.org/10.3354/meps14516.

Roman, S., Vázquez, E., Roman, M., Viejo, R. M., Weidberg, N. F., Troncoso, J. S., Woodin, S. A., Wethey, D. S., & Olabarria, C. (2023). The stress response of the seagrass *Zostera noltei* and three commercial clam species to low salinity associated with heavy rainfall. *ICES Journal of Marine Science, 0*(0)), 1–17. https://doi.org/10.1093/icesjms/fsad203.

Sartori, A. F., Mikkelsen, P. M., Bieler, R., & Printrakoon, C. (2008). Siphonal structure in the Veneridae (Bivalvia: Heterodonta) with an assessment of its phylogenetic application and a review of the Venerids of the Gulf of Thailand. *The Raffles Bulletin of Zoology Supplement, 18*, 103–125.

Stanley, S. M. (1970). In *Relation of shell form to life habits in the Bivalvia (Mollusca): 125* (pp. 1–296). Geological Society of America. https://doi.org/10.1130/MEM125-p1.

Stanley, S. M. (1972). Functional morphology and evolution of byssally attached bivalve mollusks. *Journal of Palaeontology, 46*, 165–212. https://www.jstor.org/stable/1302843.

Stanley, S. M. (1988). Adaptive morphology of the shell in bivalves and gastropods. *The Mollusca, form and function. Volume 11.* Academic Press https://doi.org/10.1016/B978-0-12-751411-6.50012-6.

Stanley, S. M. (1975). Why clams have the shape they have: An experimental analysis of burrowing. *Paleobiology, 1*(1), 48–58. https://doi.org/10.1017/S0094837300002189.

Stanley, S. M. (1981). Infaunal survival: Alternative functions of shell ornamentation in the Bivalvia (Mollusca). *Paleobiology, 7*(3), 384–393. https://doi.org/10.1017/s009483730000467x.

Takeuchi, S., Yamada, F., Shirozu, H., Ohashi, S., & Tamaki, A. (2015). Burrowing ability as a key trait in the establishment of infaunal bivalve populations following competitive release on an extensive intertidal sandflat. *Journal of Experimental Marine Biology and Ecology, 466*, 9–23. https://doi.org/10.1016/j.jembe.2015.01.011.

Wang, Z., Bai, Y., Nie, H., Xu, Q., Yin, Z., Zhang, Y., Yin, X., & Yan, X. (2021). Molecular mechanisms of wound healing and regeneration of siphon in the Manila clam *Ruditapes philippinarum* revealed by transcriptomic analysis. *Genomics, 113*(3), 1011–1025. https://doi.org/10.1016/j.ygeno.2021.02.010.

CHAPTER

7 Feeding, food sources, and assimilation

7.1 Ontogenetic progression of feeding habits

The organs involved with adult feeding are not all fully developed by the point of first larval settlement. Consequently, there is an ontogenetically linked progression of feeding habits during the early stages of benthic life. In Manila clams, the siphons begin to form at shell length around 0.4 mm, but the inhalant siphon, through which food is drawn into the adult shell, is not completed until around 2 mm length (Inoue 1980). Toba et al. 1992 observed early settled Manila clams "pedal-feeding" on organic matter at the sediment surface collected with the extended foot.

Houki et al. (2025) have determined the ontogenic progression of the postsettlement diet of clams near Kyoto, Japan, using diatom-specific DNA metabarcoding analysis to indicate the relative importance of benthic and planktonic food. They identified three phases. The first pedal-feeding stage of clams up to 2 mm shell length involved the consumption of benthic species, a second transition phase between 2 and 8 mm shell length involved consumption of both benthic and planktonic diatoms, and a third filter-feeding stage was reached at shell length greater than 8 mm, when the clams became strongly dependent on planktonic food species. At this point, both siphons and gills are fully developed, and pedal feeding ceases. Subsequently, while benthic diatoms may remain an important food source, they only become available when suspended in the water column by currents or other disturbances.

The remaining parts of this chapter address feeding and digestion in adult Manila clams.

7.2 The ctenidia and labial palps

Below, a summary of form and function in adult clams is provided. More detailed information on the structure of the bivalve gill type of adult clams can be found in early works such as Ridewood (1903), with later elucidation by Ansell (1995, 1961) and Beninger and St Jean (1997). More recently, Barnes (2004) and Gosling (2015) have also provided useful descriptions of the structures and mechanisms of filtration. Factors affecting the rates at which suspended matter is removed from ambient water in experimental conditions are considered in Chapter 8.

Because of their comb-like appearance, the gills of bivalves are termed ctenidia; they function in gas exchange but also have been greatly expanded as filter-feeding

The Manila Clam *Ruditapes philippinarum*: Science and Significance of a Globalised Species
DOI: https://doi.org/10.1016/B978-0-443-36704-5.00017-1
Copyright © 2026 Elsevier Inc. All rights are reserved, including those for text and data mining, AI training, and similar technologies.

organs. Manila clam ctenidia, along with many extant bivalves, are of the phylogenetically advanced "eulamellibranch" type. They consist of many W-shaped filaments on each side of the visceral mass. On opening a clam, these transversely orientated rows of filaments give a vertically striated comb-like appearance (Fig. 6.1). The upper center of each filament is attached to the dorsal mantle, and each filament is attached to its anterior and posterior neighbor by tissue junctions (the interfilamentar junctions), so the rows of filaments form four flat lamellae running along each side of the clam. In the Venerids, the inner half of the W (a V-shaped "demibranch") is larger and therefore visible below the ventral margin of the outer lamella when the shell is opened (Fig. 6.2A).

Further structural support for each ctenidium is provided by tissue junctions connecting the descending and ascending lamellae of each demibranch (the interlamellar junctions) and by connections between the ascending filament tips and the mantle on one side, and the visceral mass on the other. These connections and attachments create two W-shaped filtering lattices, one on either side of the visceral mass, which effectively separate the mantle cavity into ventral and dorsal chambers. Cilia on the filaments create a current of water through the inhalant siphon into the ventral inhalant mantle chamber, through the gill lamellae, into the dorsal exhalant chamber, and out of the exhalant siphon. As water flows between the gill filaments, gas exchange occurs, facilitated by a counter-current arrangement. However, the extent of the gill surface area and the rate of water flow are much greater than is needed for gas exchange alone, and the ctenidia also filter out and transport particulates.

The cilia maintaining the water flow through the system are laterally positioned on the front (upstream) edges of the filaments. In addition, "laterofrontal cirri" (consisting of fused cilia) extend across the gap between filaments and interdigitate with their opposite neighbors to form a mesh, filtering particles from the water. Mucus secreted by the gill filaments plays a role in both the capture and movement of particles, enabling the trapping of food as small as bacteria, but also forming strings of particles moving along ventral grooves. While smaller particles move along these grooves toward the mouth, larger particles remain only superficially positioned and may detach. Retained particles are moved down the filaments to connected grooves running along the ventral margins of the filaments of the demibranch, which in turn carries them anteriorly toward the labial palps.

The labial palps (Figs. 6.1 and 6.2A) are a pair of fleshy flaps, one of which envelops the anterioventral region of each ctenidium. The inside surfaces of the palps, which lie adjacent to the ctenidia, are deeply corrugated. Essentially, the front part of each ctenidium lies sandwiched between the corrugated inner surfaces of the inner and outer palp lobes. Nevertheless, the nervous and vascular supplies of the palps indicate that they are not ctenidial in origin but can rather be considered as associated with the mouth, to which they lead, and therefore as oral extensions at the beginning of the digestive system.

The size and complexity of the palps in the Veneracea tend to be larger in species likely to experience exposure to turbid water: mud-tolerant and littoral species generally have larger and more numerous palp ridges. This is indicative of the role of the labial palps as the primary active sorting sites, discriminating between particles to be rejected and ingested.

Active sorting in the Veneridae is achieved on and between the ridges of the palps based on both ciliary and muscular action, the latter controlling the erection or relaxation of the palp ridges, opening or closing off, respectively, the deep channels between them. The degree of ridge erection and, therefore, efficiency of sorting is determined by the amount of particulate material in the inhalant water. When little material arrives, the ridges relax, lying over against each other such that the deep channels between them are closed off, and little or no sorting occurs: all particles then follow "acceptance currents" running over the top of one ridge to the next and toward the mouth. Particles reaching the proximal end of the palps are transported to the mouth by cilia.

However, depending on the turbidity of the inhalant water, much or all of the filtered particles may be rejected. With moderate amounts of material in the inhalant current, the ridges undulate, allowing large particles to fall between them, where ciliary "rejection currents" carry them to the unridged palp outer surface for release into the mantle. When larger amounts of material arrive, the ridges become fully erect, exerting a strong sorting action. If necessary, the palps may adopt a protective configuration in such a way as to reject all material.

Through such responses, eulamellibranch bivalves can decouple the feeding and gas exchange functions of the gill, such that in more turbid water, ingestion of particulates can be controlled without compromising the gas exchange function of the gill. Moreover, the Venerid gill and labial palps control both the quantity and quality (size) of food particles reaching the digestive system.

Material passively detached from the ctenidia or actively rejected by the palps accumulates in the mantle cavity, from where it is channeled toward the inhalant siphon and expelled as "pseudofaeces" by muscular contractions that reverse the usual inhalant flow.

7.3 The digestive system

Food particles accepted by the ctenidia and palps pass into the mouth and are moved along ciliated tracts in a short esophagus, which connects into the anteriodorsal region of the globular stomach. Ansell (1961) and Purchon (1987) concluded that the anatomy of the stomach is more or less the same throughout the Venerid family.

Connecting with the stomach posterioventrally is the blind ended style sac within which a gelatinous, more or less rod-shaped "crystalline style" of around 5 mm length is formed. In conjunction with the eulamellibranch ctenidial gill, the structure and function of the crystalline style represent a supreme adaptation for a microphagous diet requiring the digestion and assimilation of benthic diatoms, not least as it provides a mechanism capable of crushing their hard siliceous cell walls (frustules).

The style extends from the style sac into the lumen of the stomach, where it contacts a chitinous region of the stomach wall known as the gastric shield. Cilia within the style sac cause the style to rotate, making it one of the very few structures in animal anatomy that revolves (Yonge, 1976). According to Purchon (1968), the rotation of the style has four functions. Firstly, it helps draw in the food laden mucus strings, which are effectively wound around the rotating style. Secondly, it stirs the contents of the stomach, ensuring they contact several corrugated ciliary sorting areas, which act similarly to the labial palps to separate and reject unsuitable particles. Thirdly, it rubs the stomach contents against the gastric shield, which, through mechanical attrition, breaks open cells and diatom frustules. Fourthly, as the distal end of the style itself slowly breaks down in the stomach, it liberates amylolytic enzymes. As the style end breaks down, it is pushed forward as new material, including enzymes, is added in the style sac. Additionally, it is thought that the breakdown of the style acidifies the stomach contents and, in doing so, lowers the viscosity of the incoming mucilaginous strings (Purchon 1968). Rather than being a permanent structure, the condition of the crystalline style is known to vary and can dissolve altogether in unfavorable conditions. Houki and Kawamura (2020) provide images of four style condition types. The hardest and most effective style types also exhibit an expanded region at the stomach end.

Three embayments in the stomach wall (right and left caeca and the left pouch) contain ducts which lead to extensive digestive diverticula within which phagocytosis is followed by intracellular digestion. The introduction of new material and the elimination of rejected material into and from the digestive diverticula involves various ciliary tracts and grooves. The elimination of unwanted material from both stomach and digestive diverticula is achieved via a ciliated rejection tract along an intestinal groove, which extends from the caeca to the midgut and is separated from the main stomach lumen by a flap. Once reaching the midgut, material is subsequently evacuated as feces.

Houki et al. (2018), using diatom species with frustules of different mechanical strengths, reported that the Manila clam stomach and revolving style mechanism is capable of breaking open a range of diatom sizes and hardness with over 90% efficiency. The crushing mechanism relies on a sufficient density of diatoms in the clam's stomach, such that the rotation of the crystalline style agitates densely packed diatoms, creating attrition between the frustules, causing rupture.

7.4 Food sources in nature

The anatomy and functioning of *Ruditapes philippinarum* from siphons to ctenidia and palps clearly indicate a life filtering suspended solids from the water above the sediment surface. However, this does not necessarily make them primarily reliant on planktonic prey in the wild. Breber (2002) suggested that while larval and juvenile Manila clams take their diet from the plankton, adults appear to be dependent on benthic diatoms, which grow as a film on the surface of stable sediments. In the Italian lagoons of the northern Adriatic, he observed the spatial distribution of Manila clams to coincide with those sediment flats where benthic pennate diatoms flourish to form such surface films.

Watanabe et al. (2009) used isotopes of carbon and nitrogen to investigate the relative contributions of particulate organic matter (POM) in the water column and on the sediment surface to the food assimilated by *R. philippinarum*. They showed that while phytoplankton from the general water column can be an important resource when available in large quantities, microphytobenthos (MPB) and associated detritus resuspended from the sediment surface by water movements can also be a significant food source.

Benthic mucilaginous diatom mats have been observed in Poole Harbour, UK, where Manila clams thrive, leading Franklin et al. (2012) to suggest that suspension of such mats by tidal flows provides an explanation for the relatively good clam condition index, measured when phytoplankton levels are low. Similarly, on a tidal flat near Yokohama, Japan, it is mainly MPB that maintains the high productivity of *Ruditapes philippinarum* (Watanabe et al, 2009). Similarly, in a sub-Arctic tidal lagoon, Abe et al. (2015) linked growth rates of Manila clams to current speeds and associated MPB resuspension. Moreover, Moschino et al. (2011) have suggested that the species benefits in fished areas of the seabed due to the resuspension of organic matter by the action of fishing dredges.

Zhao et al. (2013) have also demonstrated the significance of organic detritus as a food source. By comparing the stable isotope content of *R. philippinarum*, benthic microalgae, and POM, all collected from intertidal areas along the Liaodong Peninsula in northern China, it was concluded that *R. philippinarum* feeds mainly on POM (74%) but with benthic microalgae as an important supplemental food source (25.8%).

With a capacity to filter particles from 2 to 200 um (Nakamura, 2001), the food resources available to *R. philippinarum* are clearly diverse. In addition to bacteria, picocyanobacteria, and diatoms, the Manila clam is also known to exploit dinoflagellates (Li & Wang, 2001), along with organic detritus and small rotifers from the water column and sediment surface (Sorokin & Giovanardi, 1995).

In light of the widespread use of phytoplankton cultures in hatcheries and nurseries, it would be easy to overestimate the importance of phytoplankton in the diet of Manila clams in the wild. In Hichirippu Lagoon, northern Japan, phytoplankton contributes, at most, 28% of the food resources required for the Manila clam population, and particulates from degraded seagrass (*Zostera japonica*) and benthic macroalgae (*Ulva pertusa*) are also known to make significant contributions (Komorita et al., 2014). During high water on the Hichirippu tidal flats, the biomass of MPB within 0.5 cm of the sediment surface is about 100 times greater than the biomass of phytoplankton contained in 1m of water column above. Here, *R. philippinarum* is one of the dominant macrobenthic species (Komorita et al., 2010).

This wide range of food sizes and sources enables *R. philippinarum* to adjust its diet opportunistically according to the availability of different foods and is an important factor in its success in estuarine and lagoonal habitats where conditions may vary according to diurnal, lunar, and seasonal cycles. Moreover, in reporting that the efficiency with which *R. philippinarum* accepts the smaller prey groups, (2–8 um) is markedly higher than that reported by Sobral and Windows (2000) for the congeneric *Ruditapes decussatus*, Nakamura (2001) has suggested that this may be causally significant in cases where *R. philippinarum* is more successful than the native*R. decussatus* as is the case in northern Adriatic lagoons.

A capacity to accept smaller, suspended particulates may be advantageous not only in the context of prey species, since finer inorganic silt and clay particles in turbid coastal waters can accumulate organic matter on their surface, and due to their high surface-to-volume ratio, tend to have a relatively higher organic component than larger particles. Nevertheless, the capacity of *R. philippinarum* to filter smaller particles in particular also carries risks associated with their properties as physical scavengers of metal contaminants in polluted waters. The capacity of the Manila clam to process and thereby tolerate environmental metals and metalloids is discussed in Chapter 11.

Food quality is as important as food quantity for filter-feeding bivalves, and the availability of food in terms of concentration does not necessarily equate to nutritive value. Measurements of Carbon: Nitrogen ratios (C/N) indicate that senescent or nutrient-starved phytoplankton cells have lower relative carbohydrate content. Consequently, higher C/N ratios indicate higher food quality for clams. In the diatom *Thalassiosira pseudonana,* an algal C/N ratio of 8.4–10.5 is optimal for *R. philippinarum* (Gallager & Mann, 1981).

7.5 Assimilation and growth rate

Various efficiency rates are involved in determining the relationship between clearance of food from the environment and its absorption within the digestive diverticula, at which point it becomes metabolically available (see Chapter 8).

Sorokin and Giovanardi (1995) have shown that "food preferences" in the sense of assimilation efficiency are different from those of the native European species, with *R. philippinarum* favoring diatoms and *R. decussatus* cyanobacteria mixed with phytoflagellates. Breber (2002) reported that in the northern Adriatic, natural beds of Manila clams of density greater than 1000 clams m^{-2} reached shell lengths of 40mm within 2 years, in areas where benthic pennate diatoms form films on the sediment surface. Such areas are characterized as having relatively strong light intensity and weak wave action, conditions often characteristic of shallow protected bays and lagoons, and when combined with optimal salinities and warm sea temperatures, *R. philippinarum* has the potential to dominate the infaunal community.

Measures of the fraction of consumed and assimilated food used for growth vary with diet. With bacteria and mixed algae (*Chlorella, Dunaliella,* and *Oscillatoria*), these indices are comparable with those of other benthic invertebrates and clams, but double when fed on the benthic diatom *Nitzschia*, demonstrating a high affinity to diatoms as food. Investigating two sub-Arctic Japanese sites of varying hydrodynamic characteristics, Abe et al. (2015) found that Manila clam growth correlated positively with the suspension flux of MPB, which in turn was a function of the rate of tidal flow across the sediment surface. At optimal food conditions, the production coefficient (growth/biomass) of Manila clams is 1.5–2.0 times higher than that reported for mussels (*Mytilus*) and oysters (*Ostrea*), and with an optimal diet of diatoms, this index is a further three times higher.

In the Venice Lagoon, under "normal" trophic conditions of 3.5 mg of food per liter, an average-sized Manila clam could produce 5–10 g of flesh and gametes during a 6-month spring/summer period (Sorokin & Giovanardi, 1995). These authors also reported that the assimilation index was maximal with diatoms (0.92%–1.14%) and green algae (0.68%–1.02%), but minimal (0.27%) in Cyanobacteria.

A major factor in the success of *R. philippinarum* as a non-native introduction has been its growth performance in comparison to native species. In the UK and elsewhere, trials suggested the species to be a "highly lucrative" alternative to the native European clam *Ruditapes decussatus* (Spencer, 2002). In early experimental work, Laing et al. (1987) demonstrated that *R. philippinarum* grows more efficiently and faster than *R. decussatus*, and since then, many studies have confirmed this conclusion.

In summary, a combination of a wide food spectrum, strong production coefficient, and consequently high growth rate represents advantageous characteristics enhancing both the fitness of wild populations and the economic utility of the species. In shallow locations protected from strong wave action but nevertheless tidal, where benthic diatoms flourish, naturalized Manila clam populations can provide one or both of a lucrative wild fishery or an aquaculture crop.

7.6 Adaptive feeding periodicity

Houki and Kawamura (2020) have shown that the condition of the crystalline style of the Manila clam correlates significantly with feeding behavior, such that clams with extended siphons possess a harder (more effective) form of crystalline style. Conversely, the style begins to dissolve when the shell is closed. The mechanism for this dissolution is thought to be gut pH changes due to the accumulation of anaerobic metabolites when the shell closes, with consequent cessation of gas exchange. In any event, these authors have applied the correlation between style condition and siphon extension to investigate the periodicity of feeding behavior.

Since feeding necessarily involves the extension of the siphons to the sediment surface, it increases the risk of predation by visual predators. The intertidal and shallow sublittoral distribution of *R. philippinarum* exposes Manila clams to both day/night and tidal cycles. Work on Japanese clams showed that for intertidally positioned clams, the tidal cycle is predominant in determining feeding patterns (Houki & Kawamura, 2020). Here, feeding is most active at ebb and flow tides. While this may be interpreted simply in terms of tidal currents constantly replenishing phytoplankton abundance, periods of ebb and flow are also the period of maximum scour when benthic diatom mats are most likely to be resuspended just above the sediment-water interface. For intertidal clams, the light/dark cycle was not significant.

At subtidal shallow water elevations, both tidal and light/dark feeding phases were apparent and, to some extent, appear to interact. Nevertheless, feeding appeared to be predominantly at night.

These findings suggest an adaptive behavioral plasticity in *R. philippinarum* through which endogenous rhythms of feeding activity become entrained in such a way as to optimize the balance between feeding opportunity and predation risk. While feeding in the dark may protect against visual predators, intertidal clams, whose feeding periods are limited during low water, may need to feed during the daytime regardless of predator risk. However, it appears they will preferentially exploit those ebb and flow tidal periods when food is most available.

References

Abe, H., Hasegawa, N., Yoon, S., & Kishi, M. J. (2015). Evaluation of Manila clam (*Ruditapes philippinarum*) growth and microphytobenthos resuspension in a subarctic lagoon in Japan. *Hydrobiologia, 758*(1), 87–98. https://doi.org/10.1007/s10750-015-2275-4.

Ansell, A. D. (1959). The functional morphology and general biology of the British species of Veneracea, with particular reference to Venus striatula (da Costa). [Thesis, University of Glasgow]

Ansell, A. D. (1961). The functional morphology of the British species of Veneracea (Eulamellibranchia). *Journal of the Marine Biological Association of the United Kingdom, 41*(2), 489–517. https://doi.org/10.1017/S0025315400024012.

Barnes, R. D (2004). *Invertebrate zoology. A functional evolutionary approach* (seventh edition). Cengage.

Beninger, P., & St-Jean, S. (1997). The role of mucus in particle processing by suspension-feeding marine bivalves: Unifying principles. *Marine Biology, 129*, 389–397. https://doi.org/10.1007/s002270050179.

Breber, P. (2002). Introduction and acclimatisation of the Pacific carpet clam, *Tapes philippinarum*, to Italian waters. In E. Leppäkoski, S. Gollasch, & S. Olenin (Eds.), *Invasive aquatic species of Europe. Distribution, impacts, and management* (pp. 120–126). Kluwer Academic Publishers.

Franklin, D. J., Humphreys, J., Harris, M., Jensen, A. C., Herbert, R. J. H., & Purdie, D. A. (2012). An investigation into the annual cycle of phytoplankton abundance in Poole Harbour and its relationship with Manila clam nutrition. Report for. *the Marine Management Organisation*. London, UK.

Gallager, S. M., & Mann, R. (1981). The effect of varying carbon/nitrogen ratio in the phytoplankter Thalassiosira pseudonana (3H) on its food value to the bivalve Tapes japonica. *Aquaculture, 26*(1-2), 95–105. https://doi.org/10.1016/0044-8486(81)90113-7.

Gosling, E. (2015). *Marine bivalve Molluscs*. John Wiley & Sons Ltd.

Houki, S., & Kawamura, T. (2020). Feeding and digestion periodicity of Manila clam *Ruditapes philippinarum* in natural intertidal and subtidal zones estimated from the morphological condition of the crystalline style. *Journal of Molluscan Studies, 86*(4), 361–371. https://doi.org/10.1093/mollus/eyaa027.

Houki, S., Kawamura, T., Ogawa, N., & Watanabe, Y. (2018). Efficient crushing of hard benthic diatoms in the gut of the Manila clam *Ruditapes philippinarum* – Experimental and observational evidence. *Journal of Experimental Marine Biology and Ecology, 505*, 35–44. https://doi.org/10.1016/j.jembe.2018.04.007.

Houki, S., Ozaki, M., & Sano, N. (2025). Ontogenic shift of feeding habit in Manila clam *Ruditapes philippinarum* after settlement revealed by diatom-specific DNA-metabarcoding diet analysis. *Estuarine, Coastal and Shelf Science, 319*, 109244. https://doi.org/10.1016/j.ecss.2025.109244.

Inoue, Y. (1980). The ecology and environment of Manila clam in Omi Bay, Yamaguchi, Japan. *J, 16*, 22–35 ournal of Fisheries Engineering.

Komorita, T., Kajihara, R., Tsutsumi, H., Shibanuma, S., Yamada, T., Higaki, N., & Montani, S. (2010). Reevaluation of the nutrient mineralization process by infaunal bivalves (*Ruditapes philippinarum*) in a shallow lagoon in Hokkaido, Japan. *Journal of Experimental Marine Biology and Ecology, 383*(1), 8–16. https://doi.org/10.1016/j.jembe.2009.10.018.

Komorita, T., Kajihara, R., Tsutsumi, H., Shibanuma, S., Yamada, T., & Montani, S. (2014). Food sources for *Ruditapes philippinarum* in a coastal lagoon determined by mass balance and stable isotope approaches. *PLoS ONE, 9*(1), e86732. https://doi.org/10.1371/journal.pone.0086732.

Laing, I., Utting, S. D., & Kilada, R. W. S. (1987). Interactive effect of diet and temperature on the growth of juvenile clams. *Journal of Experimental Marine Biology and Ecology, 113*(1), 23–38. https://doi.org/10.1016/0022-0981(87)90080-3.

Li, S. C., & Wang, W. X. (2001). Radiotracer studies on the feeding of two marine bivalves on the toxic and nontoxic dinoflagellate Alexandrium tamarense. *Journal of Experimental Marine Biology and Ecology, 263*(1), 65–75. https://doi.org/10.1016/S0022-0981(01)00294-5.

Moschino, V., Chicharo, L., & Marin, M. G. (2011). Long-term effects of fishing on physiological performance of the Manila clam (*Ruditapes philippinarum*) in the Lagoon of Venice. *Scientia Marina, 75*(1). https://doi.org/10.3989/scimar.2011.75n1103.

Nakamura, Y. (2001). Filtration rates of the Manila clam, *Ruditapes philippinarum*: Dependence on prey items including bacteria and picocyanobacteria. *Journal of Experimental Marine Biology and Ecology, 266*(2), 181–192. https://doi.org/10.1016/S0022-0981(01)00354-9.

Purchon, R. D. (1968). *The biology of the Mollusca*. Pergamon Press.

Purchon, R. D. (1987). The stomach in the bivalvia. *Philosophical Transactions of the Royal Society of London. B, Biological Sciences, 316*(1177), 183–276. https://doi.org/10.1098/rstb.1987.0027.

Ridewood, W. G. (1903). On the structure of the Gills of the Lamellibranchia. *Philosophical Transactions of the Royal Society B., 195*, 147–284.

Sobral, P., & Widdows, J. (2000). Effects of elevated temperatures on the scope for growth and resistance to air exposure of the clam *Ruditapes decussatus (L.),* from southern Portugal. *Scientia Marina, 61*(1), 163–171.

Sorokin, Y. I., & Giovanardi, O. (1995). Trophic characteristics of the Manila clam (*Tapes philippinarum* Adams and Reeve). *ICES Journal of Marine Science, 52*(5), 853–862. https://doi.org/10.1006/jmsc.1995.0082.

Spencer, B. E. (2002). *Molluscan shellfish farming*. Fishing News Books https://doi.org/10.1002/9780470995709.

Toba, D. R., Thompson, D. S., Chew, K. K., Anderson, D. J., & Miller, M. B. (1992). *Guide to Manila Clam culture in Washington*. University of Washington.

Watanabe, S., Katayama, S., Kodama, M., Cho, N., Nakata, K., & Fukuda, M. (2009). Small-scale variation in feeding environments for the Manila clam *Ruditapes philippinarum* in a tidal flat in Tokyo Bay. *Fishery Science, 75*, 937–945. https://doi.org/10.1007/s12562-009-0113-1.

Yonge, C. M. (1976). Shallow and deep burrowing bivalves. In C. M. Yonge, & T . E. Thompson (Eds.), *Living marine Molluscs* (pp. 185–204). Collins.

Zhao, L., Yan, X., & Yang, F. (2013). Food sources of the Manila clam *Ruditapes philippinarum* in intertidal areas: Evidence from stable isotope analysis. *Chinese Journal of Oceanology and Limnology, 31*, 782–788. https://doi.org/10.1007/00343-013-2191-2.

CHAPTER

Respiration, physiological rates, and energetics

8

8.1 Gas exchange and hypoxia

In addition to their role in nutrition, the ctenidia are also the primary organs of gas exchange. In evolutionary terms, despite their effectiveness in filtering food particles, this role is secondary to the original gas exchange function for which they are also superbly adapted. Manila clam ctenidia possess the general adaptations of the eulamellibranch bivalves, in which two W-shaped ciliated lattices, one either side of the visceral mass, each established by connected individual gill filaments create a current of water through the gill lamellae (Chapter 7).

As water moves between the gill filaments, gas exchange occurs by diffusion. This is facilitated by a counter-current relationship between water flow over and blood flow within the gills, which maximizes the exchange of oxygen and carbon dioxide by maintaining concentration gradients across the full width of the gill lamellae. A constant supply of fresh water is brought in from the overlying water column through the inhalant siphon, and the separation by the gills of inhalant and exhalant chambers ensures no mixing, as does the removal of water through the anatomically distinct exhalant siphon. While the fusion of the siphons along much of their length raises the question of reinhalation of exhaled water, this possibility is reduced by the distal separation of the ends of the two siphons, which in life allows them to point in different directions (Fig. 6.3).

In filter-feeding bivalves, the nutritional function of the ctenidia has led to the evolution of surface area and water flow rates much greater than is needed for gas exchange alone. However, while Manila clams can and do adjust their depth in the sediment in response to reduced dissolved oxygen concentrations (Chapter 6), beyond such vertical migrations, unlike fin fish and other more mobile species, they cannot always avoid hypoxic conditions. Moreover, at intertidal elevations, the cessation of respiratory currents during low water and valve closures under other stresses require the initiation of metabolic responses to hypoxia, such as switching to anaerobic respiration. Consequently, Manila clams survive for many days in oxygen-deficient conditions, and the median lethal (LC_{50}) oxygen level for Manila clams over 20 days is 0.57 mg O_2 L^{-1} (Li et al., 2019).

While such adaptations ensure that in normal circumstances the Manila clam can survive episodes of minimal gas exchange and low dissolved oxygen levels, they can

The Manila Clam *Ruditapes philippinarum*. Science and Significance of a Globalised Species.
DOI: https://doi.org/10.1016/B978-0-443-36704-5.00010-9
Copyright © 2026 Elsevier Inc. All rights are reserved, including those for text and data mining, AI training, and similar technologies.

also be considered as preadaptations for anthropogenically altered eutrophic waters in which the duration and strength of oxygen depletion are intensified as a result of the aerobic decomposition of both organic pollutants and phytoplankton cells during and after blooms. Especially in warm weather conditions and at slack water of neap tides, low oxygen saturations can cause large scale mortalities (Franklin et al., 2012). Investigating causes of summer mortality in Japan, Matsuda (2022) reported that exposure to anoxia for up to 36 h did not significantly increase the mortality of *Ruditapes philippinarum* over the following 3 days, while 48 h of anoxic exposure resulted in very little survival. Moreover, conditions of severe and prolonged hypoxia in Chinese, Japanese, and Korean waters have been implicated in mass mortality events.

8.2 Energetics and scope for growth

Internal respiration describes the metabolic pathways by which energy is released in biologically useful forms from food. In aerobic conditions, this involves the consumption of oxygen and the production of carbon dioxide. Physiological energetics provides a unifying principle through which the survival, growth, and reproduction of an individual animal are linked to an energy budget determined by the rates and efficiencies of its metabolic and physiological processes. How these rates vary with environmental conditions is informative not only in terms of growth and reproductive potential but also as an indication of environmental tolerances, which in turn reflect on aspects of ecological niche and potential geographical range. Moreover, the relation between consumption and production in individual animals provides the basis of ecological energetics and the fundamentals of functional ecology.

Bayne and Newell (1983) reviewed the physiological energetics of marine mollusks from the standpoint of the established formula for steady-state conditions as follows:

$$C = P + R + F + U.$$

Where each term relates to the energy equivalent of:

C = consumption (the ingestion of food);
P = growth or reproductive products;
R = metabolic consumption/loss;
F = feces; U = excreted materials.

The energy in digested food (C–F) is lost through excretion (U), which primarily relates to the nitrogenous end products of protein metabolism, and during metabolism (R), determined through measures of oxygen consumption. Available energy remaining after all routine metabolic processes is available for somatic growth or gamete production (P). Thus, energy from consumption in excess of that required for survival has been described as the "scope for growth" (SFG) of an individual, where growth includes somatic and reproductive production (Warren & Davis, 1967). Energy equivalents of organic mass are determined using conversion factors available in the literature (e.g., Gnaiger, 1983). Clam growth is influenced mainly by water temperature and the availability of food, though other factors can also play a role (Mann & Glomb, 1978; Solidoro et al., 2000; Spillman et al., 2008).

Since C (consumption) represents ingested food, the determination of C in suspension-feeding bivalves involves the calculation of various rates and efficiencies. Filtration rate (FR) has been defined as the pumping rate, i.e., the volume of water flowing through the gills per unit time. Since the gills do not generally retain all particles, the clearance rate (CR) is defined as an equivalent volume of water completely cleared of all particulates per unit time (Riisgard, 2001). These rates are typically expressed as a volume of water per unit of clam soft body dry weight (DW) and time (e.g., L g DW^{-1} h^{-1}). The distinction between FR and CR is important as bivalves can adjust gill function in response to changing particle concentrations. When suspended particles are retained by the gill with 100% efficiency, then CR and FR are equal.

In fact, these terms are not consistently used in the literature or in any event, different technical experimental conditions and approaches to their determination, including variations in food species, can make rate comparisons between published papers difficult (see Riisgard, 2001) and interspecific comparisons among sympatric Venerids less reliable. Since the beginning of the 21st century, valuable studies on *R. philippinarum* have shed light on energy balance and the factors influencing rates. Central to these findings are the two primary limiting factors for marine ectotherms: temperature and food availability.

8.3 Factors affecting physiological rates

Han et al. (2008), using the microalga *Isochrysis galbana* reported *R. philippinarum* FR to increase with temperatures from 5°C to 25°C. Shin and Lim (2003) feeding *R. philippinarum* with the diatom *Phaedactylum tricomutum* reported a similar maximal FR at 20°C–25°C. Shin and Lim (2003) also investigated *R. philippinarum* FR in relation to salinity across a range of 15–35 PSU. FR increased with salinity, with an optimal range from 25–35 PSU. As S drops below 25, FR drops off quickly, such that even at optimum temperatures, FR at S = 15 is around 10% of the maximal rate. Gallager and Mann (1981), among others, have reported smaller specimens to have a higher FR per unit of soft body dry mass than larger specimens, although they are more sensitive to lower temperatures and salinities than older cohorts (Shin & Lim, 2003). Osmoregulation and salinity tolerances are considered further in Chapter 11.

R. philippinarum can effectively retain prey from 2–8 um (picocyanobacteria and autotrophic nanoplankton), to single-celled diatoms and dinoflagellates up to 200 um (e.g., the chain-forming diatoms *Chaetoceros* and *Skeletonema* spp). However, using monocultured food species, CR has been shown to vary with both particle size and concentration (Nakamura, 2001).

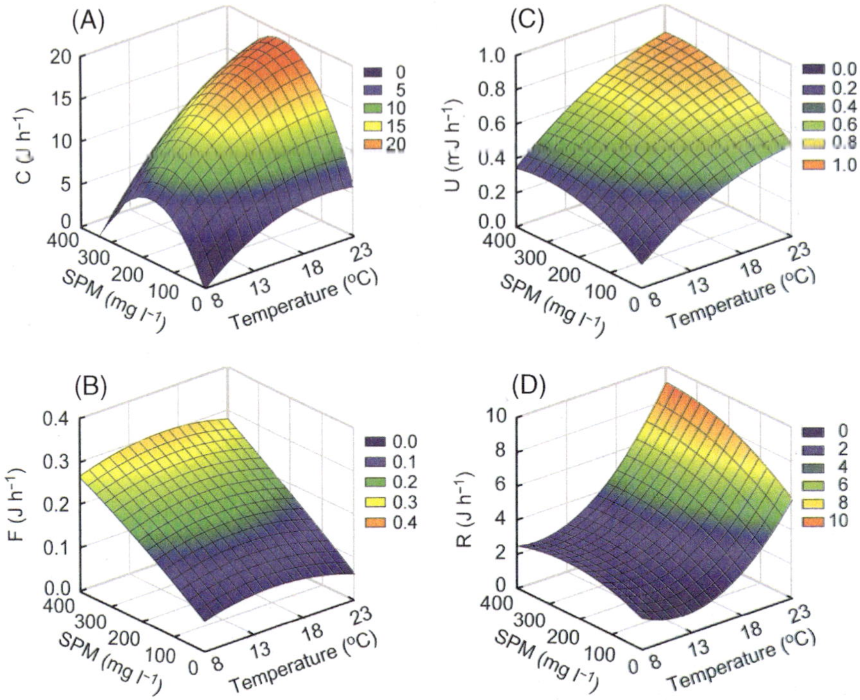

FIGURE 8.1

Mean physiological rates of *R. philippinarum* in different conditions of temperature and suspended particulate matter: (A) mean consumed energy, (B) feces energy, (C) excretion energy, (D) respiration energy. Values of C, F, U, and R (where C = consumed energy; F = feces energy; U = excreted materials energy; R = metabolic consumption/loss) represent modeled values for a 0.48 g weight-standardized Manila clam in experimental combinations of water temperatures (T) and suspended particulate matter (SPM) concentrations (SPM).

From Kang et al. (2016).

Kang et al. (2016) investigated the combined effects of temperature and food concentration using Korean *R. philippinarum* and a commercially produced mixed diet of 40% *Isochrysis* sp., 25%*Tetraselmis* sp., 20% *Thalassiosira weissflogii,* and 15% *Pavlova* sp. at concentrations from 9.5 to 350.5 mg L^{-1} and temperatures ranging from 8°C to 23°C. Fecal energy loss (F) was a function mainly of particulate matter concentration (Fig. 8.1B), and ammonia excretion (U) mainly of temperature (Fig. 8.1C). Consistent with most other studies, they found consumed energy (C) to increase with temperature but decrease with increased particulate concentration, suggesting adaptive control of CR (Fig. 8.1A). However, respiration rate increased with temperature, indicating a lack of any capacity for metabolic rate control and, in turn, suggesting high mortality risk in high temperatures with insufficient food

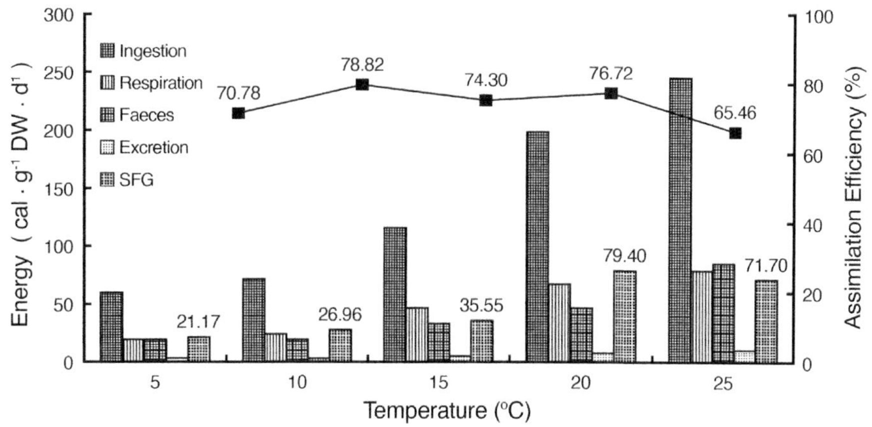

FIGURE 8.2

Energy balance of *R. philippinarum* (151 ± 12 mg dry weight) at different temperatures. Scope for growth (SFG) and assimilation efficiency is also shown.

From Han et al. (2008).

availability. Nevertheless, within the experimental conditions applied, increased consumption offset the metabolic cost of warmer temperatures.

In fact, various physiological rates are involved in determining the relationship between the clearance of food from the environment and its absorption within the digestive diverticula, when it becomes metabolically available. These include ingestion rate (IR = CR × algal concentration) and absorption or assimilation efficiency (AE: relative organic contents of food and feces), from which absorption rate (AR = IR × AE) can be determined. Q_{10} values are calculated to represent the rate of change of such relationships in response to experimental changes.

Han et al. (2008) investigated the effect of temperatures from 5°C to 25°C on Korean *R. philippinarum*. They found that, like FR, ingestion rate (IR) is also significantly related to temperature, with the highest Q^{10} between 15°C and 20°C. Absorption efficiency (AE) was found to be relatively constant across the full temperature range, although they noted a maximum mean value at 20°C. Sgro et al. (2005), with Adriatic *R. philippinarum* at salinity 24 and *Thalassiosira pseudonana* food, reported both CR and AR to increase with food concentration up to the point where IR reached a maximum, after which IR remained constant while CR declined.

Such results suggest coordination between particulate retention and the maximum capacity of the digestive system and are consistent with other Venerid bivalves, which can maintain a relatively constant IR by adjusting FR (Foster-Smith, 1975; Walne, 1972). Using *Chaetoceros neogracile* cells, Coutteau et al. (1994) reported an "incipient limiting concentration" for *R. philippinarum* beyond which valve closure occurred, restricting FR. A response thought to relate to the saturation of the alimentary canal ("saturation reduction," see Riisgård, 2001). Kang et al. (2016), in

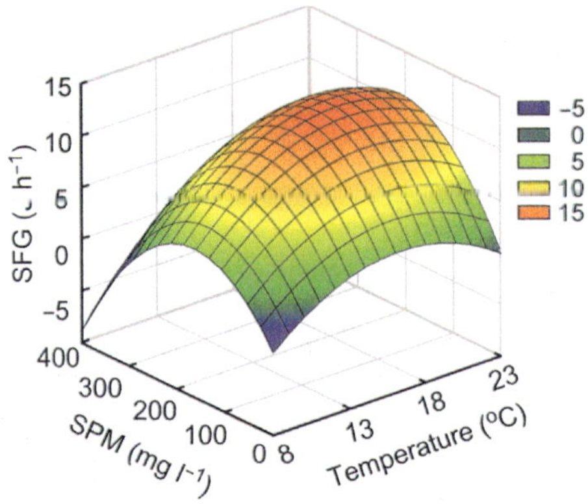

FIGURE 8.3

Mean scope for growth (SFG) of *Ruditapes philippinarum* in experimental combinations of water temperature and Suspended Particulate Matter (SPM) concentrations, derived from data and modeling represented in Fig. 8.1.

From Kang et al. (2016).

their combined temperature and food concentration experiment, reported that while consumed energy increased with both temperature and food concentration, peaking at around 100–200 mg L^{-1} and 18°C–23°C, fecal energy was largely determined by food concentration.

Fig. 8.2 by Han et al. (2008) shows the complete energy balance of *R. philippinarum* at different temperatures at a salinity of 32 PSU and an experimental food concentration of 15×10^6 *I. galbana* cells L^{-1} (corresponding to 0.45 mg DW L^{-1} and 4520 Cal g^{-1} DW). The chart shows the relationship between energy gained from consumption and that lost in feces, through respiration, and excretion. Ammonia is the excretory end product of protein catabolism. Although protein catabolism increases in starvation conditions as proteins are metabolized for survival, this is not an issue in normal conditions, where ammonia excretion represents only a minor component of the energy budget. Consistent with Kang et al. (2016), within the range from 5°C to 25°C, given sufficient available food, increased consumption more than compensates for the increased metabolic cost of warmer temperatures. By subtracting energy consumed and lost from energy consumed using the established formula, the remaining energy represents SFG, which was positive across the experimental range, with a maximum around 20°C. In the context of the temperate climatic global distribution of the species, it is of interest that both SFG and assimilation efficiency appear to be declining above 20°C.

The capacity for *R. philippinarum* to operate with positive scope for growth in a range of conditions was further revealed by Kang et al. (2016) on the interaction of temperature and food concentration. Their results are shown as surface plots in Fig. 8.3. Consistent with Han et al. (2008), SFG values were positive under almost all experimental temperature conditions. In addition the Kang et al. (2016) results show how SFG continues to increase at much higher food concentrations than were used by Han et al. (2008), (> 45 mg DW L^{-1}), with the highest SFG being recorded at mixed food concentrations of 100–200 mg DW L^{-1} at 18°C.

In temperate conditions, SFG at low temperatures is ecologically significant since, if negative, the metabolic costs of maintenance exceed the gains from food. The clam must then consume energy reserves until they are used up, at which point death occurs. Conversely, a positive SFG at low temperatures in suitable food conditions is important for early initiation of the reproductive cycle (Chapter 10).

Fig. 8.3 shows the scope for growth as derived by Kang et al. (2016) from their results in Fig. 8.1. In discussion, they observed that their findings corroborate the hypothesis that the ability of *R. philippinarum* to thrive in a broad range of environmental conditions relates, not least, to their highly flexible "physiological plasticity." Nevertheless, it is also noteworthy that SFG declines at the higher temperatures regardless of particulate concentration: a result which also suggests a temperature optimum and corresponding competitive advantage consistent with temperate rather than tropical climatic conditions (see Chapter 2).

Environmental tolerances and optima are considered further in Chapter 11 in relation to the clam's ecology.

References

Bayne, B. L., & Newell, R. C. (1983). Physiological energetics of marine Molluscs. In *The Mollusca: 4* (pp. 407–512). Academic Press.

Coutteau, P., Cure, K., & Sorgeloos, P. (1994). Effect of algal ration on feeding and growth of juvenile Manila clam *Tapes philippinarum* (Adams and Reeve). *Journal of Shellfish Research, 13*(1), 47–55.

Foster-Smith, R. L. (1975). The effect of concentration of suspension on the filtration rates and pseudofaecal production for *Mytilus edulis L., Cerastoderma edule (L.)* and *Venerupis pullastra* (Montagu). *Journal of Experimental Marine Biology and Ecology, 17*(1), 1–22. https://doi.org/10.1016/0022-0981(75)90075-1.

Franklin, D. J., Humphreys, J., Harris, M., Jensen, A. C., Herbert, R. J. H., & Purdie, D. A. (2012). *An investigation into the annual cycle of phytoplankton abundance in Poole Harbour and its relationship with Manila clam nutrition. Report for the Marine Management Organisation*. UK. Newcastle upon Tyne: MMO.

Gallager, S. M., & Mann, R. (1981). The effect of varying carbon/nitrogen ratio in the phytoplankter *Thalassiosira pseudonana* (3H) on its food value to the bivalve Tapes japonica. *Aquaculture, 26*(1–2), 95–105. https://doi.org/10.1016/0044-8486(81)90113-7.

Gnaiger, E. (1983). Appendix C. Calculation of energetic and biochemical equivalents of respiratory oxygen consumption. In E. Gnaiger, & H. Forstner (Eds.), *Polarographic oxygen sensors. Aquatic and physiological applications* (pp. 337–345). Springer-Verlag.

Han, K. N., Lee, S. W., & Wang, S. Y. (2008). The effect of temperature on the energy budget of the Manila clam, *Ruditapes philippinarum. Aquaculture International, 16*(2), 143–152. https://doi.org/10.1007/s10499-007-9133-y.

Kang, H. Y., Lee, Y. J., Choi, K. S., Park, H. J., Yun, S. G., & Kang, C. K. (2016). Combined effects of temperature and seston concentration on the physiological energetics of the Manila clam *Ruditapes philippinarum*. *PLoS ONE, 11*(3). https://doi.org/10.1371/journal.pone.0152427.

Li, Q., Sun, S., Zhang, F., Wang, M., & Li, M. (2019). Effects of hypoxia on survival, behavior, metabolism and cellular damage of Manila clam (*Ruditapes philippinarum*). *PLoS One, 14*(4), e0215158. https://doi.org/10.1371/journal.pone.0215158.

Mann, R., & Glomb, S. J. (1978). The effect of temperature on growth and ammonia excretion of the Manila clam *Tapes japonica*. *Estuarine and Coastal Marine Science, 6*, 335–339.

Matsuda, M. (2022). Effects of anoxia exposure to Manila clam *Ruditapes philippinarum*. *Bulletin of Nagasaki Prefectural Institute of Fisheries, 47*, 1–7.

Nakamura, Y. (2001). Filtration rates of the Manila clam, *Ruditapes philippinarum*: Dependence on prey items including bacteria and picocyanobacteria. *Journal of Experimental Marine Biology and Ecology, 266*(2), 181–192.

Riisgard, H. U. (2001). On measurement of filtration rates in bivalves – The stony road to reliable data: Review and interpretation. *Marine Ecology Progress Series, 211*, 275–291. https://doi.org/10.3354/meps211275.

Sgro, L., Munari, C., Angonese, A., Basso, S., & Mistri, M. (2005). Functional responses and scope for growth of two non-indigenous bivalve species in the Sacca di Goro (northern Adriatic Sea, Italy). *Italian Journal of Zoology, 72*(3), 235–239. https://doi.org/10.1080/11250000509356677.

Shin, H. C., & Lim, K. H. (2003). The influence of water temperature and salinity on the filtration rates of the short-necked clam, *Ruditapes philippinarum*. *The Korean Journal of Malacology., 19*(1), 1–8.

Solidoro, C., Pastres, R., Melaku Canu, D., Pellizzato, M., & Rossi, R. (2000). Modelling the growth of *Tapes philippinarum* in Northern Adriatic lagoons. *Marine Ecology Progress Series, 199*, 137–148.

Spillman, C. M., Hamilton, D. P., Hipsey, M. R., & Imberger, J. (2008). A spatially resolved model of seasonal variations in phytoplankton and clam (*Tapes philippinarum*) biomass in Barbamarco Lagoon, Italy. *Estuarine, Coastal and Shelf Science, 79*(2), 187–203. https://doi.org/10.1016/j.ecss.2008.03.020.

Walne, P. R. (1972). The influence of current speed, body size and water temperature on the filtration rate of five species of bivalves. *Journal of the Marine Biological Association of the United Kingdom, 52*(2), 345–374. https://doi.org/10.1017/S0025315400018737.

Warren, C. E., & Davis, G. E. (1967). Laboratory studies on the feeding bioenergetics and growth of fish. In S. D. Gerking (Ed.), *The biological basis of freshwater fish production* (pp. 175–214). Blackwell Scientific Publications.

CHAPTER

Circulation, regulation, and immunity

9

9.1 Circulation and excretion

As advanced metazoans, mollusks are triploblastic and coelomate, meaning that the basic body plan develops from three embryonic germ layers: endoderm, mesoderm, and ectoderm, with a fluid-filled cavity (coelom) within the mesodermal tissue layer. The main structural components of the circulatory system derive from the embryonic mesoderm. In particular, the bivalve heart and kidneys (nephridia) are found within a mid-dorsal coelomic cavity (Fig. 6.2B). This pericardium area is one of the few body cavities remaining from the much-reduced embryonic coelom. Together, the organs and fluid of the pericardium form a structural and functional complex responsible for both the circulation and excretion.

In the open circulatory systems of bivalves, the blood or, more accurately, hemolymph comes into direct contact with tissues and cells. Therefore, unlike in closed circulatory systems, the exchange of gases and nutrients between the hemolymph and cells does not require the crossing of capillary walls. In fact, there are no capillaries; the blood is released from arteries directly into sinuses and tissues. Consequently, there is no partitioning between blood, tissue fluids, and lymph, as is found in vertebrates, for example. For this reason, in open systems, the term hemolymph is often preferred to blood.

In bivalves, the heart itself is not the powerful organ associated with closed systems, and the open circulatory system generally is relatively inefficient compared to the closed systems of more active animals such as the cephalopod mollusks. Nevertheless, particularly during periods of higher activity, circulatory pressure and efficiency are enhanced through the contractions of somatic musculature. Similarly, it is muscular rather than heart activity that generates the levels of hydrostatic pressure and resulting turgor needed for the effective use of the foot in burrowing.

The Venerid heart consists of a single muscular ventricle, which receives oxygenated hemolymph from the ctenidia via two lateral atria. Ventricular contraction moves the hemolymph in two directions via anterior and posterior aortas with valves preventing backflow (Fig. 6.2B). From here, it passes into several arteries to be released directly into the sinuses and tissues of the mantle and visceral mass, after which it is collected into veins and returned to the heart via excretory organs (nephridia) and ctenidia.

Excretion involves ultrafiltration of the hemolymph in two places. Thickened areas of pericardial epithelium form the pericardial glands positioned adjacent to

The Manila Clam *Ruditapes philippinarum*: Science and Significance of a Globalised Species.
DOI: https://doi.org/10.1016/B978-0-443-36704-5.00016-X
Copyright © 2026 Elsevier Inc. All rights are reserved, including those for text and data mining, AI training, and similar technologies.

the atria. These release excretory products from the hemolymph into the pericardial cavity, which also receives a clear hydrostatically driven filtrate directly through the walls of the ventricles. The resulting pericardial fluid moves via ducts into paired nephridia (kidneys), where further removal of toxins along with reabsorption of valuable ions occurs. The resulting urine, with high levels of ammonia, is released via a nephridiopore into the mantle cavity and is eliminated via the exhalant siphon.

Although the pericardial complex of *Ruditapes philippinarum* has apparently not yet been properly investigated, based on studies on other bivalves, it is reasonable to suppose that the circulation of hemolymph is more variable than the above general account implies. In particular, there are likely various routes by which venous hemolymph reaches the atria, including a direct route that bypasses the ctenidia. Nevertheless, the main pathway from the viscera and pedal musculature is likely via the nephridia, and the passage of blood through the nephridia is closely associated with that of the ctenidia (Ansell, 1959; Gosling, 2015; Jennings, 1984; White, 1942) (see more general accounts in Ruppert et al. (2004)).

9.2 Hemolymph and oxygen transport

In contrast to air, which consists of 21% oxygen, available oxygen in seawater is much lower, being around 0.54% by volume at 20°C. Although the Manila clam can be perceived as a relatively sedentary animal, along with the other Venerids, it is in fact an active burrowing suspension feeder that consumes more oxygen in the tissues than could be delivered in solution by a simple aqueous hemolymph. Therefore, oxygen transport from the Manila clam's ctenidia is facilitated by the respiratory pigment hemocyanin, which is dissolved in the hemolymph plasma, rather than being contained within special corpuscles.

Hemocyanin is a globulin protein combined with copper atoms, with an affinity for molecular oxygen. In the reduced state, it is colorless, but when bound to molecular oxygen, it is light blue. Like other respiratory pigments, hemocyanin changes its affinity for oxygen under different conditions. Its level of saturation with oxygen varies according to the surrounding partial pressure of oxygen, so it takes up oxygen in the relatively high O_2 partial pressures in the ctenidia and releases it in the lower partial pressures of the tissues, showing the S-shaped oxygen dissociation curve characteristic of such pigments. CO_2 partial pressures (the Bohr effect) and temperature also shift the saturation curve in such a way as to expedite the collection and release of oxygen in the ctenidia and tissues, respectively.

9.3 Hemocytes and immune function

Along with all other invertebrate animals, bivalves do not share the "adaptive" immune systems of vertebrates, in which immunity to specific pathogens is built up after first exposure, enabling a fast and targeted response to subsequent infection. As such, they are limited in their responses to more primitive "innate" defensive systems in which resistance to foreign invasion is not a function of previous exposure. The principal components of this innate system are various types of defensive effector cells collectively known as hemocytes, which are found both in the hemolymph plasma and the tissues supplied by the open circulatory system.

In *R. philippinarum*, hemocytes have also been shown to occupy the extra-pallial fluid that lies between the Mantle's external epithelium and the inner surface of the shell, with similar concentrations to those of the hemolymph (Allam & Paillard, 1998). Hemocytes are thought to be involved in various physiological processes including tissue repair and shell production, however responses of the economically important Manila clam to pathogens such as the bacterium *Vibrio tapetis* and the protozoan parasite *Perkinsus spp,* which can include both weight loss and mass mortality, have led to a research focus on their defensive role (see Donaghy et al. (2009) for a review and Chapter 13 for further information on pathogens).

Two main types of hemocytes have been reported in *R. philippinarum*: granulocytes, named on the basis of granules within their cytoplasm, making up around 48% of hemocytes, and hyalinocytes (or agranulocytes), making up 32%. Granulocytes (dia. 6–10 um) appear larger than hemocytes (2–3 um) (Matozzo et al., 2008). In addition to these two major types (Cima et al., 2000), on the basis of staining characteristics, hemoblasts (19%) and serous cells (around 1%) have also been reported. However, differences in both stain affinities and cell size may also be the result of maturation.

Hemocyte defensive mechanisms include both cellular attacks, such as phagocytosis, and "humoral" responses involving the release of protective compounds such as antibacterial and antioxidant enzymes. For example lysins (hydrolytic enzymes which break down bacteria) and other antimicrobial peptides. Similar to various other bivalves, in the Venerid clams, including *R. philippinarum,* infection provokes an inflammatory response based on an infiltration of hemocytes. In response to *Perkinsus* infection, for example, white nodules on the surface of gill, mantle, and foot tissues (see figure 13.4) have been shown to consist of massive hemocyte infiltration into tissues surrounding the *Perkinsus* trophozoites (Choi & Park, 2005). A similar inflammatory response to the alga *Prorocentrum minimum* has been reported in the Manila clam's digestive gland and gonad (Hégaret et al., 2009).

In addition to the active movement towards infections (chemotaxis), the cellular defense mechanisms of bivalves include recognition (involving opsonization, i.e., the recognition and marking of the foreign particles by lectin-like molecules to help their identification by the defense system) and phagocytosis, followed by the intracellular digestion of foreign material. Both granulocytes and hyalinocytes appear to have amoebocyte (spreading) capabilities, although granulocytes are thought to play the major role in phagocytosis of invading pathogens.

Kang et al. (2006) have reported that hemocytes produce different isoforms of lectin according to the species of infective pathogen (e.g., Montes et al. (1995)). Alternatively, lectins can also act as agglutinins to clump together foreign particles, which become encapsulated within a layer of granular hemocytes, where they may be subject to extracellular degradation. Such encapsulation may follow failures of phagocytosis, resulting in lysis of the engulfing hemocytes, which in turn release protein and other secretions. The main product of this defensive response is a glycosylated polypeptide molecule absent in noninfected Venerid clams (Montes et al., 1995). Encapsulated pathogen aggregations take the form of the white cysts or nodules referred to above.

Comparing *R. philippinarum* and *Ruditapes decussatus*, Montes et al. (1996) noted that the percentage of dead *Perkinsus* trophozoites relative to the total encapsulated was lower (11%) for Manila clams, versus 36% in *R. decussatus*. They regarded this as consistent with the greater susceptibility of Manila clams to the spread of *Perkinsus* after infection by zoospores, as reported by Rodnguez et al. (1994).

In *R. philippinarum*, lower water temperatures, salinities, and food quality have been shown to lower the concentration of hemocytes, decrease lysozyme activity, and reduce phagocytosis capacity (Delaporte et al., 2003; Paillard et al., 2004; Reid et al., 2003). This, in turn, leads to higher susceptibility to diseases such as brown ring disease induced by *Vibrio tapetis* (see Chapter 13). Physiological status also appears to have an impact on hemocyte numbers, with minimal counts and presumably maximum susceptibility to disease in the postspawning period (Matozzo et al., 2003).

Despite these defensive responses, Manila clams can become overwhelmed with infection, sometimes resulting in mass mortalities, especially when compounded by other factors such as low food availability and thermal stress. Flye-Sainte-Marie et al. (2009) reported that heavily *Perkinsus olseni* infected Manila clams had low hemocyte phagocytosis rates. Moribund clams may surface. Nam et al. (2018) reported *Perkinsus olseni* infection rates of $1.98 \pm 1.01 \times 10^6$ cells/g flesh wet weight in moribund surfaced clams, compared to $1.16 \pm 0.49 \times 10^6$ cells/g for Manila clams from the same population in their normal buried positions. Also, the percentages of necrotic hemocytes correlated positively with infection intensity, surfaced Manila clams exhibiting a seven-fold higher percentage of necrotic hemocytes than those at normal burial depth (Nam et al., 2018).

9.4 Osmotic and ionic regulation

Salinity (S, expressed here in Practical Salinity Units or PSU)) is one of the dominant environmental factors affecting the distribution of coastal invertebrates. The salinity of offshore coastal waters is typically from 32 to 34 in temperate regions (S = 34, approximating to 34 g of salt per kilogram of water). In contrast to fully marine or freshwater environments, many coastal and all estuarine habitats experience tidal cycles and weather events, which significantly change the mix between waters of fluvial and marine origin, causing reduced salinity. Resulting fluctuations in water chemistry and salinity commonly involve increased dilution, although in enclosed areas during hot, dry seasons, especially at lower latitudes, evaporation can increase salinity above that of 100% seawater. Such variations can cause significant physiological and mechanical stresses.

In the absence of protective adaptations, freshwater dilution of seawater causes tissue swelling and cell rupture through osmotic influx, while water concentrated by evaporation causes osmotic dehydration. Physiological adaptation to such conditions in bivalves generally involves changes in tissue concentrations in line with ambient salinity (osmoconforming), while regulating cell solute composition as necessary to maintain tissue function. Animals capable of living in a wide range of salinities are termed euryhaline.

In the laboratory, *R. philippinarum* mortality is around 0% after 6 days at salinities above S = 14, below which mortality increases steeply (Carregosa et al., 2014; Dominguez et al., 2020). The reported optimum salinity range for the species falls between S = 24 and 34 (Spencer et al., 1991).

In the Ria de Aveiro, a shallow coastal system in the northwest of Portugal, Manila clams are found in areas where salinities vary up to S = 42 according to weather and season. *R. philippinarum* specimens from this estuary are capable of surviving prolonged exposure to salinities from 14 to 42. Over this range, intracellular sodium (Na) and magnesium (Mg) concentrations vary in line with those of the external medium. In terms of Na, the species can be described as osmoconforming; however, intracellular potassium (K) concentrations are maintained broadly steady across all salinities, and intracellular calcium (Ca) increases at salinities below 21 (Carregosa et al., 2014).

Clearly, Na is a factor in the achievement of an osmotic balance between intracellular and environmental salinities. At lower salinities (below 21), reduced intracellular Na may become an insufficient ion concentration for physiological cell function, a situation perhaps remedied by increasing intracellular Ca concentrations (Carregosa et al., 2014). However, in osmoconformers, intracellular organic solutes are also known to play a role in regulating cell volume during dilution (Yancy, 2005). In bivalves, including Venerids, free intracellular amino acids and small carbohydrates are actively regulated in line with salinity (Shumway et al., 1977).

Although they are categorized as osmoconformers, there is some evidence from in vivo experiments on *R. philippinarum* hemolymph that suggests the possibility of some level of osmoregulation, with hemolymph concentrations reported to be maintained about 2.5% (mg L^{-1}) above that of ambient seawater (Coughlan et al., 2009).

It is clear from the above that osmoconforming does not simply involve allowing tissues to passively dilute in low salinity conditions. In fact, maintaining tissue fluid concentrations more or less in line with ambient waters requires active adjustments in both the inorganic and organic composition within cells. Such activity is known to carry high energy costs with consequent impacts on both behavior and physiological rates during low salinity immersion.

Burrowing capability, siphon exposure, and fecal deposition are all reduced at lower salinities (Woodin et al., 2020). Domínguez et al. (2020) investigated the cumulative effects of exposure to tidal salinity changes and cycles similar to field conditions in the Ría de Arousa, Spain, using tidal cycle salinity "ramps" of 5–20, 10–25, 15–30, and 30–30. During low salinity periods (5, 10, and 15: corresponding to ebbs and low waters), *R. philippinarum* exhibited a considerable reduction in clearance rate (see Chapter 7), scope for growth, and burrowing ability. Consistent with scope for growth attenuation, lower salinities can also delay gametogenesis in *R. philippinarum* and even trigger the resorption of gametes (Vázquez et al., 2021). Carregosa et al. (2014) found that both the glucose and glycogen content of *R. philippinarum* tissues dropped considerably, below and above S = 35.

In circumstances of sudden or severe salinity fluctuation, bivalves can protect their tissues by closing their valves. Bae et al. (2021) reported valve closures of Manila clams exposed to both steady low salinity conditions and increasing and decreasing salinity fluctuations (S = 27–24 and 24–30). Valve closure is particularly useful in response to sudden tidally induced salinity changes, as it protects the tissues from immediate osmotic damage while allowing time for the necessary cellular solute adjustments. However, this behavior will induce hypoxia (Kim et al., 2001), which in turn has significant effects on cell metabolism. Decreasing oxidative phosphorylation of ATP induces the activation of alternative, less energy-efficient metabolic pathways and the consumption of the energy store, glycogen. Eventually, cells resort to protein catabolism as an energy source with the consequence that excretion of ammonia increases under osmotic stress, and cellular protein levels in *R. philippinarum* drop significantly at salinities below S = 28 (Carregosa et al., 2014).

In considering the reported salinity tolerances of *R. philippinarum*, it is important to recognize that in any one experimental situation, a number of factors will be at play. It is known, for example, that salinity and temperature act synergistically (Bae et al., 2021), but in fact each experimental treatment will involve a number of independent variables which, while they may be sufficiently consistent to elucidate particular physiological mechanisms and relationships, can only provide approximations to responses and tolerances in natural coastal habitats characterized by complex patterns of spatial and temporal volatility. In addition, the natural conditions to which different populations are acclimated vary, and beyond that, a now globally dispersed species has likely responded to selection pressures particular to their new nonnative habitats.

Nevertheless, it is useful from both economic and ecological standpoints to make generalizations relating to the range of salinities within which *R. philippinarum* can potentially thrive. Chapter 11 further considers salinity from the ecological standpoint of distribution within coastal areas to provide insights into the salinity range as part of the species' realized niche. In this respect, especially in the clam's nonnative regions, the relative distributions of related sympatric species are also significant (Chapter 15).

References

Allam, B., & Paillard, C. (1998). Defense factors in clam extrapallial fluids. *Diseases of Aquatic Organisms, 33*(2), 123–128. https://doi.org/10.3354/dao033123.

Ansell, A. D. (1959). The functional morphology and general biology of the British species of Veneracea, with particular reference to Venus striatula (da Costa). (Thesis, University of Glasgow) https://theses.gla.ac.uk/id/eprint/73626

Bae, H., Im, J., Joo, S., Cho, B., & Kim, T. (2021). The effects of temperature and salinity stressors on the survival, condition and valve closure of the Manila clam, *Venerupis philippinarum* in a holding facility. *Journal of Marine Science and Engineering, 9*(7), 754. https://doi.org/10.3390/jmse9070754.

Carregosa, V., Figueira, E., Gil, A. M., Pereira, S., Pinto, J., Soares, A. M. V. M., & Freitas, R. (2014). Tolerance of *Venerupis philippinarum* to salinity: Osmotic and metabolic aspects. *Comparative Biochemistry and Physiology Part A: Molecular and Integrative Physiology, 171*, 36–43. https://doi.org/10.1016/j.cbpa.2014.02.009.

Choi, K.-S., & Park, K.-I. (2005). Current status of *Perkinsus* infection in Korean waters. In P. Walker, R. Lester, & M. G. Bondad-Reantaso (Eds.), *Diseases in Asian aquaculture V* (pp. 263–274). Fish Health Section, Asian Fisheries Society.

Cima, F., Matozzo, V., Marin, M. G., & Ballarin, L. (2000). Haemocytes of the clam *Tapes philippinarum* (Adams & Reeve, 1850): Morphofunctional characterisation. *Fish and Shellfish Immunology, 10*(8), 677–693. https://doi.org/10.1006/fsim.2000.0282.

Coughlan, B. M., Moroney, G. A., Pelt, F. N. A. M. v., O'Brien, N. M., Davenport, J., & O'Halloran, J. (2009). The effects of salinity on the Manila clam (*Ruditapes philippinarum*) using the neutral red retention assay with adapted physiological saline solutions. *Marine Pollution Bulletin, 58*(11), 1680–1684. https://doi.org/10.1016/j.marpolbul.2009.06.020.

Delaporte, M., Soudant, P., Moal, J., Lambert, C., Quéré, C., Miner, P., Choquet, G., Paillard, C., & Samain, J. F. (2003). Effect of a mono-specific algal diet on immune functions in two bivalve species - Crassostrea gigas and *Ruditapes philippinarum*. *Journal of Experimental Biology, 206*(17), 3053–3064. https://doi.org/10.1242/jeb.00518.

Domínguez, R., Vázquez, E., Woodin, S. A., Wethey, D. S., Peteiro, L. G., Macho, G., & Olabarria, C. (2020). Sublethal responses of four commercially important bivalves to low salinity. *Ecological Indicators, 111*. https://doi.org/10.1016/j.ecolind.2019.106031.

Donaghy, L., Lambert, C., Choi, K. S., & Soudant, P. (2009). Hemocytes of the carpet shell clam (*Ruditapes decussatus*) and the Manila clam (*Ruditapes philippinarum*): Current knowledge and future prospects. *Aquaculture, 297*(1–4), 10–24. https://doi.org/10.1016/j.aquaculture.2009.09.003.

Flye-Sainte-Marie, J., Soudant, P., Lambert, C., Le Goïc, N., Goncalvez, M., Travers, M. A., Paillard, C., & Jean, F. (2009). Variability of the hemocyte parameters of *Ruditapes philippinarum* in the field during an annual cycle. *Journal of Experimental Marine Biology and Ecology, 377*(1), 1–11. https://doi.org/10.1016/j.jembe.2009.06.003.

Gosling, E. (2015). *Marine bivalve Molluscs* (2nd edition). Wiley https://doi.org/10.1002/9781119045212.

Hégaret, H., Mirella da Silva, P., Sunila, I., Shumway, S. E., Dixon, M. S., Alix, J., Wikfors, G. H., & Soudant, P. (2009). Perkinsosis in the Manila clam *Ruditapes philippinarum* affects responses to the harmful-alga, *Prorocentrum minimum*. *Journal of Experimental Marine Biology and Ecology, 371*(2), 112–120. https://doi.org/10.1016/j.jembe.2009.01.016.

Jennings, K. H. (1984). *Thesis*. Bedford College, University of London).

Kang, Y. S., Kim, Y. M., Park, K. I., Cho, S. K., Choi, K. S., & Cho, M. (2006). Analysis of EST and lectin expressions in hemocytes of Manila clams (*Ruditapes philippinarum*) (Bivalvia: Mollusca) infected with *Perkinsus olseni*. *Developmental and Comparative Immunology, 30*(12), 1119–1131. https://doi.org/10.1016/j.dci.2006.03.005.

Kim, W. S., Huh, S.-H., & Lee, T. W. (2001). Effects of salinity on endogenous rhythm of the Manila clam, *Ruditapes philippinarum* (Bivalvia: Veneridae). *Marine Biology, 138*, 157–162.

Matozzo, V., Da Ros, L., Ballarin, L., Meneghetti, F., & Marin, M. G. (2003). Functional responses of haemocytes in the clam *Tapes philippinarum* from the Lagoon of Venice: Fishing impact and seasonal variations. *Canadian Journal of Fisheries and Aquatic Sciences, 60*(8), 949–958.

Matozzo, V., Marin, M. G., Cima, F., & Ballarin, L. (2008). First evidence of cell division in circulating haemocytes from the Manila clam *Tapes philippinarum*. *Cell Biology International, 32*(7), 865–868. https://doi.org/10.1016/j.cellbi.2008.03.008.

Montes, J. F., Durfort, M., & Garcia-Valero, J. (1995). Cellular defense mechanism of the clam *Tapes semidecussatus* against infection by the protozoan *Perkinsus* sp. *Cell Tissue Research, 279*, 529–538.

Montes, J. F., Durfort, M., & Garcia-Valero, J. (1996). When the Venerid clam *Tapes decussatus* is parasitized by the protozoan *Perkinsus* sp. it synthesizes a defensive polypeptide that is closely related to p225. *Disease of Aquatic Organisms, 26*(2), 149–157. https://doi.org/10.3354/dao026149.

Nam, K. W., Jeung, H. D., Song, J. H., Park, K. H., Choi, K. S., & Park, K. I. (2018). High parasite burden increases the surfacing and mortality of the Manila clam (*Ruditapes philippinarum*) in intertidal sandy mudflats on the west coast of Korea during hot summer. *Parasites and Vectors, 11*(1). https://doi.org/10.1186/s13071-018-2620-3.

Paillard, C., Allam, B., & Oubella, R. (2004). Effect of temperature on defense parameters in Manila clam *Ruditapes philippinarum* challenged with *Vibrio tapetis*. *Diseases of Aquatic Organisms, 59*(3), 249–262. https://doi.org/10.3354/dao059249.

Reid, H. I., Soudant, P., Lambert, C., Paillard, C., & Birkbeck, T. H. (2003). Salinity effects on immune parameters of *Ruditapes philippinarum* challenged with *Vibrio tapetis*. *Diseases of Aquatic Organisms, 56*(3), 249–258. https://doi.org/10.3354/dao056249.

Rodnguez, F., Godoy, T., & Navas, J. I. (1994). Cross-infection with *Perkinsus atlanticus* in *Ruditapes decussatus, Ruditapes philippinarum* and *Venerupis pullastra*. *Bulletin of the European Association of Fish Pathologists, 14*(1), 24–27.

Ruppert, E. E., Fox, R. S., & Barnes (2004). Invertebrate zoology. *A functional evolutionary approach* (7th edition). Cengage.

Shumway, S. E., Gabbott, P. A., & Youngson, A. (1977). The effect of fluctuating salinity on the concentrations of free amino acids and ninhydrine positive substances in the adductor muscles of eight species of bivalve molluscs. *Journal of Experimental Marine Biology and Ecology, 29*, 131–150.

Spencer, B. E., Edwards, D. B., & Millican, P. F. (1991). Cultivation of Manila clams (Laboratory Leaflet No. 65). Ministry of Agriculture, Fisheries and Food (MAFF), Directorate of Fisheries Research, Lowestoft, UK.

Vázquez, E., Woodin, S. A., Wethey, D. S., Peteiro, L. G., & Olabarria, C. (2021). Reproduction under stress: Acute effect of low salinities and heat waves on reproductive cycle of four ecologically and commercially important bivalves. *Frontiers in Marine Science, 8*. https://doi.org/10.3389/fmars.2021.685282.

White, K. M. (1942). The pericardial cavity and pericardial gland of the Lamellibranchia. *Journal of Molluscan Studies, 25*, 37–88. https://doi.org/10.1093/oxfordjournals.mollus.a064426.

Woodin, S. A., Wethey, D. S., Olabarria, C., Vázquez, E., Domínguez, R., Macho, G., & Peteiro, L. (2020). Behavioural responses of three venerid bivalves to fluctuating salinity stress. *Journal of Experimental Marine Biology and Ecology, 522*, 151256.

Yancey, P. H. (2005). Organic osmolytes as compatible, metabolic and counteracting cytoprotectants in high osmolarity and other stresses. *Journal of Experimental Biology, 208*, 2819–2830.

CHAPTER

Reproduction, larval dispersal, and settlement 10

10.1 Gonads and gametogenesis

Ruditapes philippinarum is a broadcast spawner with external fertilization. Therefore, the reproductive system is simple, consisting of little more than gonads from which ducts lead to the mantle cavity. The gonads are not anatomically discrete, instead being an integral part of the visceral mass surrounding the loops of gut ventral to the pericardium (Fig. 6.2B) but penetrating also into the muscular tissue of the foot (Drummond et al., 2006). Although paired, the two are difficult to distinguish as, when active, they form a ramifying mass of microscopic branched ciliated ducts ending in numerous follicles. In suitable conditions, the gonads become active sites of gametogenesis by which cell division and differentiation result ultimately in haploid oocytes and spermatozoa.

Release of gametes is achieved via ciliated gonoducts leading to the posterio-ventral surface of the visceral mass. Here, gametes enter the exhalant chamber of the mantle cavity, from where they are ejected into the water column.

R. philippinarum is generally gonochorous (dioecious) but not sexually dimorphic, so the sex of an individual clam is determined by histological examination of the gonad (Chung et al., 1994; Devauchelle, 1990; Eversole, 1989). However, there is evidence that they are capable of sequential (i.e., asynchronous) hermaphroditism, changing sex, possibly more than once during their lifetime.

As is generally the case in bivalves, gametogenesis begins within a mitotic proliferation of undifferentiated primordial germ cells in the epithelial lining of gonadal follicles (acini). The resulting cells become differentiated into spermatogonia or oogonia as they move toward the lumen. In male gonads, further mitotic division produces spermatogonia, then spermatids, before meiotic division results in spermatozoa with flagella oriented toward the lumen.

Female gametogenesis must ensure that each egg is provided with an energy store for its postfertilization development. This involves yolk deposition (vitellogenesis), which continues until full maturation, facilitated by associated follicular cells that provide nutrition to developing oocytes. In both male and female gonads, vesicular cells are present within the gonadal follicles. Also known as storage or Leydig cells, these are rich in glycogen and gradually release glucose during gametogenesis. Rodríguez-Moscoso and Arnaiz (1998) have reported that in the European congeneric *Ruditapes decussatus*, the volume of vesicular cells declines as gametogenesis

The Manila Clam *Ruditapes philippinarum*: Science and Significance of a Globalised Species
DOI: https://doi.org/10.1016/B978-0-443-36704-5.00011-0
Copyright © 2026 Elsevier Inc. All rights are reserved, including those for text and data mining, AI training, and similar technologies.

commences, eventually disappearing. Medhioub & Lubet, 1988 found that vesicular cells disappear more quickly in female than male *R. philippinarum.*

While it is possible to obtain a general idea of gonad condition from the size of the visceral mass, various methods have been employed to more sensitively evaluate the reproductive condition of Manila clams. The most accurate assessment is achieved by histological examination of gonad sections. This then allows statistics such as the ratio of follicular tissue to total gonad area and/or mean oocyte diameter to be determined, using image analysis software (Meneghetti et al., 2004). Many histological studies on Manila clam gonads have now been published to determine the factors affecting aspects of the reproductive cycle. For such a purpose, although the process of gametogenesis is a continuum, descriptions of the sequence of events are usually expressed in terms of histological stages, based primarily on the degree of maturation of the gametogenic cells in the follicles.

Holland and Chew (1974) provided the earliest gonadal stage descriptions for *R. philippinarum*. Since then, among the (largely similar) published stage descriptions for *R. philippinarum*, those of Beninger and Lucas (1984) and Valli et al. (1996) have been applied with minor amendments in various studies, such as Tumnoi (2012). As oogenesis progresses, the oocytes grow to a maximum, which may vary according to conditions but can reach diameters of more than 35 μm. In practice, in any one reproductively active individual, follicles can be found at various stages, and the most abundant follicular stage may be used to represent the reproductive condition of the animal (Drummond et al., 2006).

Once the gonad stage in a sample of clams has been determined, a general population-level statistic can be calculated. Examples include a "gonadic index" based on the number of clams at each stage multiplied by the numerical score attributed to that stage, the products added, then divided by the total number of individuals sampled. As applied by Meneghetti et al. (2004), the gonadal index ranged from 0 (all individuals in the sample are in resting or spent stages) to 3 (all individuals ripe). However, other similar calculations have also been applied with Manila clams (Baek et al., 2014).

Gametogenesis is energetically demanding and requires resources that would otherwise be used for growth. Some bivalve species store reserves such as glycogen before gametogenesis when food is plentiful, and metabolic energy demand is at a minimum. Such reserves then become available for gametogenesis. In contrast, *R. philippinarum* appears to adopt an opportunistic strategy in which gamete production and accumulation of reserves occur simultaneously (Bayne, 1976). Baek et al. (2014) report no evidence of the mobilization of stored glycogen to lipids for the formation of oocytes during gametogenesis. These results suggest that a sufficient contemporaneous food supply is essential for gametogenesis in *R. philippinarum.*

10.2 Temperature and spawning

The required minimum temperatures for each stage of the reproductive cycle have been established in various studies, and the achievement of each stage is dependent on both absolute temperature and duration of exposure (Mann, 1979). In the laboratory, Toba and Miyama (1995) found the rate of sexual development to decrease with lower water temperatures in the range between 10°C and 21°C and estimated that no sexual activity occurred below 5°C.

Mann (1979), using Californian Manila clams, found that after an experimental period with abundant food supply, somatic growth was greatest in clams kept at 12°C and progressively lower with a series of temperatures up to 21°C. He attributed this to increased gametogenic activity at the higher temperatures, noting a temperature-related sequence of carbohydrate accumulation and gametogenesis coincident with carbohydrate depletion; and weight loss through spawning. Moreover, increased ammonia excretion postspawning (indicating protein metabolism) suggested high metabolic demand at a time of carbohydrate depletion, presumably for tissue regeneration postspawning. At this time, mortality may be high, especially at temperatures sufficient to accelerate gametogenesis to an energetically stressful level, even with an abundant food supply (Mann, 1979; Mann & Glomb, 1978).

Whereas temperature is considered the main exogenous factor determining the onset of gamete production in many temperate coastal bivalves, including *R. philippinarum*, the process itself is influenced by a complex interaction of those environmental factors (not least food availability) which determine the physiological condition and energetic "scope for growth" of the individual (see Chapter 8).

The processes from gonad activation to gametogenesis and spawning require progressively higher temperatures. The initiation of gonad activity is variously reported as 4.8°C (Tumnoi, 2012), 8°C (Laing & Utting, 1994), and 10°C (Millican & Williams, 1985). Laing and Utting (1994) report gamete ripening at 12°C with spawning and larval growth at 14°C. Mann (1979) found ripe gonads at all temperatures between 12°C and 21°C, but spawning occurred only at the higher temperatures, between 15°C and 21°C. Others report spawning to require 14°C (Holland & Chew, 1974; Ohba, 1959), 13.3°C (Tumnoi, 2012), and 13.1°C (Drummond et al., 2006).

Sustainable populations require spring and summer seasons not only warm enough but also long enough for all the phases of reproduction to be completed. Beyond process minima, lower temperatures will extend the period from the commencement of gametogenesis to successful spawning. In temperate conditions, this will depend on the rate of spring warming, so the achievement of a successful spatfall will be determined by a combination of time and temperature. Consequently, Mann (1979) suggested the calculation of "day-degrees" as a method of capturing this combined time-temperature effect. A day-degree (D°) is calculated as:

$$D^\circ = d(t_1 - t_o)$$

Where t_o = the minimum temperature to initiate the process; t_1 = ambient temperature; and d = number of days more than t_o.

Table 10.1 Reported spawning months of *Ruditapes philippinarum* by region and latitude.

Site													Latitude ° North	Sources
East Asia (Native distribution)	J	F	M	A	M	J	J	A	S	O	N	D		
Tikhaya Zavod, Vostok Bay, Russia					*	*	*	*	*				55.05	Ponurovsky and Yakovlev (1992)
Litovka estuary, Vostok Bay, Russia					*	*	*						53.64	Denisenko (1978)
Slavyansky and Possjet Bays, Russia						*	*						51.16	Rakov (1986)
Srednyaya Bay, Vostok Bay, Russia						*	*	*					48.72	Denisenko (1978)
Busse Lagoon, Sakhalin, Russia							*	*					46.63	Kulikova (1979)

(continued on next page)

Hokkaido Island, Japan						*	*	*	*				43.38	Cahn (1951)
Mutsu Bay, Japan						*	*	*	*	*			41.05	Sugiura and Kikuya (2022)
Tsugaru Strait, Moheji, Japan						*	*	*	*				40.86	Shimizu et al. (2006)
Simpo, Jollabuk-do, Korea (west coast)						*	*	*	*	*			37.57	Chung et al. (2013)
Inchon Bay, Korea					*	*	*	*					37.48	Choi (1965)
Jinjiang District, China									*	*	*	*	36.00	Qi and Yang (1988)
Tokyo Bay, Japan		*	*	*					*	*			35.69	Naito (1931)
Tokyo Bay Japan				*	*	*		*	*	*	*		35.69	Kikuchi (1984)

(continued on next page)

Table 10.1 Reported spawning months of *Ruditapes philippinarum* by region and latitude—cont'd

Site													Latitude ° North	Sources
Ariakenuomi, Japan			*	*	*	*		*	*	*	*		35.69	Kikuchi (1984)
Bisan-Seto Channel, Inland Sea, Japan					*				*	*	*	*	35.22	Ohba (1959) (14°C–21°C during May spawning)
Hiroshima Bay, Japan		*	*	*	*				*	*	*		34.39	Fisheries Experimental Station (1952)
Kasaoka Bay, Japan				*	*	*				*	*	*	34.50	Yasuda, Hamai, & Hatta (1954)
Jeju Island, East coast, Korea							*	*					33.30	Ngo and Choi (2004)
Ariake Bay, Japan		*	*	*				*	*	*	*		32.82	Tanaka (1954)
Fujian Province, China									*	*			26.07	Qiu et al. (1983), Shi et al. (1984)

(continued on next page)

Western Europe	J	F	M	A	M	J	J	A	S	O	N	D		
Drumcliffe Bay, Sligo, Ireland					*	*	*	*	*				54.55	Drummond et al. (2006)
Dungloe Bay, Donegal, Ireland					*	*	*	*	*				54.95	Drummond et al. (2006)
Cork Harbour, Ireland						*	*	*	*				51.83	Xie and Burnell (1994)
Poole Harbour, UK				*	*	*	*		*	*			50.69	Humphreys et al. (2007)
Normandy France						*	*	*					48.88	Medhioub anc Lubet (1988)
Brittany (North), France								*					48.58	Paillard (1992)
Ile Tudy, Finistere, Brittany, France					*	*	*						47.84	Beninger and Lucas (1984)
Bay of Brest, Brittany, France					*		*	*	*				41.92	Laruelle et al. (1994)

(continued on next page)

Table 10.1 Reported spawning months of *Ruditapes philippinarum* by region and latitude—cont'd

Site													Latitude ° North	Sources
Etel Ria, Brittany, France						*	*		*	*			47.65	Laruelle et al. (1994)
Morbihan Bay, Brittany, France						*	*		*	*			47.29	Laruelle et al. (1994)
Bay of Arcachon									*	*			44.40	Robert et al. (1993)
Northern Spain							*	*	*	*			42.75	Carballo et al. (1992)
Southern Spain				*	*	*	*	*	*	*			36.32	Sarasquete et al. (1990)
Mediterranean														
Thau Lagoon, Herault, France				*	*	*	*	*	*	*			43.63	Maitre-Allain (1985)

(continued on next page)

	J	F	M	A	M	J	J	A	S	O	N	D		
North America														
Str of Georgia, B. Columbia, Canada						*	*	*	*				48.67	Bourne (1982)
Misery Point, Hood Canal, USA			*	*	*	*	*	*	*	*			47.65	Holland and Chew (1974)
Big Beef Harbour, Hood Canal, USA				*	*	*	*	*	*				47.60	Nosho and Chew (1974)
Big Beef Harbour, Hood Canal, USA					*	*	*	*	*				47.60	Holland and Chew (1974)

(continued on next page)

Table 10.1 Reported spawning months of *Ruditapes philippinarum* by region and latitude—cont'd

Site													Latitude ° North	Sources
Other Oceans and Seas														
	J	F	M	A	M	J	J	A	S	O	N	D		
Eilat (Red Sea) Israel	*	*	*	*	*	*	*	*	*	*	*	*	29.55	Shpigel and Fridman (1990)
Oahu, Hawaiian Islands, USA	*	*	*	*	*	*	*	*	*	*	*	*	21.44	Yap (1977)
St. Croix, Virgin Islands, USA		*	*	*				*	*	*			17.75	Rodde et al. (1976)
St. Croix, Virgin Islands, USA	*	*									*	*	17.75	Rodde et al. (1976)
Bandirma Bay, Sea of Marmara, Turkey					*	*	*	*					40.24	Colakoglu and Palaz (2014)

All months within which spawning is reported are marked. The chart does not discriminate between low-level spawning and intense spawning pulses.

Beninger and Lucas (1984) found Manila clams in a rearing pond in northern France to require 500–600 D°, similar to the 480 D° over 6 weeks reported by Utting and Doyou (1992).

The idea of day-degrees assumes that other exogenous factors (such as food availability) affect the quality of gametes and spawn, rather than the rate of development, and that while the temperature regime and exact pattern of warming will vary by locality, the number of day-degrees for successful spawning will be a limiting constant. However, these assumptions are problematic, not least as food quality and thermal adaptation can all be influential on rates. Nevertheless, such a measure can be useful in informing the optimal ambient conditions in clam hatcheries. Also, D° can provide some tentative appreciation of the chances of successful spawning and naturalization on the northern boundaries of distribution, where summer temperature maxima may be short-lived.

Since, in temperate latitudes, temperature restricts sexual maturation, the spawning periods of the Manila clam are seasonal. Where temperatures are conducive, such as in Ariake Bay, southern Japan (water temperatures range from 12°C to 28°C throughout the year), an extended breeding season is reported with multiple cycles of gonad maturation and spawning and three peaks of juvenile recruitment between April and November (Tsutsumi et al., 2002). Further north in Tokyo Bay (water temperatures range from 8°C to 28°C), only two major spawning events occur in May and November (Toba et al., 2007). Still further north than Tokyo, in the Tsugaru Strait, first spawning has been reported as commencing in June, but with most clams spawning between the end of July and the beginning of August. Here, all clams were in a spent condition by the end of September (Shimizu et al., 2006). Therefore, a latitudinal effect along the Japanese coast is apparent.

In even more northerly East Asian latitudes, a single spawning period tends to be compressed into a relatively short period. Table 10.1 shows how spawning events vary according to latitude and location.

10.3 Fecundity

As an indication of an organism's ability to produce offspring, measures of fecundity provide insights into reproductive strategies and the population dynamics of a species. Technical definitions of fecundity vary, as do methods of estimation. In bivalves such *as R. philippinarum*, in which broadcast spawning precedes external fertilization, individual fecundity is typically defined as the estimated number of gametes produced by a female during an induced spawning event, although it may be calculated by counting mature oocytes in the ripe gonad. Since the estimated number does not reflect fertilization rates, an alternative method involves counting eggs released or viable larvae in a closed chamber during a single spawning event.

Yap (1977) reported fecundity to increase from 4.32×10^5 eggs per individual with shell size. Park and Choi (2004) reported an estimated fecundity of 0.94 to 11×10^6 for clams of shell length 21–43 mm with a mean of 4.1×10^6. They further estimated that Gomso Bay (Korea) Manila clams released eggs equivalent to 14.3% of the total dry-tissue weight. Toba and Miyama (1991), using Tokyo Bay Manila clams of shell length 34–36 mm, reported fecundity of $0.24–1.35 \times 10^6$. Chung et al. (2001), using Korean clams (Gomso Bay) of length 20–47 mm, reported $0.20–1.79 \times 10^6$. Moreover, depending on conditions, consecutive spawning events may occur. In each of two such spawning events, 15 days apart, Chung et al. (2001), using Gomso Bay clams of length 20–47 mm, reported a first spawning of $0.20–1.79 \times 10^5$ and a second spawning of $3.2–17.2 \times 10^5$.

Despite prodigious predation, settled spat abundance can also provide some indication of fecundity at the population level. Ikematsu (1957) reported densities in Japan ranging from 20,000/m^2 to 1.8 million/m^2.

Clearly, many variables are at work, including different methods of estimation, environmental conditions, clam sizes, and spawning patterns. In any event, the fecundity of broadcast-spawning bivalves is clearly prodigious, and *R. philippinarum* is no exception. Nevertheless, there is no reason to suppose that fecundity is a key determinant in the success of *R. philippinarum* relative to its congeneric sympatrics.

10.4 Sex ratio and size at sexual maturity

A change in the sex ratio of an age cohort as animals increase in size provides indirect evidence of sex reversal, especially in a sexually nondimorphic species where differential predation is unlikely. In a sample of Korean Manila clams (n = 390). Lee et al. (2013) reported an overall sex ratio of 1:1, but the proportion of females increased with shell size. Clams of length 15.5–20 mm had a ratio of 1:1, whereas in larger clams the percentage of females increased. In their study, the overall sex reversal ratio found was 19%. Among these, 12.9% displayed a reversal from female to male, while the reversal from male to female was 25.9%.

They further suggested that sex reversal probably occurs during the inactive period after spawning. Various studies have reported occasional simultaneous hermaphrodite individuals. Drummond et al. (2006) reported around 0%–0.1% hermaphrodites among over 1000 individuals, and Ponurovsky and Yakovlev (1992) reported up to 2% synchronous hermaphrodite individuals on the Russian coast of the Sea of Japan. Delgado and Perez-Camacho (2007) found one such specimen in over 40 clams. Since reproductively inactive clams cannot be histologically distinguished, it seems likely that these individuals are in transition or have become fixed in this condition during the active reproductive period.

Size at sexual maturity is a crucial factor in the conservation of sustainable fisheries and has therefore been a subject of great interest among fishery regulators, not least for informing minimum landing sizes. The size of *R. philippinarum* and *Ruditapes decussatus* at first reproduction has been reported as 20 and 23 mm, respectively (Moura et al., 2018). However, size at sexual maturity can vary with conditions both directly and as a consequence of selection pressure on a population. On the Russian coast of the Sea of Japan, Ponurovsky and Yakovlev (1992) reported that age at sexual maturity varied with location and habitat. Around 60% of subtidal Manila clams were sexually mature in their first year and occasionally at shell lengths as small as 8 mm in males and 7.5 mm in females.

Heavily exploited fish populations are also subject to selection pressure toward smaller sizes at sexual maturity, exerted artificially by the fishery itself (Caill-Milly et al., 2025). This subject and the question of minimum landing sizes are considered further in Chapter 19 with reference to the Arcachon Bay population in France.

10.5 Larval development

In contrast to somatic cytoplasm, the germplasm consists of cells that form the repository of hereditary genetic resources involved in the formation of sex cells in the gonads. As such, it is intimately involved in meiotic cell division, which produces haploid eggs and sperm. During meiosis, primary diploid oogonia and spermatogonia divide asymmetrically twice to form mature haploid gametes and two haploid polar bodies. The latter can be seen as repositories for the homologous chromosomes not retained in the sex cells, while also leaving the greater part of the cytoplasm conserved, especially in the developing oocyte. Reunov et al (2018) provided a detailed account of these germplasm processes in *R. philippinarum.*

After spawning and external fertilization, the zygote of *R. philippinarum* follows a largely conventional molluscan bivalve embryology. Table 10.2 shows the embryonic stages with time elapsed for *R. philippinarum.* At 22°C a fertilized egg of diameter 64.2 μm commences rapid mitotic cleavage after just over an hour becoming a 32-cell embryo in 2 h 50 m. Subsequent embryonic stages shown in Fig. 10.1 include a compact solid ball of cells, the morula (at 3 h 40 m), before forming a cavity (the blastocoel) to become the hollow blastula (at 5 h 05 m). By 7 h 15 m, the embryo is reorganized into the three embryonic germ layers (endoderm, mesoderm, and ectoderm) of the gastrula, which provide the organizational foundation of the triploblastic phyla. However, various environmental factors will affect the rate of Manila clam development, not least temperature. Qiuzheng and Mingyue (1988) reported a much faster 3 h 45 m to gastrula formation at 25°C compared with the 7 h 15 m reported at 22°C by Hur et al (2005).

After around 20 h at 22°C, the veliger larva develops, during which time a second visible shell increment grows (prodissoconch II). Early veligers resemble small bivalves with calcified valves that completely enclose the soft tissues. However, they differ in having a ciliated velum, which protrudes from the shell to provide both motility and the collection of food. They also possess a pronounced straight hinge line and are consequently commonly referred to as D-shaped larvae or D-shaped veligers. The more familiar clam shape emerges as the umbo grows (Fig. 10.1). Finally, before settlement, the pediveliger stage includes both a foot and siphons, and a final metamorphosis corresponds to settlement and the adoption of the benthic habit. Veligers are planktonic and sufficiently motile to maintain a position and depth in

Table 10.2 Stages and mean time intervals in the embryological development of *Ruditapes philippinarum*.

Stage	Elapsed time	Elapsed time	Mean dimensions	Notes
	Hur et al. (2005) (22°C)	Quizhong and Mingyue (1988) (25°C)		
Egg			64.2 μm	Diameter
First polar body	0 h 35 m	0 h 15 m		First meiotic division
Second polar body	1 h 00 m	0 h 20 m		Second meiotic division
2 cells	1 h 10 m	0 h 35 m		
4 cells	1 h 40 m	1 h 00 m		
8 cells	2 h 05 m	1 h 20 m		
16 cells	2 h 25 m	1 h 45 m		
32 cells	2 h 50 m	2 h 05 m		
Morula	3 h 40 m	2 h 20 m		Solid cell mass. Approx. 64 cells
Blastula	5 h 05 m	2 h 38 m		Hollow single-layer cell sphere
Gastrula	7 h 15 m	3 h 45 m		Three germ layers
Trochophore	10 h 00 m			Ciliated, free-swimming larva
Early D-shaped larva	20 h 00 m		85 μm	Shell length
Late D-shaped larva	5 d			
Early umbonal veliger	6 d		172 μm	With two ciliated lobes (velum)
Late umbonal veliger	12 d			With foot
Pediveliger	13 d		209 μm	Settlement with adult body plan

After around 10 hours at 22°C, a fully organized top-shaped motile trochophore larva emerges with a functioning gut, a conspicuous girdle of cilia (prototroch) near the middle, and apical flagella. This stage also marks the development of a bivalved appearance in the form of the early "prodissoconch I" shell, whose length of approximately 100 μm corresponds to the final length of the trochophore.

the water column. Due to their relatively long duration, veligers have a key role in the dispersal of the species. Fig. 10.2 shows a scanning electron micrograph of early juvenile *R. philippinarum*.

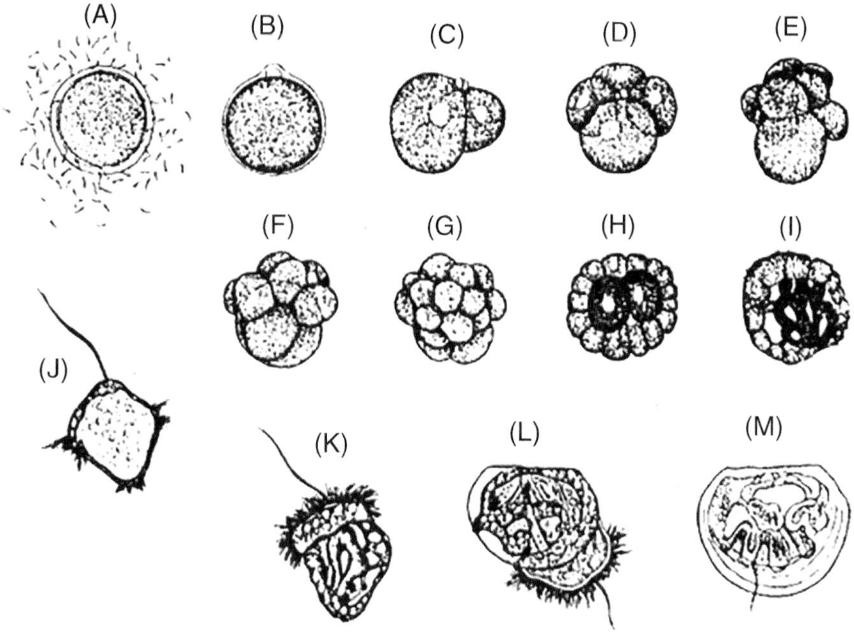

FIGURE 10.1

Development of the Manila clam embryo and larva. A-B depict the initial fertilization of the clam oocyte, with C–I depicting cell division and J and K showing the trochophore larval stages. L and M depict the D-larva veliger stages. D-larvae have a shell length of about 90–95 μm.

From Utting and Spencer (1991).

Zhang et al. (2022) have elucidated gene groups controlling larval development in the Manila clam through transcriptomic analysis of gene expression.

10.6 Larval dispersal and settlement

In suitable locations and conditions, there is a clear positive relationship between newly settled Manila clams ("settlers") and larval density. Working in Puget Sound, United States, Williams (1980) reported the average length of newly settled Manila clams to range from 0.197 to 0.207 mm and average density from 18,600 to 31,200/m^2. He also noted more settlement in areas with moderate or low adult clam densities in comparison to areas with abundant adult clams. Williams attributed this effect in part to the ingestion of larvae by adults, while also recognizing that even in dense areas, larvae are not entirely prevented from settling.

FIGURE 10.2

Scanning electron micrograph of early juvenile *Ruditapes philippinarum*. The prodissoconch shell is the smoother shell area to the left. The postmetamorphosis dissoconch shell is the more ridged, larger portion of the shell. The protruding soft tissue is likely extended siphons.

From Matt Harris and Simon Cragg.

Not all spatfalls contribute equally to adult populations. At Ise Bay, Japan, Miyawaki and Sekiguchi (2000) found that density peaks of planktonic larvae, new settlers (shell length less than 0.3 mm), and small clams (0.3–1.0 mm), which occurred in early winter, contributed to established benthic populations, whereas those that fell in spring and early summer did not. Conversely, in Ariake Bay, density peaks of new settlers and small clams in autumn and early winter failed to contribute, in contrast to those in spring (Ishii et al., 2001). Location and season-specific environmental factors, therefore, appear to influence the success of a spatfall event in contributing to an established benthic population.

Larval dispersal depends on an interaction between the duration of the planktonic stages, hydrodynamic conditions including tidal current patterns, and the behavior of the larvae in the water column. Although planktonic Manila clam larvae are free-swimming, the scale of motility is such that dispersal is largely accomplished by advection within water currents into new suitable habitats. However, this does not mean that larval motility is necessarily a trivial factor in dispersal since they can change their position in the water column in response to factors, including salinity, temperature, current strength, and light level (Ishida et al., 2005).

Swimming speeds vary with larval stage from about 0.5 mm/s for the trochophore, increasing gradually to 3 mm/s in the pediveliger. The prevailing direction of movement similarly varies with the trochophore showing a 90% probability of upward movement, and the veliger around 50% in each vertical direction. Such movements tend to maintain the larvae in preferred conditions. For example, in the presettlement stage, there is a high probability of the pediveliger remaining within 1 m of the sediment surface (Bidegain et al., 2013; Ishida et al. 2005).

In contrast to coastal salinities typically around 34PSU, estuarine salinities tend to vary more with depth, and Manila clam larvae appear to target lower salinities (Ishida et al., 2005). Behavioral data have been used in conjunction with empirically determined hydrodynamic current and salinity patterns to model dispersal from four sites in Poole Harbour, UK. The model estimates that the resulting distribution of larvae at settlement can be highly influenced by vertical migrations in response to reduced salinity preferences of S = 17 and 23, in comparison to neutral buoyancy. For larvae "released" at two out of the five sites modeled, the estimated proportion being retained within the harbor was greatly increased at both these sites with lower salinity preferences. These sites appeared to resupply the whole harbor, ensuring the persistence of populations (Herbert et al., 2011). Fig. 10.3A and B show a good correspondence between modeled larval dispersal patterns and actual Manila clam distribution in the harbor.

Bidegain et al. (2013) have predicted Manila clam dispersal in the Bay of Santander, Spain, including habitat suitability in their model. This enabled spatial predictions of recruitment success, which corresponded well with observed distribution data. This correlation was not evident when larval behavior was removed from the model, which resulted in an underestimate of larval recruitment within the Bay. This further supports the contention that larval behavior favors retention within an estuary, as it also does within Poole Harbour (Herbert et al., 2011). Both Poole Harbour and the Bay of Santander are complex areas containing many smaller bays and inlets. Bidegain et al. (2013) results also support a combination of largely isolated "self-recruiting nursery grounds" with more generally connected areas, as suggested by Herbert et al. (2011) in Poole Harbour. It is also clear from these models that larval duration in the plankton is significant for retention and recruitment within an estuary. Larval dispersal along a coast from one estuary to the next is considered further in Chapter 20.

Tezuka et al. (2013) also reported that Manila clam larvae prefer lower salinity water, and that settlement rate increased as salinity decreased, up to a low of S = 13. Such studies demonstrate that hydrodynamic conditions are far from being the only determinants of larval settlement locations, and that the veliger behavior plays a

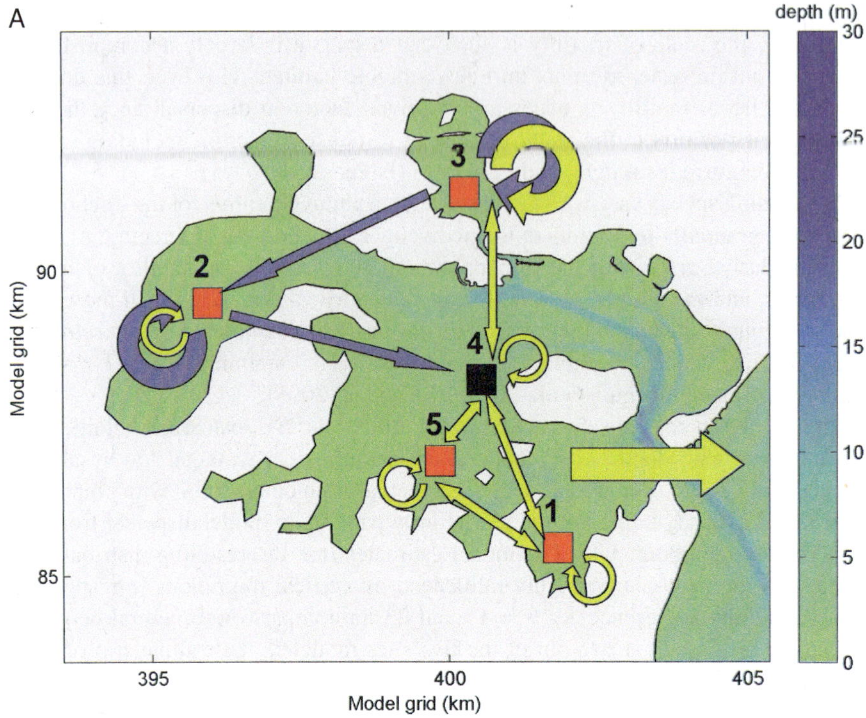

FIGURE 10.3

(A) Manila clam larval dispersal in Poole Harbour, UK, indicated by hydrodynamic and behavioral modeling. Yellow arrows demonstrate connectivity between zones with larval swimming behavior targeting a salinity of 23PSU. Blue arrows show connectivity when a lower salinity of 17PSU is targeted. The size of the arrows predicts the relative strength of the supply, indicating that areas 2 and 3 resupply strongly. (The prospect of successful dispersal out of the harbor to adjacent estuaries is considered in Chapter 20). (B) Biomass (ash-free dry weight) of Manila clams sampled in 2002 and 2009 shows approximate correspondence with the modeled predictions.

From Herbert et al. (2011).

significant role in retention within estuarine areas. Tezuka et al. (2013) also reported that larval settlement does not differ by sediment grain size, although they observed that larvae failed to dig in with sediment grain sizes of 0.125 and 0.063 mm, suggesting an unsuitability of homogeneous fine sediments and the possibility of secondary settlement in such circumstances.

10.7 The role of byssus

Attachment to a substrate by the secretion of byssus filaments is unique to bivalves. Secreted by byssal glands on the foot, these filaments function to secure the position

B

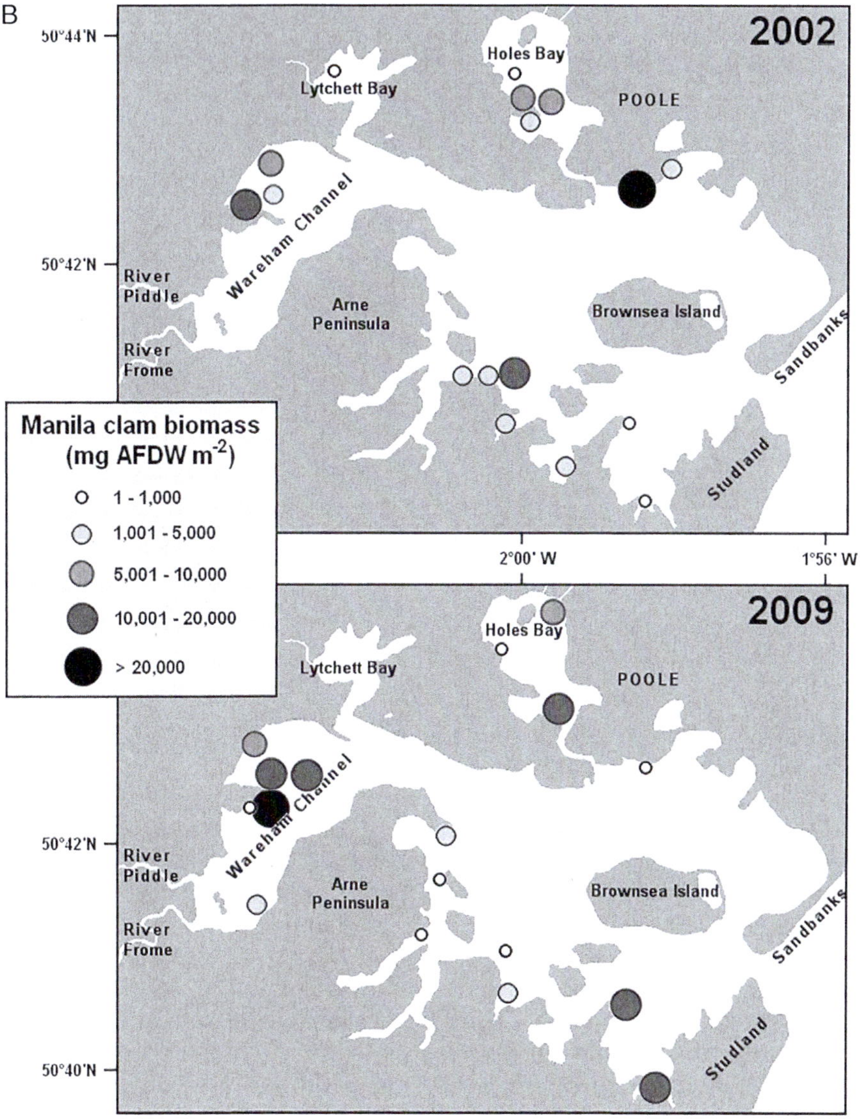

FIGURE 10.3, cont'd.

of the animal against disturbance from tidal water currents and waves. For epibenthic bivalves such as mussels, attachment is essential throughout life, and movement of adults to more favorable conditions requires the secretion of new threads. Infaunal bivalves vary in the extent to which they retain a capacity for byssal secretion. Although adult *R. philippinarum* retain a byssal gland and capacity to secrete byssus,

FIGURE 10.4

Manila clam spat showing byssal attachment to a large pebble in the Yagami Estuary, Yokohama, Japan (finger indicates scale). Photo credit: the author.

the behavior is most associated with newly settled spat, and byssus production is reduced with growth (Takeuchi et al., 2019).

Once settled in suitable conditions, Manila clam larvae can be observed quickly embedding into the sediment, but they may also secrete byssus filaments and select sediment grain sizes conducive to byssal attachment (Fig. 10.4).

Takeuchi et al. (2026) conducted field surveys in areas of Ariake Bay, Japan, which differed with regard to sediment characteristics. No attachment preference was observed between shell debris and mineral components of the sediment.

In laboratory experiments, these authors used sediment mixtures of sand grains, small pebbles, and large pebbles. In mixed sediments, clams preferentially attached to small or large pebbles with no preference for pebble size. However, when presented with sand grains alone, attachment probability was not reduced, indicating a preference for attachment rather than nonattachment even when pebbles were not available. They reported zero tendency for the clams to attach to each other in the laboratory experiments, although this is sometimes observed in the field.

Takeuchi et al. (2026) also observed byssally attached clams from the field, which, when moved to sluggish flow or still water environments, self-detach and burrow into the sediment. This raises the possibility that byssal attachment may sometimes function as a temporary anchor mechanism before secondary dispersal and settlement. Moreover, rather than securing a position, byssal attachment may sometimes be instrumental in facilitating passive secondary settlement when currents cause general bedload sediment transport with clams attached.

10.8 Secondary settlement

Despite veliger behavior and preferences, the relative power of hydrodynamic processes is such as to limit their influence on settlement sites. In many cases, larvae will settle in suboptimal locations, and there is no doubt that postsettlement mortality plays a large part in establishing the distribution of clam populations in the benthos. However, in bivalves, settlement, rather than being a single event, may involve secondary migrations through active movement across or within the sediment, or resuspension at the sediment-water boundary layer to passively exploit waves or currents. Species can either limit such secondary dispersal by burial or attachment, or they can facilitate passive migration by staying on the sediment surface or detaching to promote resuspension.

The ability to drift in the boundary layer is a function of the drag force, which varies with shell shape. However, the secretion of byssus threads by bivalves, aside from securing a position, can also significantly enhance drag, resulting in "byssopelagic drifting," even at low flow velocities. Since Manila clam settlers do produce byssal filaments for initial attachment, the question arises as to whether these may also be involved in resuspension and horizontal relocation before secondary settlement.

To answer this question, Montaudouin (1997) performed a comparative study of the settlement behavior of *R. philippinarum* and the cockle *Cerastoderma edule*. In a laboratory flume experiment, Manila clams of 1.1 mm mean length were placed on an unsuitable surface (plexiglass). They were reported to never resuspend, but to crawl to a distance of up to 42 cm in 4 h. In contrast, the cockle mostly moved by resuspension. Montaudouin (1997) interpreted these differences as alternative strategies for movement to a more suitable location. However, the Manila clams' movement showed no apparent tropisms, for example, in response to the direction of water flow.

Foret et al. (2018) have also examined the potential for secondary migrations in settled *R. philippinarum*, along with three other bivalves of varying shell shape: the warty venus, *Venus verrucosa*; the scallop, *Pecten maximus*; and the mussel, *Mytilus edulis*. They assessed postsettlement sinking velocities and sediment-water boundary shear stresses due to water movements on tidal flats where the four species are found. The datasets were then combined to estimate the secondary migration potential and the temporal windows within which secondary migration in suspension might be possible. As might be expected, they found that secondary migration potential varied with shell shape, byssus secretion, and behavior, up to a threshold shell size above which drifting with water movement was no longer possible. For *R. philippinarum*, the maximum shell length compatible with resuspension was 462 μm. Therefore, opportunities for resuspension decline as the animal grows.

Of these four species, *R. philippinarum* demonstrated the highest sinking velocity and lowest estimated resuspension threshold size. In prevailing hydrodynamic conditions at the study site in Normandy, France, Foret et al. (2018) calculated the temporal window within which these species could plausibly resuspend. This was the shortest for *R. philippinarum*, ranging between 5 and 16 days depending on the growth rate. Both the spatial and temporal scales of potential relocation are also heavily dependent on the timing of initial settlement in relation to the spring/neap tidal cycle and consequent current speed.

There is little doubt that current speeds and waves will sometimes resuspend and relocate Manila clam settlers, and such hydrodynamic processes may account for some observed settler distributions, perhaps from subtidal to low intertidal areas. However, it seems that, for Manila clams, active crawling is also a significant horizontal migration mechanism. Moreover, active crawling may not be confined to early benthic stages. In a journal note, Sai (1963) reported "Wakagu" clams (*R. philippinarum)* to be capable of crawling up to 570 cm over 30 days.

Therefore, we cannot rule out significant postsettlement movements in Manila clams either passively in response to hydrodynamic effects on early settlers or actively in response to less suitable initial substrates. Passive migrations have been conjectured to account for mismatches between larval supply and new settlers in circumstances of heavy river discharge and tidal currents, combined with the absence of evidence of mass settler mortality, such as dead shells (Miyawaki & Sekiguchi, 2000).

On tidal flats in Tokyo Bay, Toba et al. (2011) investigated the physical transport of newly settled Manila clam larvae up to 1 mm shell length using sediment traps. They found that the mobilization of juvenile clams was caused by wave-generated oscillatory turbulence, especially when wave shear stress exceeded 0.3 N $m{-}^2$, and once resuspended, they would be carried inshore by advection currents. Although sediment particles and clams differed in size and specific gravity, they were both observed to be transported in a similar way, suggesting bedload transport included clams adhered to sediment grains as one possible explanation. Using a release/recovery method, these authors found clams moving several hundred meters as a consequence of these combined wave and current effects. They concluded that variation in physical conditions could account for changing patterns of juvenile abundance.

References

Baek, M. J., Lee, Y.-J., Choi, K.-S., Lee, W. C., Park, H. J., Kwak, J. H., & Kang, C.-K. (2014). Physiological disturbance of the Manila clam, *Ruditapes philippinarum*, by altered environmental conditions in a tidal flat on the west coast of Korea. *Marine Pollution Bulletin, 78*(1–2), 137–145. https://doi.org/10.1016/j.marpolbul.2013.10.050.

Bayne, B. L. (1976). Aspects of reproduction in bivalve molluscs. In M. L. Wiley (Ed.), *Estuarine processes* (pp. 432–448). Academic Press.

Beninger, P. G., & Lucas, A. (1984). Seasonal variations in condition, reproductive activity, and gross biochemical composition of two species of adult clam reared in a common habitat: *Tapes decussatus* L. (Jeffreys) and *Tapes philippinarum* (Adams & Reeve). *Journal of Experimental Marine Biology and Ecology, 79*(1), 19–37. https://doi.org/10.1016/0022-0981(84)90028-5.

Bidegain, G., Bárcena, J. F., García, A., & Juanes, J. A. (2013). LARVAHS: Predicting clam larval dispersal and recruitment using habitat suitability-based particle tracking model. *Ecological Modelling, 268*, 78–92. https://doi.org/10.1016/j.ecolmodel.2013.07.020.

Bourne, N. (1982). Distribution, reproduction and growth of Manila clam *Tapes philippinarum* (Adams & Reeve) in British Columbia. *Journal of Shellfish Research, 2*, 47–54.

Cahn, A. R. (1951). *Clam culture in Japan. Leaflet FL-399* (p. 103). U.S. Fish and Wildlife Service.

Caill-Milly, N., Bru, N., de Montaudouin, X., Dang, C., Briaudeau, T., & Sanchez, F. (2025). Biological features for a minimum conservation reference size revision- case of *Ruditapes philippinarum* in Arcachon Bay (French Atlantic coast). *Estuarine, Coastal and Shelf Science, 320*, 109288.

Carballo, R. Q., Selas, F. J. V., Carballo, S. R., Ferreiro, L. L., & Pascual López, M., de la C. (1992). Evaluation of the reproductive cycle of Ruditapes philippinarum and Tapes decussatus in the Ria de Muros-Noya. (Evaluacion del ciclo reproductor de *Ruditapes philippinarum,* y *Tapes decussatus* en la Ria de Muros-Noya). *Alimentaria: Revista de tecnología e higiene de los alimentos, 236*, 37–44.

Choi, K. C. (1965). Ecological studies on early stages of the bivalve *Tapes philippinarum*. *College Education Review, 7*(1), 161–234.

Chung, E. Y., Ryou, D. K., & Lee, J. H. (1994). Gonadal development, age and growth of the short-necked clam *Ruditapes philippinarum* (Pelecypoda: Veneridae) on the coast of Kimje, Korea. *Korean Journal of Malacology, 10*, 38–54.

Chung, E.-Y., Chung, J. S., & Ki-Young Lee, K.-Y. (2013). Gametogenic cycle, the spawning season, first sexual maturity, and the biological minimum size in male *Ruditapes philippinarum* (Bivalvia: Veneridae) in Western Korea. *Journal of Life Sciences, 7*(6), 613–622.

Chung, E.-Y., Hur, S. B., Hur, Y.-B., & Lee, J. S. (2001). Gonadal maturation and artificial spawning of the Manila clam *Ruditapes philippinarum* (Bivalvia: Veneridae), in Komso Bay, Korea. *Korean Journal of Fish Science and Technology, 4*, 208–218.

Colakoglu, S., & Palaz, M. (2014). Some population parameters of *Ruditapes philippinarum* (Bivalvia, Veneridae) on the southern coast of the Marmara Sea, Turkey. *Helgoland Marine Research, 68*, 539–548.

Delgado, M., & Perez-Camacho, A. (2007). Comparative study of gonadal development of *Ruditapes philippinarum* (Adams and Reeve) and *Ruditapes decussatus* (L.) (Mollusca: Bivalvia): Influence of temperature. *Scientia Marina, 71*(3), 471–484.

Denisenko, S. G. (1978). Population structure of the bivalve *Venerupis japonica* (Deshayes) in Vostok Bay of the Sea of Japan. Problems of applied and regional ecology of the shelf. In *The 2nd all-union conference on biology of the shelf. Abstracts of papers Part 2,* (pp. 33–35). (in Russian).

Devauchelle, N. (1990). Sexual development and maturity of *Tapes philippinarum*. In ESVA (Ed.), *Tapes philippinarum, Biologia e sperimentazione* (pp. 47–62). ESVA.

Drummond, L., Mulcahy, M., & Culloty, S. (2006). The reproductive biology of the Manila clam, *Ruditapes philippinarum*, from the North-West of Ireland. *Aquaculture, 254*(1–4), 326–340. https://doi.org/10.1016/j.aquaculture.2005.10.052.

Eversole, A. G. (1989). Gametogenesis and spawning in North American clam populations: Implications for culture. In J. J. Manzi, & M. Castagna (Eds.), *Clam mariculture in North America*. Elsevier.

Forêt, M., Tremblay, R., Neumeier, U., & Olivier, F. (2018). Temporal variation of secondary migrations potential: Concept of temporal windows in four commercial bivalve species. *Aquatic Living Resources, 31*, 19.

Goshima, S., Ide, N., Fujiyoshi, Y., Noda, T., & Nakao, S. (1996). Reproductive cycle and shell growth of transplanted Manila clam *Ruditapes philippinarum* in Saroma Lagoon. *Nihon Suisan Gakkai, 62*(2), 195–200. https://doi.org/10.2331/suisan.62.195.

Herbert, R. J. H., Willis, J., Jones, E., Ross, K., Hübner, R., Humphreys, J., Jensen, A., & Baugh, J. (2011). Invasion in tidal zones on complex coastlines: Modelling larvae of the non-native Manila clam, *Ruditapes philippinarum*, in the UK. *Journal of Biogeography, 39*(3), 585–597. https://doi.org/10.1111/j.1365-2699.2011.02626.x.

Holland, D. A., & Chew, K. K. (1974). Reproductive cycle of the manila clam (*Venerupis japonica*) from hood canal, Washington. *Proceedings of the National Shellfish Association, 64*, 53–58.

Humphreys, J., Caldow, R. W. G., McGrorty, S., West, A. D., & Jensen, A. C. (2007). Population dynamics of naturalised manila clams *Ruditapes philippinarum* in British coastal waters. *Marine Biology, 151*(6), 2255–2270. https://doi.org/10.1007/s00227-007-0660-x.

Hur, Y. B., Bae, J. H., & Hur, S. B. (2005). Comparison of development and larval growth of four venerid clams. *Journal of the World Aquaculture Society, 36*(2), 179–187. https://doi.org/10.1111/j.1749-7345.2005.tb00383.x.

Ikematsu, W. (1957). Ecological studies on the clam, *Tapes japónica* (Reeve)—II. On the setting season and the growth in early young stage. *Nippon Suisan Gakkaishi, 22*(12), 736–741. https://doi.org/10.2331/suisan.22.736.

Ishida, M., Ogasawara, M., Murakami, C., Momoi, M., Ichikawa, T., & Suzuki, T. (2005). Changes in the behavioural characteristics in relation to salinity selection and vertical movement at different growth stages of the planktonic larvae of the Japanese littleneck clam *Ruditapes philippinarum. Bulletin of the Japanese Society of Fisheries Oceanography, 69*, 73–82.

Ishii, R., Sekiguchi, H., Nakahara, Y., & Jinnai, Y. (2001). Larval recruitment of the Manila clam *Ruditapes philippinarum* in Ariake Sound, southern Japan. *Fisheries Science, 67*(4), 579–591. https://doi.org/10.1046/j.1444-2906.2001.00293.x.

Kikuchi, T. (1984). Reproductive ecology and life history traits in the marine invertebrates: XII introductory notes on life history traits. *Contributions from the Amakusa Marine Biological Laboratory, Kyushu University, 6*(4), 53–58.

Kulikova, V. A. (1979). Some peculiarities in reproduction of Bivalvia species in Busse lagoon (Sea of Okhotsk) with reference to water temperatures. *Biologiya Morya Vladivostok, 1*, 34–38.

Laing, I., & Utting, S. D. (1994). The physiology and biochemistry of diploid and triploid Manila clam (*Tapes philippinarum* Adams & Reeve) larvae and juveniles. *Journal of Experimental Marine Biology and Ecology, 184*(2), 159–169. https://doi.org/10.1016/0022-0981(94)90002-7.

Laruelle, F. J., Guillou, J., & Paulet, Y. M. (1994). Reproductive pattern of the clams, *Ruditapes decussatus* and *Ruditapes philippinarum* on intertidal flats in Brittany. *Journal of the Marine Biological Association of the UK, 74*(2), 351–366.

Lee, J. S., Park, J. S., Shin, Y. K., Lee, Y. G., & Park, J. J. (2013). Sequential hermaphroditism in Manila clam *Ruditapes philippinarum* (Bivalvia: Veneridae). *Invertebrate Reproduction and Development, 57*(3), 185–188. https://doi.org/10.1080/07924259.2012.717109.

Hiroshima Fisheries Experimental Station, (1952), Suishi-dayori, No, 24, 1 G. (cited in Ohba, 1959).

Maitre-Allain, T. (1985) Data on the reproduction of the Japanese clam *Ruditapes philippinarum* in the Thau pond (Heault, France) (Donnees sur la reproduction de la palourde japonaise *Ruditapes philippinarum* dans l'etang de Thau (Heault, France). Rapports et

Procès-Verbaux des Réunions de la Commission. Internationale pour l'Exploration Scientifique de la Mer Méditerranée. 29(4) 109–110.

Mann, R. (1979). The effect of temperature on growth, physiology, and gametogenesis in the Manila clam *Tapes philippinarum* (Adams & Reeve, 1850). *Journal of Experimental Marine Biology and Ecology, 38*(2), 121–133. https://doi.org/10.1016/0022-0981(79)90016-9.

Mann, R., & Glomb, S. J. (1978). The effect of temperature on growth and ammonia excretion of the Manila clam *Tapes japonica*. *Estuarine and Coastal Marine Science, 6*(3), 335–339. https://doi.org/10.1016/0302-3524(78)90021-X.

Medhioub, N. M., & Lubet, P. E. (1988). Cytological research on the cellular environment (reverse tissue) of the gonads of the clam Ruditapes philippinarum (Recherches cytologiques sue l'environment cellulaire (tissue de reserve) des gonads de la Palourde (*Ruditapes philippinarum* Adams et Reeve), Mollusque bivalve. *Annales des Sciences Naturelles. Zoologie, 9*, 87–102.

Meneghetti, F., Moschino, V., & Da Ros, L. (2004). Gametogenic cycle and variations in oocyte size of *Tapes philippinarum* from the Lagoon of Venice. *Aquaculture, 240*(1–4), 473–488. https://doi.org/10.1016/j.aquaculture.2004.04.011.

Miyawaki, D., & Sekiguchi, H. (2000). Long-term observations on larval recruitment processes of bivalve assemblages on temperate tidal flats. *Benthos Research, 55*(1), 1–16. https://doi.org/10.5179/benthos1996.55.1_1.

Montaudouin de, X. (1997). Potential of bivalves' secondary settlement differs with species: A comparison between cockle (*Cerastoderma edule*) and clam (*Ruditapes philippinarum*) juvenile resuspension. *Marine Biology, 128*, 639–648. https://doi.org/10.1007/s002270050130.

Moura, P., Vasconcelos, P., Pereira, F. á, Chainho, P., Costa, J. éL., & Gaspar, M. B. (2018). Reproductive cycle of the Manila clam (*Ruditapes philippinarum*): An intensively harvested invasive species in the Tagus Estuary (Portugal). *Journal of the Marine Biological Association of the United Kingdom, 98*(7), 1645–1657. https://doi.org/10.1017/s0025315417001382.

Millican, P. F. and Williams, D. R. (1985). The seasonal variation in the level of meat content, lipid and carbohydrate in Mercenaria mercenaria and Tapes semidecussatus grown in fertilised and unfertilised water CM2985/K.51, International Council for the Exploration of the Sea.

Naito, S. (1931). Kio-jigyo-hokolrn, Naiwan Branch, Chiba Fish. Exp. Stat., yearbook, 10–19. 1930-1931. https://www.pref.chiba.lg.jp/english/introduction/industry.html

Ngo, T. T. T., & Choi, K. S. (2004). Seasonal changes of *Perkinsus* and *Cercaria* infections in the Manila clam *Ruditapes philippinarum* from Jeju, Korea. *Aquaculture, 239*(1–4), 57–68. https://doi.org/10.1016/j.aquaculture.2004.06.026.

Nosho, T. Y., & Chew, K. K. (1974). Reproductive cycle of the Manila clam (*Venerupis japonica*) from Hood Canal, Washington. *Proceedings of the National Shellfisheries Association, 64*, 53–58.

Ohba, S. (1959). Ecological studies in the natural population of a clam, *Tapes japonica*, with special reference to seasonal variations in the size and structure of the population and to individual growth. *Biological Journal of Okayama University, 5*, 13–47.

Paillard, C. (1992). Etiology and characterisation of brown ring disease in breeding clams *Ruditapes philippinarum* (Etiologie et caracterisation de la maladie de la anneau brun chez palourde de'elevage *Ruditapes philippinarum)* Thesis. Universitie de Bretange Occidentale.

Park, K. I., & Choi, K. S. (2004). Application of enzyme-linked immunosorbent assay for studying of reproduction in the Manila clam *Ruditapes philippinarum* (Mollusca: Bivalvia): I. Quantifying eggs. *Aquaculture, 241*(1–4), 667–687. https://doi.org/10.1016/j.aquaculture.2004.08.017.

Ponurovsky, S. K., & Yakovlev, Y. M. (1992). The reproductive biology of the Japanese littleneck, *Tapes philippinarum* (Adams and Reeve, 1850) (Bivalvia: Veneridae). *Journal of Shellfish Research, 11*(2), 265–277.

Qi, Q., & Yang, M. (1988). The growth and development of the clam Ruditapes philippinarum. *Journal of Fisheries of China, 12*(1), 1–11.

Qiuzheng, Q., & Mingyue, Y. (1988). The growth and development of the clam *Ruditapes philippinarum. Journal of Fisheries of China, 12*(1), 1–11.

Rakov, V. A. (1986). Biological basis for the cultivation of the Pacific cockle in Peter the Great Bay. The fifth congress of the all-union hydro-biological society. In *Abstracts of papers. Part 1* (pp. 114–116). Volzhskaya Kommuna Press.

Reunov, A., Alexandrova, Y., Reunova, Y. A., Komkova, A., & Milani, L. (2018). Germ plasm provides clues on meiosis: The concerted action of germ plasm granules and mitochondria in gametogenesis of the clam *Ruditapes philippinarum. Zygote.*, 1–11. https://doi.org/10.1017/S0967199418000588.

Robert, R., Trut, G., & Laborde, J. L. (1993). Growth, reproduction and gross biochemical composition of the Manila clam *Ruditapes philippinarum* in the Bay of Arcachon, France. *Marine Biology, 116*, 291–299.

Rodde, K. M., Sanderlin, J. B., & Roles, O. A. (1976). Experimental cultivation of *Tapes japonica* (Deshayes) (Bivalvia: Veneridae) in artificial upwelling culture systems. *Aquaculture, 9*(3), 203–215.

Rodríguez-Moscoso, E., & Arnaiz, R. (1998). Gametogenesis and energy storage in a population of the grooved carpet-shell clam, *Tapes decussatus* (Linne, 1787), in northwest Spain. *Aquaculture, 162*(1–2), 125–139. https://doi.org/10.1016/S0044-8486(98)00170-7.

Sai, S. (1963). Movement of a clam, Ruditapes philippinarum. *Aquaculture (Japan), 11*, 13–24.

Sarasquete, M. C., Gimeno, S., & Canales, M. D (1990). Reproductive cycle of the palourde *Ruditapes philippinarum* (Adams and Reeve) from the south-west Atlantic coast (Spain) Cycle reproductor de la palourde *Ruditapes philippinarum* (Adams & Reeve) de la cote sud-quest Atlantique (Espagne). *Revue Internationale d'Oceanographie Medicale, Nice, 97–98*, 90–99.

Shi, B, Wu, Q., Ma, C., & Zhu, Y. (1984). Studies on the spawning of the Philippine clam *Ruditapes philippinarum* (Adams and Reeve) in artificial rearing earth ponds. *Journal of Xiamen University (Natural Science), 23*(2), 211–216.

Shimizu, Y., Ootsu, H., Ebisu, A., & Tada, M. (2006). Reproductive cycle of the Japanese littleneck clam in the waters off Moheji (Japan), Tsugaru Strait. *Scientific Reports of Hokkaido Fisheries Experimental Station, 70*, 99–104.

Shpigel, M., & Fridman, R. (1990). Propagation of the Manila clam (*Tapes semidecussatus*) in the effluent of fish aquaculture ponds in Eilat. *Aquaculture, 90*(2), 113–122. https://doi.org/10.1016/0044-8486(90)90335-K.

Sugiura, D., & Kikuya, N. (2022). Cycle of reproduction, oocyte atresia, and maturation size of the Manila clam *Ruditapes philippinarum* on a pebble beach in Mutsu Bay, Northern Japan. *Invertebrate Reproduction & Development, 66*(1). https://doi.org/10.1080/07924259.2021.1999337.

Takeuchi, S., Honma, K., & Hirasaka, K. (2019). Byssus production ability and degree of byssal-gland development in the infaunal clam *Ruditapes philippinarum*. *Acta Zoologica, 101*(4), 392–403.

Takeuchi, S., Nakachi, M., Matsubara, M., & Sankoda, S. (2026). Substrate selectivity in the byssal attachment behavior of *Ruditapes philippinarum* (Bivalvia: Veneridae). *Journal of Experimental Marine Biology and Ecology, 594*, 152158. https://doi.org/10.1016/j.jembe.2025.152158.

Tanaka, Y. (1954). Spawning season of important bivalves in Ariake Bay—III. T*apes philippinarum*. *Bulletin of the Japanese Society of Scientific Fisheries, 19*, 1165–1167.

Tezuka, N., Kanematsu, M., Asami, K., Sakiyama, K., Hamaguchi, M., & Usuki, H. (2013). Effect of salinity and substrate grain size on larval settlement of the Asari clam (Manila clam, *Ruditapes philippinarum*). *Journal of Experimental Marine Biology and Ecology, 439*, 108–112. https://doi.org/10.1016/j.jembe.2012.10.020.

Toba, M., Ito, M., & Kobayashi, Y. (2011). Bedload transport of newly-settled juveniles of the Manila clam *Ruditapes philippinarum* observed in situ at Banzu tidal flat, Tokyo Bay. *Journal of Shellfish Research, 30*(3), 777–789. https://doi.org/10.2983/035.030.0318.

Toba, M., & Miyama, Y. (1991). Gonadal development and spawning induction in artificially conditioned Manila clams *Ruditapes philippinarum*. *Nippon Suisan Gakkaishi, 57*(7), 1269–1275. https://doi.org/10.2331/suisan.57.1269.

Toba, M., & Miyama, Y. (1995). Influence of temperature on the sexual maturation in Manila clam, *Ruditapes philippinarum*. *Japan Aquaculture Society., 43*(3), 305–314.

Toba, M., Yamakawa, H., Kobayashi, Y., Sugiura, Y., Honma, K., & Yamada, H. (2007). Observations on the maintenance mechanisms of metapopulations, with special reference to the early reproductive process of the Manila clam *Ruditapes philippinarum* (Adams & Reeve) in Tokyo Bay. *Journal of Shellfish Research, 26*(1), 121–130. https://doi.org/10.2983/0730-8000(2007)26[121:ootmmo]2.0.co;2.

Tsutsumi, H., Ishizawa, K., Tomishige, M., Moriyama, M., Sakamoto, K., & Montani, S. (2002). Population dynamics of a Clam, *Ruditapes philippinarum*, on an Artificially-created sand cover on the tidal flats at the river mouth of Midorikawa River. *Japanese Journal of Benthology, 57*, 177–187. https://doi.org/10.5179/benthos.57.177.

Tumnoi, W. (2012). *The autecology of* Tapes philippinarum *(Adams and Reeve, 1850) in Southampton Water, UK. School of Ocean and Earth Sciences*. University of Southampton.

Utting, S. D., & Doyou, J. (1992). The increased utilization of egg lipid reserves following induction of triploidy in the Manila clam (*Tapes philippinarum*). *Aquaculture, 103*(1), 17–28. https://doi.org/10.1016/0044-8486(92)90274-O.

Utting, S. D., & Spencer, B. E. (1991). *The hatchery culture of bivalve mollusc larvae and juveniles*. Ministry of Agriculture Fisheries and Food, UK Laboratory Leaflet N0 68.

Valli, G., Mazzolini, D., & Raimondi, V. (1996). Reproductive cycle and biometrics in *Tapes philippinarum* of the upper Adriatic during an annual cycle (Ciclo riproduttivo e biometria in *Tapes philippinarum* (Adams and Reeve, 1850) dell'Alto Adriatico durante un ciclo annuale. *Annales de l'Hydrobiologie, 12*(13), 41–53.

Williams, J. G. (1980). The influence of adults on the settlement of spat of the clam, Tapes japonica. *Journal of Marine Research, 38*(4). https://elischolar.library.yale.edu/journal_of_marine_research/1527.

Xie, Q., & Burnell. , G. M. (1994). A comparative study of the gametogenic cycles of the clams *Tapes philippinarum* (Adams and Reeve, 1850) and *Tapes decussatus* (Linnaeus) on the south coast of Ireland. *Journal of Shellfish Research, 13*(2), 467–472.

Yamamoto, K., & Iwata, F. (1956). Studies on the bivalve *Venerupis japonica* in Akkeshi Lake II: Growth rate and biological minimum size. *Bulletin of the Regional Fisheries Research Laboratory, 14*, 57–63.

Yap, W. G. (1977). Population biology of the Japanese littleneck clam *Tapes philippinarum* in Kaneohe Bay, Oahu, Hawaiian Islands. *Pacific Science, 31*, 223–244.

Yasuda, J., Hamai, I., & Hatta, H. (1954). A note on the spawning season in *Venerupis philippinarum. Bulletin of the Japanese Society of Scientific Fisheries, 20*, 277–279.

Zhang, Y., Nie, H., Yin, Z., & Yan, X. (2022). Comparative transcriptomic analysis revealed dynamic changes of distinct classes of genes during development of the Manila clam (*Ruditapes philippinarum*). *BMC Genomics, 23*(1), 1471–2164. https://doi.org/10.1186/s12864-022-08813-0.

CHAPTER

Physicochemical tolerances and distribution 11

11.1 Introduction

The geographical limits of *Ruditapes philippinarum*, as described in Chapters 2 and 3, provide insights into its broad climatic tolerances as a temperate species. This chapter examines physical and chemical tolerances as determinants of local distribution. While some comparisons with other species are described here, further such comparisons are included in Chapter 15.

Attempts to account for patterns of local distribution based on the Manila clam's laboratory-determined physiological responses make important contributions to understanding distribution, not least in the context of physico-chemically volatile estuarine and intertidal habitats. However, in practice, local distribution is determined by a complex interaction of physical and biological factors. Therefore, Chapter 12 addresses biotic factors of distribution and abundance, such as predation and competition.

11.2 Temperature

In benthic ectotherms such as the Manila clam, temperature influences distribution and abundance through its direct effect on physiological processes as described in Chapters 8 and 10. At constant temperature, with increasing departure from physiologically optimal conditions, growth will decline, and eventually, mortality will increase. However, in northerly temperate tidal habitats, considerable tidal and seasonal fluctuations in temperature are the norm, and the sediment surface can fall below freezing.

Although the relatively thick shell, when closed, may theoretically provide some direct temporary refuge from environmental temperatures, in fact, the thermal properties of the aragonite-based valves (Chapter 5) provide less effective thermal insulation than the calcite shells of oysters (Gomez-Martinez et al., 2022). However, shell closure also involves the cessation of cold water flow through the mantle cavity, and acute, potentially lethal extreme temperatures may provoke avoidance mechanisms, including valve closure and deeper burial.

Conversely, during rising temperatures, valve closure with cessation of respiratory currents and therefore oxygen exchange will suppress an otherwise escalating metabolic rate. Valve closure has therefore been considered a mechanism for active

The Manila Clam *Ruditapes philippinarum*: Science and Significance of a Globalised Species
DOI: https://doi.org/10.1016/B978-0-443-36704-5.00014-6
Copyright © 2026 Elsevier Inc. All rights are reserved, including those for text and data mining, AI training, and similar technologies.

metabolic depression as a strategy for energy conservation during rising temperatures, a response which in other mollusks has been regarded as an adaptation for upper shore life (Anacleto et al., 2014).

In most circumstances, an infaunal position will provide some protection from extreme temperatures since intertidal deposits typically exhibit a marked vertical temperature gradient in the top 10 cm, which moderates extremes at the surface (Perkins, 1974). However, the shallow burrowing habit of Manila clams suggests that, especially for recently settled small specimens, with correspondingly short siphons, other mechanisms are involved in extreme low-temperature tolerance.

In Hichirippu Lagoon, Hokkaido, a subarctic region of Japan, toward the northern limit of the Manila clam's indigenous distribution, the tidal flats in winter can be covered by ice. Yet a wild clam population persists with a stable density fluctuating around 8000/m^2 (Komorita et al., 2006). In light of this, Komorita et al. (2009) investigated the effect of extreme winter temperatures on the mortality of newly settled Manila clams in the lagoon. Sediment surface temperature in winter months fluctuated between −9.5°C and +8.0°C, with a lowest daily mean in January 2007 around −1.8°C. Here, spawning occurs once at the end of the summer, meaning that the postsettlement juvenile cohort experiences these severe temperatures. Juvenile mortality in the 9 months following settlement was calculated as 97.8%. Since high postsettlement mortality is a general phenomenon, these authors compared their results with postsettlement mortality at the Kikuchi River tidal flat, adjacent to Ariake Bay in Southern Japan (as reported by Tsukuda (2008), where the daily mean temperature was approximately 6°C higher in the postsettlement period of juveniles. While recognizing the possible influence of other variables, it was suggested that extremely low temperatures do not cause additional large-scale mortalities of the postsettlement clams. Such evidence from the Hichirippu Lagoon indicates some level of tolerance of persistent seasonal extreme low temperatures, even in Manila clams at the most vulnerable stage of their postsettlement life cycle.

In Canada, non-native Manila clams have naturalized along the coast of Vancouver Island (around 50 °N), where it occupies mid to upper intertidal shores. Bower (1992) conducted field and laboratory investigations in the spring of 1989 and 1991 following severe winters in which the beaches, when exposed, experienced air temperatures lower than −10°C. Their laboratory experiments sought to replicate beach conditions of around 8 h of exposure to freezer air temperatures between −4°C and −16°C. In experiments involving two consecutive days of freezing treatment, 5%–7% were dead after 1 day and 76% after 17days. Nevertheless, both field and laboratory evidence indicated that while severe winters such as those of 1989 and 1991 can cause mass mortalities in the Manila clam populations of Vancouver Island, a sufficient number of clams survive to perpetuate sustainable populations in the region.

Bower (1992) examined the histological effects of freezing conditions, finding varying degrees of damage to gill tissues ranging from adhesion of adjacent filaments and detachment of the ciliated epithelium to severe necrosis of both epithelium and connective tissue. Although intracellular ice formation is invariably lethal, as temperature falls significantly below subzero, the formation of extracellular ice may provide some protection since, as pure water separates as the solid phase, the remaining liquid phase becomes more concentrated, consequently drawing out intracellular water, increasing its concentration and reducing its freezing point.

In the temperate infaunal Venerid *Mercenaria,* tissue damage does not occur until the temperature in the tissues reaches −6°C and until 64% of intracellular water has been removed from the cells (Williams, 1970). Other bivalves, such as epibenthic mussels, can tolerate even lower temperatures. Since such freezing requires tolerance of tissue dehydration as pure water is collected in the ice phase, Newell (1979) has observed that the same tolerance of dehydration during freezing must also enhance survival in conditions on upper intertidal shores.

In interpreting patterns of local distribution, it must be expected that other variables, such as food availability and salinity, will interact with temperature to affect the resilience and sustainability of Manila clam populations. Moreover, some care must be taken in the interpretation of distributions where overlying water or sediment surface temperatures are taken rather than direct field measurements of temperatures within the sediments as actually experienced by clams. Also, even in laboratory conditions, experimental procedure is important for conclusions: For example, Bower's (1992) work on tissue damage and mortality indicated that death due to extremely low temperatures may occur long after a freezing episode, such that, for example, a mortality of 0% after 1 day becomes 61% after 22 days. Nevertheless, on the basis of a combination of field and laboratory studies, it seems clear that Manila clam populations are capable of remaining viable in environments which experience episodes of sub-zero temperatures.

At cold temperatures above freezing, factors related to physiological energetics rather than direct tissue damage are involved in mortality rates. Laing and Child (1996) investigated the tolerance of juvenile Manila clams to temperatures of 3°C, 6°C, and 9°C. for 11 weeks. At 3°C, 42% had died by 11 weeks, but at 6°C and 9°C, 100% of clams survived, showing slight or good growth, respectively. Even without food at 3°C, 88% of small Manila clams survived for 3 weeks, although losing condition. This confirms that below 3°C, thermal tolerance is a function of a combination of both the intensity and duration of exposure.

Other variables contributing to tolerances are acclimation and adaptation. The first involves metabolic and physiological adjustments in response to conditions experienced, and the second is a consequence of natural selection. The capacity to acclimate requires adjustment in biochemical pathways to produce new stable physiological states in response to conditions. For example, a switch to a different isoform of an enzyme with a different thermal optimum or shifts in energy storage and allocation (Angilletta et al., 2003).

Hongtao et al. (2023) studied the effects of low-temperature stress (5°C and −2°C) on gene expression in wild northern and southern Chinese Manila clam populations. They reported specific fatty acid metabolism-related genes to be highly expressed in the northern population. The existence of such alternative pathways arises through adaptive genetic mutations fixed by natural selection. The capacity to acclimate is itself a consequence of adaptation, which, over longer timescales, may permanently expand temperature tolerance and the distribution of species at both high and low temperature extremes.

Acclimation during experimental pretreatment is also known to affect the thermal resilience of the species (Shin et al., 2000). Consequently, the area of origin of collected animals may affect reported tolerances. In NW Spain, Gonzalo et al. (2016) studied the effects of high-temperature episodes simulating low tide exposures for 3 days on clams buried at a depth of 2 cm. *R. philippinarum* showed no mortality at 21°C, 27°C, and 32°C, but suffered 33% mortality after 2 days of tidal exposure to 36°C. In the laboratory, Isono et al. (1998) subjected Manila clams of various sizes to constant temperatures between 10°C and 34°C for 1 month. Their results suggested an optimum in terms of growth rate and oxygen consumption (as Q_{10}) of 25°C. At 34°C, all clams died within 10 days. Bishui et al. (1983) also reported an optimum temperature of 25°C for spat (year 0 clams recently settled), a high tolerance up to 35°C, and a lethal critical temperature at about 40°C. Optimum temperatures can also be estimated on the basis of physiological energetics. For Korean clams, Han et al. (2008) estimated a thermal optimum for the species as 20°C.

Examining thermal tolerances in the context of climate change heatwave scenarios (Anacleto et al., 2014) subjected Manila clams to temperature increases of 1°C per 30 min in aerated seawater of salinity S = 35, after 7 days of acclimation at 22°C. They reported an L50 (the temperature corresponding to 50% mortality) of 42°C and a Critical Thermal Maximum (CT_{max} – approximating to 100% mortality) of 46°C. Bertolini and Pastres (2021) developed a "tolerance landscape model" for *R. philippinarum* based on parameters including CT_{max}, which they defined as the lethal temperature after 1 min exposure (54.5°C) and a temperature sensitivity estimate (Q10). The model suggested that in some northern Adriatic lagoons, the mortality threshold was exceeded in three of the years from 2015 to 2018.

It is thought that metabolic rates generated by warm spring or summer temperatures may trigger high mortalities when energy intake requirements are greater than available food supplies can deliver. This phenomenon has been attributed to warming water temperatures preceding the spring phytoplankton bloom (Franklin et al., 2012) and/or postspawning energy demand by depleted clams during food scarcity (Goulletquer, 1988).

Whatever the temperature tolerances of settled clams, the thermal niche dimensions of the species are limited by the reproductive cycle. Since successful reproduction requires higher temperatures than those sufficient for the survival and growth of juveniles and adults, reproductive rather than somatic thermal minima will establish the limits of sustainable populations. In particular, there is a necessary minimum temperature below which gametogenesis will not commence, and in temperate winters, the Manila clam gonad is quiescent. Above this minimum, the rate of gonadal development is directly related to temperature, although food supply will affect the quality of the resulting gametes (Eckelbarger & Watling, 1995). In a naturalized UK population, when early spring temperatures sufficient for gonad activation coincide with the spring phytoplankton bloom, good recruitment is likely (Tumnoi, 2012).

The lowest temperatures compatible with reproductive activity vary by population and geography, and the processes from gonad activation to gametogenesis and spawning each require a progressively higher temperature. Consequently, the timing and number of spawning events will vary by region according to prevailing climatic conditions, with consequent implications for the total number of propagules released and recruitment of new age cohorts. These aspects of reproduction have been described in Chapter 10.

In summary, it appears that, notwithstanding a physiological thermal optimum around 20°C, viable Manila clam populations can: exist in seasonal temperature ranges from less than 2°C to greater than 35°C; commence gonad maturation at less than 4°C; and spawn from around 13°C. Such temperature tolerances are indicative of ectotherms well adapted to temperate seasonal climates requiring a capacity for acclimation not manifest in the stenothermal invertebrates found in truly polar or tropical regions (Bates & Morley, 2019).

11.3 Salinity

The salt composition of seawater includes magnesium, calcium, and potassium ions among many others, but mostly sodium and chloride. The salinity of water (S) is traditionally expressed as a number equivalent to grams of salt per kilogram of water (g/Kg or ‰). With a small number of exceptions, the range of salinities that exist in the world's oceans and seas is S32–S38. With S = 35 being considered an oceanic norm. Nowadays, salinity is measured using electrical conductivity, and such determinations technically do not have formal units, although the letters PSU (for practical salinity units) are often added to the figure. For many practical purposes. PSU can be considered to correspond to g/Kg or ‰.

In contrast to strictly marine invertebrates whose body fluids are generally in osmotic equilibrium with seawater, truly estuarine species must be euryhaline: able to tolerate the significant dilution of waters of mixed river and marine origin, along with fluctuations of salinity related to the tidal cycle and seasonal patterns of rainfall and river discharge. In intertidal habitats, evaporation can also raise salinity above that of seawater. However, the more typical challenge of dilution relates to swelling of tissues due to osmosis and reduction of important intracellular ions by dilution and diffusion.

In Spain, Roman et al. (2023) detected little change in oxygen consumption, clearance rate (see Chapter 8), growth, and mortality in buried juvenile Manila clams (shell length = 15–18 mm) exposed to 6 days of cyclic salinity changes between S = 20 and 5 in the overlying water. This work mimicked real exposures during heavy rainfall on Galician shellfish beds. In contrast, all four rates were affected in the sympatric natives *R. decussatus* and *Venerupis corrugata*.

With a sudden salinity change, bivalves can respond behaviorally by valve closure. The tendency of *R. philippinarum* to effectively seal off the environmental medium at around S = 15 is an essential feature of the ability to survive acute hyposaline inundation. This response is not available to all Venerids, such as the more osmotically sensitive *Venerupis corrugata*, whose valves cannot form a seal (Carregosa et al., 2014; Parada et al., 2012). Moreover, it appears that the biochemical response to osmotic stress is more pronounced in *V. corrugata* than in *R. philippinarum*, a finding that points to the significance of complete valve closure as part of an adaptive strategy for hyposaline episodes.

A second mitigation of hyposalinity is burrowing. Breber (1996) reported that in a northern Adriatic lagoon during exceptional spring conditions, no *R. philippinarum* deaths were observed when the salinity fell to less than S = 5 for a "few days." He attributed this to the clams going deeper in the sediment, where environmental stresses are buffered. Nevertheless, the need for gas exchange and feeding means that the principal responses to dilution in euryhaline animals must be physiological (Lockwood, 1976).

Many estuarine crustaceans are capable of osmoregulation, meaning they can maintain the concentration of their extracellular body fluids (hemolymph) above that of the surrounding medium, often assisted in this by an exoskeleton that can be significantly reduced in permeability, thus largely restricting osmosis and diffusion to the gills. Bivalves, however, face a greater challenge in that once the valves are open, the integument is relatively permeable.

Although Coughlan et al. (2009) have reported the salinity of Manila clam hemolymph to be consistently 2.5 ± 0.5 higher than that of the ambient medium, this is inadequate as protection from estuarine salinity variations. Therefore, being unable to sufficiently regulate the concentration of the hemolymph, *R. philippinarum* is generally described as an osmoconformer. Consequently, to prevent potentially catastrophic osmotic cellular swelling and the effects of intracellular dilution, bivalves regulate at the cellular level. Quantification of ionic content at various salinities reveals *R. philippinarum* to be capable of maintaining cellular ions, including calcium, magnesium, and especially sodium, all above that which would be expected from a passive equilibrium (Carregosa et al., 2014).

However, in contrast to extracellular fluids dominated by NaCl, the major osmotic components within the cells of invertebrates are typically small organic compounds (osmolytes), which can be quantitatively adjusted according to external osmotic conditions. In *R. philippinarum*, intracellular proteins decrease while glycogen and glucose increase with salinity (Carregosa et al., 2014). In the clams *Saxidomus giganteus* and *Calyptrogen spp.* organic osmolytes are dominated by taurine, betaine, and α-amino acids such as glycine. Such osmolytes are also known to have other properties beneficial in stress conditions, such as protecting metabolic reactions, counteracting destabilizing forces on macromolecules, and acting as antioxidants, all of which help prevent cellular damage (Yancey, 2005).

Blanco et al. (2022) have studied intracellular proteins of the mantle edge of *R. philippinarum* during hyposaline conditions designed to resemble local clam bed habitats in NW Spain. Here, during rainfall events, salinity can remain at around 10 for prolonged periods and can reach values close to S = 0 for several days (Parada et al., 2012). Proteins are significant in intracellular responses to osmotic stress in their role as enzymes regulating many protective responses, some of which, including those involved in protein folding under stress conditions (chaperones), osmoprotectants, and DNA repair factors, are highly demanding of energy. Consequently, low salinity episodes decrease the energy available for other activities. For example, low salinities can inhibit burial and reduce filtration rates in adult Manila clams. At salinities below 10.8, Nakamura et al. (2005) reported that Manila clams are no longer able to filter or actively bury.

Moreover, the extent to which these protections are deployed in *R. philippinarum* appears to be affected by reproductive phase, such that, in contrast to the European sympatric species *Ruditapes decussatus,* energetic priority may be given to gametic reabsorption (atresia) rather than simply delaying gonad development. According to Blanco et al. (2022), this difference in their osmotic strategy may explain why *R. philippinarum* is more prone to mass mortalities when acute hyposaline episodes coincide with reproductive activity. However, in near-optimal conditions, the priority given to gonad tissue may also explain the greater reproductive output of *R. philippinarum.*

By subjecting Manila clams to various salinities, Elston et al. (2003) found no significant mortality after long-term exposure to S = 15 and higher. Clams withstood salinities of 5–10 for 6 to 8 days, but 14 days proved lethal, with no recovery even when subsequently placed, still alive, into S = 31. A salinity of 12.5 was reported as marginal with variable mortality. At these salinities temperatures of 6°C, 12°C, and 18°C made difference in mortality. Histological examination of the digestive gland revealed that clams exposed to salinities of 10 and 12.5 from 2 to 14 days exhibited a sequential series of changes which progressed from loss of granulation of the tubular absorptive cells, which then suffered swelling to occlude the tubule lumen, followed by the shedding of the epithelium into the lumen.

Dominguez et al. (2020) exposed *R. philippinarum* to simulated tidal cycles involving four salinity "ramps" applied during different seasons. Temporary valve closures were exhibited at salinities less than 15, and with the lowest salinity ramp (5–20), reduced burrowing coincided with a negative scope for growth (SFG). SFG became positive with the 10–25 ramp applied in spring, albeit still with reduced burrowing (for SFG, see Chapter 8). Such work demonstrates the energetic costs of low salinities. Bae et al. (2021) exposed Manila clams to six salinities for 15 days. Mortality of clams exposed to a salinity fluctuation of 24–30 was higher than that of clams exposed to a constant salinity of 30. Valve closures of clams exposed to constant low salinity conditions (S = 24) and two salinity fluctuation conditions (24–30 and 27–24) were higher than those exposed to a constant 30.

In reflecting on the relation between distribution and salinity at the sediment surface, it is necessary, as with temperature, to recognize the increase of relative stability of conditions with sediment depth. While this benefit will vary according to the grade and porosity of sediment, it can apply equally well to hypo- and hypersaline conditions. The latter, typically caused by evaporation from intertidal or lagoonal habitats, will result in cell shrinkage. In the laboratory, *R. philippinarum* has been reported as tolerant of salinities up to 42 (Carregosa et al., 2014), and clams grown in mariculture pond effluent in Israel reportedly survived well and grew in a salinity of 41 PSU (Shpigel & Fridman, 1990). But, since vulnerability is a function of severity and duration, in the field, lower shore habitats, being exposed to hypersaline conditions for lesser periods, are more tolerable.

Clearly, salinity variation represents a challenge for *R. philippinarum*, and the extent of this challenge depends on a combination of absolute salinity, the rate of change, and the duration of exposure (Parada et al., 2012). Nevertheless, the various euryhaline adaptations of the Manila clam underpin its successful occupation of extensive areas of estuarine and intertidal habitats not available to stenohaline marine species.

11.4 Sediments

The energy of water movements, such as wave action and current flow, combined with sediment supply and biogenic processes, determines the nature of nearshore substrates. Manila clams can themselves affect various aspects of sediment character, both directly and indirectly through fishing and aquaculture impacts. Such impacts are considered in Chapter 21. Here, the range of sediment types within which Manila clams can be found is considered.

Generally, sediment particles are classified into categories of increasing size. On the Wentworth Scale, these range from clay and silt (< 63 μm), through sands of various coarseness (63–2000 μm), to granules, pebbles, and cobbles. The term mud is often used for mixtures of clay and silt. In coastal areas, finer particle sizes and flatter shore gradients are associated with lower energy environments, such as many estuaries and protected bays, which are sheltered from the degree of wave action typical of open coasts. Here, relatively fine particles of marine or fluvial origin can often settle from the water column (Newell, 1979), producing homogeneous sediments. However, where local geomorphological or anthropogenic conditions provide coarser materials such as pebbles or larger stones, low-energy conditions exert little sorting effect, resulting in heterogeneous sediments. Accumulation of shell fragments also contributes to substrate heterogeneity.

The shape and size of sediment particles determine ecologically important properties of benthic deposits, including the volume of the interstitial space (porosity). Muds, due to their high silt and clay content, along with compacted sands, have low porosity, which, along with beach gradient, generally determines sediment permeability. Lower-permeability sediments retain more water at low tide, a property that, due to the thermal properties of water and the slow rate of diffusion of NaCl, can generate vertical gradients of temperature and salinity, especially in the top 10 cm (Perkins, 1974). This property moderates the effects of acute variations at the sediment surface through rainfall and diurnal temperature variations and also keeps sediments moist at low tide, even to the extent of maintaining a thin surface film of water, which may be beneficial for the survival of microphytobenthos (see Chapter 7 on food sources).

With water content above about 25% by weight, sediments exhibit thixotropic properties in which applied force reduces the sediment's resistance to burrowing. Finer sediments tend to have higher organic content derived mostly from sedimentation of planktonic and other detritus (Grey & Elliott, 2009). Consequently, oxygen level and redox potential within a sediment will also vary with particle size. However, a direct connection with the water column via siphons normally makes clams largely independent of interstitial oxygen.

Manila clams are capable of living in a wide range of sediment types. It appears that, if the sediment is relatively stable and there is some sand fraction, then other components from mud (up to at least 30% and probably more) (Vincenzi et al., 2006), pebbles, and stones are compatible with sustainable wild populations. Consequently, in any one estuary or bay, sediment type is not necessarily a good predictor of distribution.

In the Hood Canal, Puget Sound, Washington, Nosho and Chew (1971), investigating natural recruitment rates, were unable to demonstrate a preferred substratum and concluded the distribution of spat to be largely dependent on water currents. Similarly, in Poole Harbour, UK, Herbert et al. (2011) showed the distribution of adult clams to align well with tidal hydrodynamic influence on larval dispersal (Snelgrove & Butman, 1994, and Chapter 10). This dependence on hydrodynamics can also explain areas that, although ostensibly suitable for Manila clam occurrence, are not occupied. Despite this tolerance of various sediment grades, the mortality, condition, and growth of Manila clams do appear to vary with sediment characteristics, suggesting a narrower optimal range.

Several studies have shown *R. philippinarum* to exhibit a higher growth rate and successful juvenile settlement in sandy sediments compared to those with a higher fraction of silt (Barillari et al., 1990; Melià et al., 2003). Cigarria and Fernandez (2000) reported a significant influence of substratum on mortality, which was lowest in "sand–gravel" beds. Similarly, Soobin et al. (2021) working with juvenile Manila clams, found that mortality was lower in medium sand compared to finer sediment (< 0.09 mm). While mortality may be a function of differences in the ease of predation in different sediments, these authors also reported a higher condition index in medium sand than in either coarser or finer sediment, although particle size did not affect shell growth. Shell growth rate was significantly higher in poorly sorted sediment, although sorting appeared not to affect mortality or CI. They concluded that poorly-sorted sediment with an average grain size of medium sand is optimal for the survival and growth of Manila clams. In an investigation of a wild population of Manila clams in the Spanish Bay of Santander. Bidegain et al. (2015) concluded on the basis of distribution that the percentage of sand and organic matter was positively associated with high habitat suitability.

In Washington State, United States, and elsewhere, it is recognized that adding gravel and stones onto natural settlement areas greatly enhances settlement (Chew, 1989). When the larvae of *R. philippinarum* settle on the seabed, they stabilize their position by secreting byssus threads for attachment to a solid object such as a small pebble, larger stone, or shell fragment (Chapter 10). In contrast, for cultivated populations that rely on seeding with juveniles, even firm mud substrates have been considered "ideal spots" for clam culture (Anderson et al., 1982). This suggests that while well-sorted mud sediments will provide little opportunity for spat to attach and settle, when established, they might subsequently grow well.

So, the range of suitable habitats for adult clam growth is wider than that required for successful natural settlement and survival of larvae. While the determination of optimal conditions for growth is useful for deciding where to locate aquaculture beds, it is of more limited use in understanding and predicting the distribution of wild populations. Therefore, when seeking to determine the sediment niche range of the Manila clam, the requirements of larval clams and spat around the time of settlement are as important as the optimal conditions for juvenile and adult clams laid for aquaculture. Natural wild populations need conditions suitable for all life cycle stages.

In general, the evidence is consistent with the observation that thriving wild populations of Manila clams are most often associated with stable, predominantly sandy sediments, typically relatively unsorted, including pebbles, stones, or shell debris, which allow for secure attachment at settlement while also providing a suitable habitat for burrowing, protection, and growth. Despite this, the borders of natural clam beds are not always attributable to a change in substrate, indicating that other factors, not least larval dispersal on prevailing water currents. They are important in determining patterns of distribution.

Finally, it is worth emphasizing that not all the beneficial properties of particular sediments are a direct consequence of clam and substrate interactions. Although often associated with bare sediment, work by Galvan and Puente (2023) and Roman (2023, 2024) on the Atlantic coast of Spain has provided evidence consistent with their hypothesis that seagrass meadows (*Zostera noltei*) can provide "nursery" areas that protect juvenile Manila clams by decreasing the visibility and/or efficiency of access for predators. Which of these two possibilities is unclear, and both may be related to a tendency for Manila clams to bury more deeply in coarser deposits. In Poole Harbour, UK, after widespread spatfalls, clam fishers attribute a greater relative decline in abundance of juvenile clams in sorted sandy substrates to greater predation pressure from the shore crab (*Carcinus maenas*), in comparison to finer but heterogeneous sediments.

Experiments in Japan indicate that gravel substrates of grain size similar to the length of 20 mm clams have a strong protective effect through the reduction of predation by the naticid snail *Glossaulax didyma* (Takeyama et al., 2020). Predation is further addressed in Chapter 12.

11.5 Elevation and exposure

Manila clams can live in both intertidal and subtidal habitats. This range represents a gradation of environmental conditions due to the semidiurnal rise and fall of the tide, which creates patterns of exposure, immersion, depth, and current speed that vary over the 14-day interval between high-amplitude (spring) and low-amplitude (neap) tidal phases (along with equinoctial and other annual cycles). Elevations above and below the mean tide level are influential on distribution. Intertidal distribution presents challenges during aerial exposure, including cessation of feeding, exposure to terrestrial predators, and risk of desiccation. In addition, exposure can increase the range of temperatures and salinities to which the species must be tolerant. However, an infaunal intertidal position within the sediment will buffer environmental extremes at the surface during aerial exposure and also offer periodic protection from aquatic predators.

With regard to short-term physical tolerance, commercially motivated studies have shown Manila clams to be capable of a 100% survival rate after 58 h of exposure even in unburied fully aerial conditions of 15°C and low relative humidity, as may be experienced during shipment to market (Ali & Nakamura, 1999). Moreover, the species is capable of gas exchange with air, which, although less effective when measured in terms of aerobic metabolism, may be part of the clam's strategy for surviving aerial exposure (Ali et al., 1999).

Day (1977), working in Hong Kong, reported a time to 50% mortality (L_{50}) in fully aerial conditions of 60 h, noting that Manila clams would gape, thereby increasing desiccation. Comparing *R. philippinarum* with two other intertidal bivalve species, he was able to correlate position on the shore with $L_{50,}$ which in turn correlated with gaping: the Manila clam being restricted to the lower half of the shore and being the only species to gape during exposure. This suggests desiccation may be a significant factor affecting the upper shore limit of *R. philippinarum.*

Since the amplitude of semi-diurnal tides varies over the spring-neap cycle, only organisms living around the mean tide level (the eulittoral zone) will constantly experience a twice-daily exposure/submersion cycle. Around the lower-amplitude neap tide phase of the cycle, higher elevations will remain exposed (and lower-elevation areas will remain submerged) for much more prolonged periods. Therefore, the mean tidal height on a shore marks the center of a transition area that limits the distribution of many species of marine origin. While the risk of desiccation will also vary with the porosity of the sediment, humidity, and temperature, the eulittoral transition zone will also mark a significant change in available feeding time.

The total cumulative duration of exposure across a full 14-day tidal cycle can be expressed as a percentage. Experimental work in the Menai Strait, Wales, UK, suggested an upper limit of viable commercial Manila clam laying around mean low water of neap tides (MLWN), a zone corresponding to an approximate exposure to air of only about 12% of the full tidal cycle (Spencer, 2002). Nevertheless, populations can be found at higher elevations than this, although typically with reduced growth rates in comparison to lower elevations, especially on shores with large tidal amplitudes (Glock, 1978; Glock & Chew, 1978). Above 15%–30% exposure time, Manila clam growth reportedly drops off significantly, with zero growth being recorded at levels with 50% exposure (Spencer, 2002). Nevertheless, in different locations, adult clams can be laid and survive with low mortality above even this level, although growth rates suggest such high shore elevations are beyond the natural ecological niche of the species.

In practice, the upper limit of distribution of the species is likely determined by a combination of regular cessation of feeding during low water and the risk of desiccation during aerial exposure. Since food availability and sediment factors affecting desiccation will vary by site, so will the upper limit of intertidal distribution.

Despite reduced growth rates, higher elevations may nevertheless convey some advantage in terms of mortality. In commercial terms, Quayle and Bourne (1972) and Glock and Chew (1978) reported that higher shore positions result in increased survival. Based on observations in Washington State, USA, Toba et al. (1965) reported wild Manila clams to be most abundant at around 1–1.8 m above means low water of spring tides (MLWS), and Miller et al. (1978) comment that although Manila clams may occur between around 1–2.4 m above MLWS, better survival rates occur above MLWS + 1 m. Both reports attribute this apparent intertidal optimum to increased exposure to marine predation at lower intertidal levels, which correspond to increased periods of immersion.

These observations suggest that predation is a significant factor influencing shore distribution, and greater predation impact from aquatic rather than terrestrial predators in the postsettlement size range. It therefore appears that the Manila clam can potentially maintain viable populations across a wide range of elevations. However, in any particular locality, while an upper shore limit may be determined by physical conditions and feeding duration, marine predator pressure becomes increasingly influential with progressively lower shore elevations. In many cases, optimal elevations are often found somewhere in the lower shore area.

Shore gradient also appears to correlate with Manila clam mortality. In addition to finer sediments, low-energy shores are also associated with flatter shore gradients. Such places are typical of estuaries, enclosed bays, and coastal lagoons, where extensive intertidal flats with gradients as low as 1:600 can often be measured in kilometers. On the coasts of Oregon, USA, planted seed clams had the highest recovery (harvest) rates on relatively level beaches when compared to steeply sloped beaches (Lukas & Gaumer, 1974). Miller (1982), working in Puget Sound, Washington, reported good Manila clam recoveries from gradients less than 15:1, with much reduced recoveries with slopes in excess of 10:1. Whether this is a direct effect of the gradient seems unlikely, and other variables that correlate with the gradient may well be at work.

Indeed, caution is required with many such generalizations, as different shores, even along the same coast, can vary considerably in ways that will impact distribution. These include tidal amplitude, aquatic and terrestrial predator communities, food availability, abundance of competitor species, and the extremes of temperature when exposed. In the microtidal Poole Harbour, UK, for example, a prolonged high-water stand, or "double high water," greatly reduces percentage exposure high up the shore (Humphreys, 2005; Humphreys et al., 2007). Conversely, Breber (2002) attributed the success of the Manila clam in the Northern Adriatic to the "pronounced tidal range" as the flushing created conditions favorable to benthic pennate diatom production, a food source for the Manila clam (Chapter 7). He reports that in the Adriatic lagoons, the

Manila clams are usually concentrated by the sea mouth of the lagoons and considers the natural habitat to be on mud flats on the inside of the channels which link the lagoons to the sea.

Despite the common general assertion that Manila clams most commonly find optimal conditions on intertidal shores between MLWS and MLWN, not much is known about the extent to which distribution extends below intertidal shores and the relative abundance of such populations. Nevertheless, the species is recognized as extant on shallow sublittoral seabeds (e.g., Zhuang 1964), but not in deeper waters. Moreover, it is also well known from aquaculture practices such as pond and suspended culture that Manila clams can thrive in conditions of constant immersion (Chapter 17). In the Shandong and Liaoning provinces of China, increasing demand has resulted in farmed clam grounds reaching subtidal depths of 10 m. While this improves growth rate, the main problem is regarded as increased predation from crabs, starfish, and gastropod mollusks (Fang, 2015). The question of biological factors affecting the lower limit of Manila clam distribution, in particular predation, is discussed further in Chapter 12.

11.6 Coastal pollution

Coastal waters and estuaries in particular are subject to anthropogenic pressures in the form of many types of contaminants. Among these are hazardous organic compounds such as polychlorinated biphenyls, polycyclic aromatic hydrocarbons, petroleum hydrocarbons, as well as endocrine-disrupting chemicals and inorganic compounds such as metals and metalloids. Such pollutants are hazardous due to their toxic and persistent properties, and benthic sediments are known to be major sinks for contaminants. Moreover, many are known to bioaccumulate in benthic species. Aside from the risk to human health from the consumption of fishery and aquaculture production, such pollutants can also present direct physiological challenges, especially to benthic filter feeders such as *R. philippinarum*.

Regoli and Giuliani (2014) have elucidated the interactions between different classes of pollutants and the production of reactive oxygen species, which in turn cause oxidative stress in marine mollusks. Oxidative stress involves a surfeit of unstable free radicals, which damage important cell components, including enzymes and DNA. While reactive oxygen species are naturally produced during cell metabolism, their damaging impacts are mitigated by antioxidant defenses. However, this balance can be detrimentally affected by various classes of environmental pollutants, including trace metals and organic xenobiotics such as synthetic drugs, pesticides, and industrial chemicals

Sfriso et al. (2018) determined the accumulation of inorganic contaminants in bivalves from the Venice Lagoon, Italy, using a Biota-Sediment Accumulation Factor (BSAF). BSAF for a given metal(loid) is calculated as the total concentration in the soft tissue divided by the concentration in the sediment: bioaccumulation being indicated by factors greater than 1. BSAF values for *R. philippinarum* were Cadmium

(Cd) 21; Zinc (Zn) 18.6; Copper (Cu) 13.77; Arsenic (As) 5.71; Mercury (Hg) 1.79. Metals with BSAF values below 1 were: Lead (Pb); Chromium (Cr); Manganese (Mn); Nickel (Ni); Vanadium (V), and Iron (Fe).

Similar BSAF values were reported for the lagoon cockle *Cerastoderma glaucum* and the razor clam *Solen marginatus*. Although *R. philippinarum* generally exhibited somewhat higher bioaccumulation levels, it nevertheless had the highest maximum abundance (up to 200 $m-^{2}$ in Venice Lagoon). Bivalve species responses vary as toxicity is not simply related to bioaccumulation. Metals can be excreted or made metabolically unavailable through cellular biochemical responses. In *R. philippinarum,* detoxifying protective responses include the production of biotransformation enzymes such as glutathione S-transferase and metal-binding proteins such as metallothionein, which protect against oxidative stress (Chiesa et al., 2018). In each species, toxicity only occurs when detoxification and excretion are exceeded by metal uptake (Rainbow, 2006).

Similarly, in Laizhou Bay, China, responses to organic contaminants in *R. philippinarum* have been shown to include production of a wide range of protective compounds, including superoxide dismutase, catalase, glutathione peroxidase, glutathione S-transferase, glutathione, etc., in both gills and digestive glands (Wang et al., 2021). Using genomic techniques involving transcriptome (RNA) sequencing, as an indication of gene expression in response to environmental contaminants, Milan et al. (2011) have reported over 100 Glutathione S-transferase encoding transcripts in *R. philippinarum.*

Chiesa et al. (2018), investigating clams from the Tagus estuary, Portugal, found that *R. philippinarum* did not exhibit a reduction of overall physiological condition at high environmental contamination levels and tissue content. In fact, the highest values of the condition index were recorded in areas characterized by higher metals and Arsenic concentrations. Similarly, Moschino et al. (2012) investigated *R. philippinarum* responses to both organic and inorganic contaminants in the Venice Lagoon, and reported relatively weak biological sensitivity to such pollutants, including relatively low bioaccumulation of organic contaminants compared to other species.

Moreover, it appears that *R. philippinarum* can adapt to pollution, becoming increasingly tolerant when subjected to chronic contamination, thereby exhibiting a "remarkable resistance" to various contaminants (Paul-Pont et al., 2010). Moschino et al. (2012) suggest that overall weak responsiveness and low sensitivity to pollution are the basis for the Manila clam's competitive ability to live and reproduce in the North Adriatic transitional areas and at the most contaminated sites within the Venice Lagoon. Characteristics also facilitate its rapid spread in the northern Adriatic more generally.

Similar studies in Portugal, Italy, China, and Japan demonstrate the Manila clam's ability to cope with complex mixtures of organic and inorganic pollution, which at levels found in many industrialized coastal environments rarely directly constitute the primary factor determining the local distribution of the species. Of course, this is not to say that extreme levels of contamination are not ultimately lethal. Tsutsumi et al. (2003) investigated metal (Pb, Cd, Cu, Zn, and Mn) impacts on *R. philippinarum* biomass and abundance in an effort to understand the collapse of populations on the tidal flats of Ariake Bay, Japan. They established a relationship between both density and abundance and the concentration of manganese in the sediment. While all other metal concentrations were consistently less than 63 μg/g, manganese varied up to 2900 μg/g. Stations where the clam populations collapsed had 1700–2900 μg/g of manganese Their results indicated that levels of manganese in excess of 1000 μg/g was incompatible with otherwise dense clam populations and, based on earlier work, attributed the collapse of the Ariake Bay fisheries to the impact of this toxin on juvenile clams (less than 5 mm) which suffered high mortality after settlement (Tsutsumi et al., 2000, 2002). Cases of mass mortality are considered further in Chapter 14.

Because *R. philippinarum* takes its food from the bottom of the water column, a zone intimately connected to the underlying sediment, the species has been recommended as having utility as a possible bioindicator of pollution in both seawater and sediment (e.g., Wang et al., 2010). This suggestion rests on a reportedly strong correlation between environmental and tissue metal content. However, in contrast, Moschino et al. (2012) report only a weak correlation and, noting the ability of the species to adapt to extant metal levels, suggest caution in adopting the species as a pollution "sentinel" at least in the northern Adriatic lagoons.

References

Ali, F., & Nakamura, K. (1999). Effect of temperature and relative humidity on the tolerance of the Japanese clam, *Ruditapes philippinarum* (Adams & Reeve), to air exposure. *Aquaculture Research, 30*(9), 629–636. https://doi.org/10.1046/j.1365-2109.1999.00303.x.

Ali, F., Nakamura, S., & Yamasaki, S. (1999). Physiological characteristics of aerial survival in Japanese clam, *Ruditapes philippinarum* (Adams & Reeve). *Aquaculture Science, 47*(3), 381–389.

Anacleto, P., Maulvault, A. L., Lopes, V. M., Repolho, T., Diniz, M., Nunes, M. L., Marques, A., & Rosa, R. (2014). Ecophysiology of native and alien-invasive clams in an ocean warming context. *Comparative Biochemistry and Physiology Part A: Molecular & Integrative Physiology, 175*(1), 28–37. https://doi.org/10.1016/j.cbpa.2014.05.003.

Anderson, J., Miller, M. B., & Chew, K. K. (1982). *Technical Report, Washington Sea Grant Program 82-4* (p. 45). Seattle, WA: University of Washington.

Angilletta, M. J., Wilson, R. S., Navas, C. A., & James, R. S. (2003). Tradeoffs and the evolution of thermal reaction norms. *Trends in Ecology and Evolution, 18*, 234–240.

Bae, H., Im, J., Joo, S., Cho, B., & Kim, T. (2021). The effects of temperature and salinity stressors on the survival, condition and valve closure of the Manila Clam, *Venerupis philippinarum* in a holding facility. *Journal of Marine Science and Engineering, 9*, 754. https://doi.org/10.3390/jmse9070754.

Barillari, A., Boldrin, A., Pellizzato, M., & Turchetto, M. (1990). Environmental conditions in the breeding of Tapes philippinarum. (Condizioni ambientali nell'allevamento di Tapes philippinarum). In Tapes philippinarum *biologia e sperimentazione* (pp. 183–195). Veneto: E.S.A.V. Regione.

Bates, A., & Morley, S. A. (2019). Interpreting empirical estimates of experimentally derived physiological and biological thermal limits in ectotherms. *Canadian Journal of Zoology, 98*(4), 237–244. https://doi.org/10.1139/cjz-2018-0276.

Bertolini, C., & Pastres, R. (2021). Tolerance landscapes can be used to predict species-specific responses to climate change beyond the marine heatwave concept: Using tolerance landscape models for an ecologically meaningful classification of extreme climate events. *Estuarine, Coastal and Shelf Science, 252*, 107284.

Bidegain, G., Bárcena, J. F., García, A., & Juanes, J. A. (2015). Predicting coexistence and predominance patterns between the introduced Manila clam (*Ruditapes philippinarum*) and the European native clam (*Ruditapes decussatus*). *Estuarine. Coastal and Shelf Science, 152*, 162–172.

Bishui, L., Wu Tianming, W., & Bingzhang, H. (1983). The effects of temperature and salinity on growth and development of spats of the clam *(Ruditapes philippinarum*). *Journal of Fisheries of China, 7*, 15–23.

Blanco, S., Morán, P., Diz, A. P., Olabarria, C., & Vázquez, E. (2022). Effects of short-term hyposalinity stress on four commercially important bivalves: A proteomic perspective. *Environmental Research, 215*(3), 114371.

Bower, S. M. (1992). Winter mortalities and histopathology in Japanese Littlenecks *(Tapes philippinarum* (A. Adams and Reeve, 1850) in British Columbia due to freezing temperatures. *Journal of Shellfish Research, 11*(2), 255–263.

Breber, P. (1996). An account of the acclimatisation of the Manila clam *Ruditapes philippinarum* (Adams and Reeve) syn. *Tapes semidecussatus*. Reeve (Mollusca; Bivalvia; Venerinae) in Italian waters. *Lavori della Societa Italiana di Malacologia, 24*, 47–52.

Breber, P. (2002). Introduction and acclimatisation of the Pacific carpet clam, *Tapes philippinarum*, to Italian waters. In E. Leppäkoski, S. Gollasch, & S. Olenin (Eds.), *Invasive aquatic species of Europe. Distribution, impacts, and management* (pp. 529–547). Kluwer Academic Publishers.

Carregosa, V., Figueira, E., Gil, A. M., Pereira, S., Pinto, J., Soares, A. M., & Freitas, R. (2014). Tolerance of *Venerupis philippinarum* to salinity: Osmotic and metabolic aspects. *Comparative Biochemistry and Physiology Part A: Molecular & Integrative Physiology, 171*, 36–43. https://doi.org/10.1016/j.cbpa.2014.02.009.

Chew, K. K. (1989). Manila clam biology and fishery development in western North America. In J. J. Manzi, & M. Castagna (Eds.), *Clam mariculture in North America*. Elsevier.

Chiesa, S., Chainhob, P., Almeida, A., Figueira, E., Amadeu, S. A. M. V. M., & Freitas, R. (2018). Metals and as content in sediments and Manila clam *Ruditapes philippinarum* in the Tagus estuary (Portugal): Impacts and risk for human consumption. *Marine Pollution Bulletin, 126*, 281–292.

Cigarria J and Fernandez J M (2000). Management of Manila clam beds: I. Influence of seed size, *type of substratum and protection on initial mortality. Aquaculture 182*, (1–2), 173–182

Coughlan, B. M., Moroney, G. A., van Pelt, F. N. A. M., O'Brien, N. M., Davenport, J., & O'Halloran, J. (2009). The effects of salinity on the Manila clam (*Ruditapes Philippinarum*) using the neutral red retention assay with adapted physiological saline solutions. *Marine Pollution Bulletin, 58*, 1680–1684.

Day, J. E. (1977). *The malacofauna of Hong Kong and Southern China*. In B. Morton, & R. A. Dudgeon (Eds.). University of Hong Kong. Department of Zoology Correlation of gill physiology, emersion survival, and intertidal distribution of three bivalves from Hong Kong.

Dominguez, R., Vázquez, E., Woodin, S. A., Wethey, D. S., Peteiro, L. G., Macho, G., & Olabarria, C. (2020). Sublethal responses of four commercially important bivalves to low salinity. *Ecological Indicators, 111*, 106031. https://doi.org/10.1016/j.ecolind.2019.106031.

Eckelbarger, K. J., & Watling, L. (1995). Role of the digestive gland in the reproductive cycle of the Manila clam, *Ruditapes philippinarum*. *Invertebrate Biology, 114*(4), 256–269.

Elston, R. A., Cheney, D. P., MacDonald, B. F., & Suhrbier, A. D. (2003). Tolerance and response of Manila clams, *Venerupis philippinarum* (A. Adams and Reeve, 1850) to low salinity. *Journal of Shellfish Research., 22*, 667–674.

Fang, J. (2015). Development of Manila clam industry in China. *The third international symposium on Manila (Asari) clam*. https://nria.fra.affrc.go.jp/asari/booklet.pdf.

Franklin, D. J., Humphreys, J., Harris, M., Jensen, A. C., Herbert, R. J. H., & Purdie, D. A. (2012). *An investigation into the annual cycle of phytoplankton abundance in Poole Harbour and its relationship with Manila clam nutrition*. Marine Management Organization.

Galvan, C., & Puente, A. (2023). Exploring the newly emerging effects of native seagrasses on survival and growth of nonnative juvenile clams. *Frontiers in Marine Science, 10*, 1160734. https://doi.org/10.3389/fmars.2023.1160734.

Glock, J., & Chew, K. K. (1978). *A report prepared for the Squaxin Island Tribe* p. 69.

Glock, J. W. (1978). *Growth, recovery and movement of Manila clams* Venerupis japonica *planted under protective devices and on open beaches at Squaxin Island*. University of Washington.

Gomez-Martinez, O., Aldana-Aranda, D., Quintana, P., Pichardo, J. L., & Alvarado-Gil, J. J (2022). Photoacoustic determination of the thermal properties of bivalve mollusk shells. *Marine Biology, 141*, 911–914. https://doi.org/10.1007/s00227-002-0880-z.

Gonzalo, M., Woodin, S. A., Wethey, D. S., & Vázquez, E. (2016). Impacts of sublethal and lethal high temperatures on clams exploited in European fisheries. *Journal of Shellfish Research, 35*(2), 405–419. https://doi.org/10.2983/035.035.0215.

Goulletquer, P. (1988). Winter mortality of Japanese and Philippine Ruditapes on the Atlantic coast: Aspects of biochemistry and ecophysiology. *Haliotis, 17*, 152–163.

Grey, J. S., & Elliott, M. (2009). Ecology of marine sediments: From science to management. *Ecology of marine sediments*. Oxford University Press. https://doi.org/10.1093/oso/9780198569015.001.0001.

Han, K. N., Lee, S. W., & SY, W. (2008). The effect of temperature on the energy budget of the Manila clam, *Ruditapes philippinarum*. *Aquaculture. International, 16*, 143–152.

Herbert, R J H, et al. (2011). Invasion in tidal zones on complex coastlines: modelling larvae of the non-native Manila clam, *Ruditapes philippinarum*, in the UK. *Journal of Biogeography, 39*(3), 585–597.

Hongtao, N., Lu, Z., Li, D., Dong, S., Li, N., Chen, W., Huo, Z., & Yan, X. (2023). The potential roles of fatty acid metabolism-related genes in Manila clam *Ruditapes philippinarum* under cold stress. *Aquaculture, 562*, 738750.

Humphreys, J., Caldow, R. W. G., McGrorty, S., West, A. D., & Jensen, A. C. (2007). Population dynamics of naturalised Manila clam *Ruditapes philippinarum* in British coastal waters. *Marine Biology, 151*, 2255–2270. https://doi.org/10.1007/s00227-007-0660-x.

Humphreys, J. (2005). Salinity and tides in Poole Harbour. Estuary or lagoon? In J. Humphreys, & V. May (Eds.), *The ecology of Poole Harbour*. Elsevier.

Isono, R. S., Kita, J., & Kishida, C. (1998). Upper temperature effects on rates of growth and oxygen consumption of the Japanese littleneck clam *Ruditapes philippinarum*. *Nippon Suisan Gakkaishi, 64*(3), 373–376.

Komorita, T., Shibanuma, S., Yamada, T., Kajihara, R., Tsukuda, M., & Montani, S. (2009). Impact of low temperature during the winter on the mortality in the post-settlement period of the juvenile of short-neck clam, *Ruditapes philippinarum*, on the tidal flats in Hichirippu lagoon, Hokkaido, Japan. *Plankton Benthos Research, 4*(1), 31–37.

Komorita, T., Tsutsumi, H., Inaba, Y., Kajihara, R., Shibanuma, S., Yamada, T., & Montani, S. (2006). The characteristics of productivity on a subarctic shallow water (Hichirippu Lagoon, Hokkaido): Studies on the seasonal fluctuations of environmental factors and macro-benthic community. In S. Tanabe, H. Takeoka, T. Isobe, & Y. Nishibe (Eds.), *Chemical pollution and environmental changes* (pp. 397–400). Universal Academy Press.

Laing, I., & Child, A. R. (1996). Comparative tolerance of small juvenile palourdes (*Tapes decussatus* L.) and Manila clams *Tapes philippinarum* (Adams & Reeve) to low temperature. *Journal of Experimental Marine Biology and Ecology, 195*, 267–285.

Lockwood, A. P. M. (1976). Physiological adaptation to life in estuaries. In R. C. Newell (Ed.), *Adaptation to environment* (pp. 315–392). Butterworths.

Lucas, G., & Gaumer, T. F. (1974). Clam-abalone stock supplementation feasibility Study. Fish Commission of Oregon. In J. J. Manzi, & M. Castagna (Eds.), *Clam mariculture in North America* (pp. 243–261). Elsevier. Cited in Chew K K. Manila clam biology and fishery development in western North America.

Melia, P., Nizzoli, D., Bartoli, M., Naldi, M., Gatto, M., & Viaroli, P. (2003). Assessing the potential impact of clam rearing in dystrophic lagoons: An integrated oxygen balance. *Chemical Ecology, 19*, 129–146.

Milan, M., Coppe, A., Reinhardt, R., Cancela, L. M., Leite, R. B., Saavedra, C., Ciofi, C., Chelazzi, G., Patarnello, T., Bortoluzzi, S., & Bargelloni, L. (2011). Transcriptome sequencing and microarray development for the Manila clam, *Ruditapes philippinarum*: Genomic tools for environmental monitoring. *BMC Genomics, 12*, 234.

Miller, M. B. (1978). *Washington Sea Grant Publication (WSG) 78-2* (p. 18). Seattle, Washington.

Miller, M. B. (1982). *Recovery and growth of hatchery-produced juvenile Manila clams* Venerupis japonica *(Deshayes) planted on several beaches in Puget Sound*. University of Washington.

Moschino, V., Delaney, E., & Da, R. L. (2012). Assessing the significance of *Ruditapes philippinarum* as a sentinel for sediment pollution: Bioaccumulation and biomarker responses. *Environmental Pollution, 171*, 52–60.

Nakamura, Y., Hashizume, K., Koyama, K., & Tamaki, A. (2005). Effects of salinity on sand burrowing activity, feeding, and growth of the clams *Mactra veneriformis, Ruditapes philippinarum,* and *Meretrix lusoria*. *Journal of Shellfish Research, 24*(4), 1053–1059.

Newell, R. C. (1979). *Biology of intertidal animals*. Faversham: Marine Ecological Surveys.

Nosho, T., & Chew, K. K. (1971). The setting and growth of the Manila clam *Venerupis japonica* (Deshayes), in the Hood Canal, Washington. *Proceedings of the National Shellfisheries Association, 62*, 50–58.

Parada, J. M., Molares, J., & Otero, X. (2012). Multispecies mortality patterns of commercial bivalves in relation to estuarine salinity fluctuation. *Estuaries and Coasts, 35*, 132–142.

Paul-Pont, I., de Montaudouin, X., Gonzalez, P., Soudant, P., & Baudrimont, M (2010). How life history contributes to stress response in the Manila clam *Ruditapes philippinarum*. *Environmental Science Pollution Research, 17*, 987–998. https://doi.org/10.1007/s11356-009-0283-5.

Perkins, E. J. (1974). *The biology of estuaries and coasts*. Academic Press.

Quayle, D. B., & Bourne, N. (1972). The clam fisheries of British Columbia. *Fisheries Research Board of Canada, 179*, 70.

Rainbow, P. S. (2006). Trace metal bioaccumulation: Models, metabolic availability and toxicity. *Environment International, 33*, 576–582.

Regoli, F., & Giuliani, M. E. (2014). Oxidative pathways of chemical toxicity and oxidative stress biomarkers in marine organisms. *Marine Environmental Research, 2014*, 106–117. https://doi.org/10.1016/j.marenvres.2013.07.006.

Roman, M., Vázquez, E., Viejo, R. M., Woodin, S., Wethey, D. S., Román, S., Weidberg, N. F., Troncoso, J. S., Mendez, M. M., & Olabarria, C. (2024). Context-dependent resilience of intertidal seagrass and Venerid clams after hyposalinity stress. *Marine Ecology Progress Series, 729*, 117–133. https://doi.org/10.3354/meps14516.

Roman, S., Vázquez, E., Román, M., Viejo, R. M., Weidberg, N., Troncoso, J. S., Woodin, S. A., Wethey, D. S., & Olabarria, C (2023). The stress response of the seagrass *Zostera noltei* and three commercial clam species to low salinity associated with heavy rainfall. *ICES Journal of Marine Science, 0*(0)), 1–17. https://doi.org/10.1093/icesjms/fsad203.

Sfriso, A. A., Chiesa, S., Sfriso, A., Buosi, A., Gobbo, L., Gnolo, A. B., & Argese, E. (2018). Spatial distribution, bioaccumulation profiles and risk for consumption of edible bivalves: A comparison among razor clam, Manila clam and cockles in the Venice lagoon. *Science of the Total Environment, 643*, 579–591.

Shin, Y.-K., Kim, Y., Chung, E.-Y., & Hurl, S.-B. (2000). Temperature and salinity tolerance of the Manila clam, *Ruditapes philippinarum*. *Journal of the Korean Fisheries Society, 33*(3), 313–318.

Shpigel, M., & Fridman, R. (1990). Propagation of the Manila clam (*Tapes semidecussatus*) in the effluent of fish aquaculture ponds in Eilat, Israel. *Aquaculture, 90*, 113–122.

Snelgrove, P. V. R., & Butman, C. A. (1994). Animal-sediment relationships revisited: Cause versus effect. *Oceanography and Marine Biology: An Annual Review, 32*, 111–177.

Soobin, J., Jo, K., Bae, H., Seo, H., & Kim, T. (2021). Optimal sediment grain size and sorting for survival and growth of juvenile Manila clams, *Venerupis philippinarum*. *Aquaculture, 543*, 737010. https://doi.org/10.1016/j.aquaculture.2021.737010.

Spencer, B. E. (2002). *Molluscan shellfish farming*. Fishing New Books/Blackwell Science.

Takeyama, K., Sassa, S., & Kohno, H. (2020). Development and verification of a new maintenance and management method considering the geoenvironmental dynamics for a habitat of the Manila clam *Ruditapes philippinarum*. *Global Ecology and Conservation, 24*, e01367.

Toba, D. R., Thompson, D. S., Chew, K. K., Anderson, G. J., & Mill, M. B. (1965). *Washington Sea Grant Program*. Seattle: University of Washington.

Tsukuda, M. (2008). *Mechanism of population persistence of shortneck clam (*Ruditapes philippinarum*) on the sandy tidal flats in the eastern coast of Ariake Bay* (p. 108). Graduate School of Environmental and Symbiotic Sciences, Prefectural University of Kumamoto.

Tsutsumi, H., Takeguchi, T., Maruyama, A., & Nakahara, Y. (2000). Seasonal fluctuations in the benthic community after the population collapse of a clam, *Ruditapes philippinarum*, on the Midori River tidal flats in Kumamoto. *Japanese Journal of Benthology, 54*, 1–8.

Tsutsumi, H., Ishizawa, K., Tomishige, M., Moriyama, M., Sakamoto, K., & Montani, S. (2002). Population dynamics of a clam, *Ruditapes philippinarum* on an artificially-created sand cover on the tidal flats at the river mouth of Midorikawa River. *Japanese Journal of Benthology, 57*, 177–187.

Tsutsumi, H., Tsukuda, M., Yoshioka, M., Koga, M., Shinohara, R., Nomura, Y., Choi, K. A., Cho, H. S., & Hong, J. (2003). Heavy metal contamination in the sediment and its effect on the occurrence of the most dominant bivalve, *Ruditapes philippinarum*, on the tidal flats of Ariake Bay in Kumamoto prefecture, the west coast of Kyushu, Japan. *Benthos Research, 58*(2), 121–130.

Tumnoi, W. (2012). *The autecology of* Tapes philippinarum *(Adams and Reeve, 1850) in Southampton Water, UK*. School of Ocean and Earth Sciences. University of Southampton.

Vincenzi, S., Caramori, G., Rossi, R., & De Leo, G. A. (2006). A GIS-based habitat suitability model for commercial yield estimation of *Tapes philippinarum* in a Mediterranean coastal lagoon (Sacca di Goro, Italy). *Ecological Modelling, 193*, 90–104.

Wang, X., Zhou, Y., Yang, H., Wang, Q., & Liu, S (2010). Investigation of heavy metals in sediments and Manila clams *Ruditapes philippinarum* from Jiaozhou Bay, China. *Environmental Monitoring and Assessment, 170*(1–4), 631–643. https://doi.org/10.1007/s10661-009-1262-5.

Wang, Z., Bai, Y., Nie, H., Xu, Q., Yin, Z., Zhang, Y., Yin, X., & Yan, X. (2021). Molecular mechanisms of wound healing and regeneration of siphon in the Manila clam *Ruditapes philippinarum* revealed by transcriptomic analysis. *Genomics, 113*(3), 1011–1025.

Williams, R. J. (1970). Freezing tolerance in *Mytilus edulis*. *Comparative Biochemistry and Physiology, 35*, 145–161.

Yancey, P. H. (2005). Organic osmolytes as compatible, metabolic, and counteracting cytoprotectants in high osmolarity and other stresses. *Journal of Experimental Biology, 208*, 2819–2830.

Zhuang, Q. Q. (1964). Studies of Chinese species of Veneridae (Class *Lamellibranchia*). *Studia Marina Sinica*, 43–106.

CHAPTER

12 Biological constraints on distribution and abundance

12.1 Introduction

In the microtidal Poole Harbour on the central south coast of England, a double high water phenomenon requires wild Manila clam fishers to dredge clams from boats (Chapter 17). This technique also enables the exploitation of subtidal clams from the adjacent bay beyond the Harbour entrance. However, due to low abundance, no fishers bother with Manila clams from outside the harbor. While some Manila clams can be found subtidally in these bays, especially in the vicinity of freshwater sources, there is evidence to suggest that these are dependent on larval supply from within the harbor (Herbert et al., 2011), and in any event, economically viable catches are only found in the main intertidal fishery grounds in the moderately reduced salinity areas of the harbor. The fact that, in various respects, subtidal environments in protected bays are less physico-chemically challenging for species of typically marine taxa, such as the Manila clam, suggests that biological explanations limiting the species' subtidal abundance seem likely.

On rocky shores, the generalization that upper shore species limits are due to physical factors, whereas lower limits are determined by biological factors, has been described as "paradigmatic" by Kaiser et al. (2005). This concept may also apply to the Manila clam on stable sediment shores, though the causal factors—particularly those defining its lower limits—remain poorly researched. Nevertheless, studies on endogenous activity rhythms and their alignment with the tidal and day/light cycles of Manila clams of Akkeshi Bay, Japan, indicate that feeding periodicity can adjust to either intertidal or subtidal cycles. Whereas in the intertidal zone an entrained endogenous feeding cycle is aligned with regular patterns of emersion and exposure, acclimation to subtidal conditions brings a light/dark endogenous activity rhythm in which night feeding is thought to reduce susceptibility to more intense visual predation during daylight hours (Houki & Kawamura, 2020).

The role of a number of biological factors, which influence distribution and abundance, is discussed in this Chapter. Pathogens and parasites, which are also capable of affecting distribution and abundance, are considered separately in Chapter 13.

The Manila Clam *Ruditapes philippinarum:* Science and Significance of a Globalised Species.
DOI: https://doi.org/10.1016/B978-0-443-36704-5.00012-2
Copyright © 2026 Elsevier Inc. All rights are reserved, including those for text and data mining, AI training, and similar technologies.

12.2 Food availability

Many authorities have reported on the relationship between food availability and growth in *Ruditapes philippinarum*. Through modeling, Yoon et al. (2013) have estimated that Manila clam growth rate is dependent on variations in food availability in different parts of a brackish subarctic lagoon in northern Japan. In warming scenarios, a predicted decrease in phytoplankton due to ingestion increases among other grazers, coincided with improved Manila clam growth. Therefore, while technically the different filter feeders can be seen as competing for phytoplankton, this suggests that, in subarctic waters at least, temperature rather than food competition is the limiting factor.

On the other hand, clearance of phytoplankton from the water column by filter-feeding infauna is such that in a static water layer, visible reductions in turbidity immediately above the seabed can be observed. Nevertheless, although technically such infauna, which includes other bivalves, can be said to compete for this food, the frequency and extent to which this can become a limiting factor on distribution in the field is not clear and may vary by location.

Light intensity is a factor determining the lower elevation limits of benthic photosynthetic algae and seagrasses. The Manila clam is known to consume microphytobenthos resuspended during ebb and flow tidal water movement (Chapter 7), and it is likely that in many locations, the availability of this particular photosynthetic food source may quickly decline with depth. Notwithstanding the fact that *R. philippinarum* can thrive exclusively on phytoplankton food, it has been reported that Manila clam distribution responds to microphytobenthic algal availability.

12.3 Competition

Competition is rarely cited as a significant contributing factor in explanations for patterns of *R. philippinarum* distribution and abundance, although it does sometimes feature in general terms among hypothesized reasons for population declines (Chapter 14). On balance, though secure field evidence is difficult to obtain, the infrequency of references to competition as a major factor suggests that in most locations it is not.

Nevertheless, between 2004 and 2008, Takeuchi et al. (2015) observed a higher recruitment of three bivalve species, including *R. philippinarum* in Ariake Sound, Japan, following a decline, almost to extirpation, in the populations of two bioturbating shrimp species (*Upogebia major* and *Nihonotrypaea japonica*), which had previously dominated in the high to mid tide zones. This suggests that, rather than simple direct competition for space, species that significantly modify the environment (ecosystem engineers), in this case by bioturbation of the sediment, may effectively exclude the Manila clam from otherwise suitable locations.

However, even with the removal of the shrimp populations, in contrast to the two other bivalve species (*Meretrix lusoria* and the surf clam, *Mactra veneriformis*), *R. philippinarum* failed to establish an adult population in these mid to high zones, despite initial recruitment, a result which Takeuchi et al. (2015) attributed to the greater physical disturbance in the first winter after recruitment. In contrast, in the lower tidal zone, *R. philippinarum* was successful where the surface sediment was less subject to hydrodynamic disturbance.

Laboratory tests on the burrowing performance of these three bivalve species revealed that *R. philippinarum* was less capable of repetitive reburrowing than the other two species. Therefore, while this case indicates strong spatial competition between the shrimps and the two other bivalve species, this appears not to be a straightforward explanation for *R. philippinarum* distribution. It seems that pelagic dispersal, in taking larvae to mid and high-intertidal zones, had carried them beyond the physical dimensions of the species' ecological niche. A phenomenon not unusual in benthic species with pelagic larvae.

The extent to which competition for physical space is a common limiting factor in sediment communities in estuaries can also be questioned based on the significant physiological challenges posed by estuarine environments, which have resulted in a relative paucity of species (Chapter 20), which, all else being equal, present relatively low direct competitive challenge. Consequently, the impact of interspecific competition for food and space on the growth and survival of Manila clams is normally considered less significant than predation.

However, beyond its native range, *R. philippinarum* encounters new and potentially competitive sympatric venerid species. Chapter 15 compares some of the tolerances and distributions of the Manila clam with those of the European sympatric *R. decussatus* with a view to elucidating aspects of the realised niche

12.4 Predation

Especially since the late Paleozoic era and the following "Mesozoic revolution," during which an "infaunalization" of soft-bottom benthic animals occurred (Vermeij, 1977), predation on bivalves has evolved in a wide variety of taxa. Today, predation is a significant factor in abundance, distribution, and population dynamics, and occurs at all life stages from released gametes to pelagic larvae, settled spat, juveniles, and mature clams. Comparing biotic and abiotic factors on postsettlement Manila clams in Washington State, United States. Dethier et al. (2019) concluded that, albeit context-dependent, predator density was more important to Manila clam survival than salinity, pH, and elevation. Similarly, on the basis of experimental field work in enclosed and exposed areas around Maizuru Bay, Japan, Sun et al. (2017) concluded that predation was more effective in depleting clams than other environmental factors.

Each life cycle stage attracts the attention of a different community of predators. Gametes in the water column are subject to predation by filter feeders. The planktonic larval stages of bivalves are subject to pelagic predators ranging from scyphozoan jellyfish to larval polychaete worms and small fish species (Esteves et al., 2000; Johnson & Brink, 1998; Short et al., 2013). Predators among benthic invertebrates include adult bivalves (Lehane & Davenport, 2004), which have been observed to filter out bivalve larvae from the water column with the same efficiency as algal cells (Troost et al., 2008). Such predation of bivalve pelagic larval stages can be prodigious, accounting for mortalities as high as 99% (Gosling, 2015).

Once settled, Manila clams are predated by infaunal and epibenthic predators, including polychaetes, crabs, gastropods, cephalopod mollusks, starfish, and fish. Table 12.1 provides examples of Manila clam predators. When exposed by the tide, both maritime and terrestrial birds also become significant consumers. While introductions in nonnative areas have greatly expanded the number of predator species, the taxa and niches of these are often broadly similar.

Many coasts and estuaries serve as over-wintering sites for migratory shorebirds. These depend on food resources sufficient to survive the winter in good enough condition to return to their more northerly summer breeding grounds in spring. Based on field observations, Caldow et al. (2007) reported on winter Manila clam consumption by the Eurasian oystercatcher (*Haematopus ostralegus*) in Poole Harbour, UK. Oystercatchers consumed Manila clams within the length range 16–50 mm, and since the birds ignored clams smaller than 15 mm shell length, the size distribution of Manila clams consumed differed significantly from that of the clams available in the sediment.

Some individuals tended to "specialize" in particular prey items. Fig. 12.1 shows that, by mid-January,the coldest period of winter, the proportion of oystercatchers consuming Manila clams was recorded at 40%–50%, though the actual percentage was likely higher. The Manila clam intake rate measured as ash-free dry mass (mean 2.02 mg AFDM s^{-1}) did not significantly differ from that of oystercatchers specializing in native cockles (*Cerastoderma edule*) or sand gapers (*Mya arenaria*). Modeling estimated that the presence of the non-native Manila clam population at a mean density of only 5 clams m^{-2} (Caldow et al., 2007; Jensen et al., 2004) reduced overwintering oystercatcher mortality from 4.6% to 1.1%. Crows and gulls are also known to predate Manila clams, the latter by dropping them from flight onto hard surfaces to fracture the shells for access.

Predators of juvenile and mature Manila clams must extract them from the sediment and then either swallow them whole or drill, break, or otherwise open the valves before consumption. Consequently, successful predation has been described as a sequence of stages from search to encounter, followed by capture, handling, and consumption. The achievement of the full sequence contributes to the predation rate. Optimal foraging theory (Hughes, 1980) predicts that the behavior of a predator will be such as to maximize the net rate of energy intake. It therefore follows that clam characteristics which impede, delay, or prevent any of these stages are of selective value, since they represent increased energy cost or time for predation, and therefore, all else being equal, reduced net benefit. Similarly, the rejection by oystercatchers of smaller Manila clams in Poole Harbour described above may be interpreted in terms of optimal foraging theory.

Incompatibility between an infaunal habit and elaborate predator-protective shell sculpture in bivalves informed Stanley's (1977) opinion that the burrowing habit,

Table 12.1 Range of taxa predating *Ruditapes philippinarum* demonstrated by example species.

Class	Species	Vernacular	Region	Sources
Polychaeta	*Halla okudai*	Marine worm	NE Pacific	Saito et al. (2000)
Crustacea				
	Thalamita crenata	Crab		Yap (1977)
	Cancer productus	Red crab	NE Pacific	Quayle and Bourne (1972)
	Carcinus maenas	Shore crab	NE Atlantic	Cigarría and Fernández (2000)
	Portunus trituberculatus			Sun et al. (2017)
	Charybdis japonica	Blue crab		Takahashi et al. (2016)
Gastropoda				
	Euspira lewisii	Moon snail		Anderson et al. (1982)
	Euspira fortunei			Hiromi and Shin'ichi (2009)
	Ocenebra erinacea	Oyster drill	NE Atlantic	Dang et al. (2010)
	Glossaulax didyma		NW Atlantic	Takeyama et al. (2020)
	Paratectonatica tigrina		NW Pacific	Takeshita and Maekawa (2020)
	Neverita didyma		NW Pacific	Rodrigues et al. (1987)

(continued on next page)

Table 12.1 Range of taxa predating *Ruditapes philippinarum* demonstrated by example species—cont'd

Class	Species	Vernacular	Region	Sources
Cephalopoda	*Octopus minor*	Octopus	NW Pacific	Song et al. (2019)
	Amphioctopus fangsiao		NW Pacific	Ebisawa et al. (2011)
Chondrichthyes	*Hemitrygon akajei*	Ray	NW Pacific	Tsutsumi et al (2018)
	Aetobatus narutobiei	Eagle ray	NW Pacific	Tezuka et al. (2021)
Osteichthyes	*Balistes capricus*	Trigger fish		Cigarría and Fernández (2000)
	Sparus auratus	Sea bream		Cigarría and Fernández (2000)
	Tagifugu rubripes	Puffer fish	NW Pacific	Takahashi et a . (2016)
Aves	*Melanilla deglani*	Diving duck		Bourne (1984)
	Melanilla perspincillata			
	Melanilla nigra			
	Haematopus ostralegus	Oystercatcher	Europe	Caldow et al. (2007)
	Corvus corone	Crow		Richardson (1985)
	Larus argentatus	Gull		Herbert et al. (2011)
Mammalia	*Homo sapiens*	Human	Cosmopolitan	Numerous (Chapter 17)

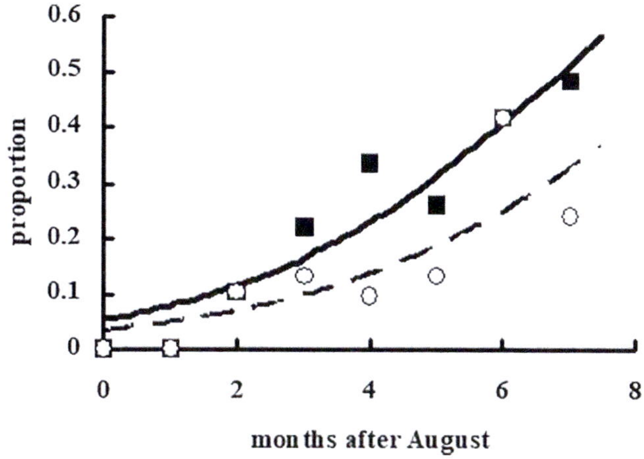

FIGURE 12.1

Seasonal variation in the proportion of over-wintering Eurasian oystercatchers (*Haematopus ostralegus*) that specialized in feeding on Manila clams (open circles) and those that ever included Manila clams in their diet (filled squares) in Poole Harbour, UK.

From Caldow et al. (2007).

rather than predation, is the major selective factor in bivalve shell morphology. However, other aspects of shell morphology have been experimentally demonstrated to have a bearing on vulnerability to predation. Relative to some other bivalves, the NE Pacific Venerid *Leukoma staminea,* when exposed to the red rock crab *Cancer productus*, benefits from greater shell thickness and more pronounced shell inflation (Boulding, 1984). These features, which are shared by the sympatric Manila clam, prolong the shell-breaking time to an extent that even clams that could be opened are sometimes abandoned in favor of easier prey. Evidence of failed crab attacks can be seen as chipped shells.

Notable among Manila clam predators are gastropod mollusks. Rodrigues et al. (1987) investigated the naticid *Neverita didyma* when predating *R. philippinarum.* They showed that this predator is unable to assess shell thickness and relies only on prey size based on ease of handling. Prey size limits are determined by the predator's foot size, this being the organ used in handling prey. Takeshita and Maekawa (2020) established that in predation of *R. philippinarum* by another Naticid gastropod, *Paratectonatica tigrina*, the probability of predator success varied with the relative shell weights of the protagonists (shell weight ratio). Moreover, they observed three different tactics of attack: wall drilling, edge drilling, and no drilling (which involved suffocation). They determined that the entry tactic employed varied with shell weight ratio, being 6/25 (24%) for suffocation, 8/25 (32%) for edge drilling, and 11/25 (44%) for wall drilling.

Generally, the depth to which an individual is buried also influences the time required by predators to excavate it and increases the chances of it being discarded (Haddon et al., 1987, Zwarts & Wanink, 1989). Although burial in the sediment is protective, the shallow depth to which Manila clams burrow is less advantageous in this respect. Dudas et al. (2005) have demonstrated that the crab *Cancer productus*, when presented with two bivalve species of different shell thickness (*R. philippinarum* and the thin-shelled but deeper burrowing North American Varnish clam *Nuttallia obscurata)*, will select the latter for predation when both are presented on the sediment surface, but the shallower burrowing Manila clam, when both are allowed to bury.

Since *R. philippinarum* food includes suspended and resuspended matter, the depth to which it can bury is determined by the length of the extended siphons. The relatively shallow depths, typically around 5 cm for Manila clams, will provide some protective advantage, but evidence that a deeper position would further reduce predation risk raises the question of whether any countervailing selective advantage accrues from a shallow position. Richardson (1985), who investigated Manila clam predation by crows, suggested that the shallower burrowing habit minimizes the costs that would otherwise be associated with longer and heavier siphons. However, since deep-living bivalves tend to have lighter shells, in so far as the heavy and presumably more energetically demanding shell of *R. philippinarum* is a protective countermeasure to shallow burrowing predation risk, the overall cost-benefit equation may favor a shallower habitat. In any event, and since most epibenthic predators occupy the sediment surface, it should be assumed that the combination of a heavy tumid shell and shallower burrowing habit represents an effective combined contribution to the fitness of this species.

Predatory mortality is generally size-specific, and this has been shown to be the case with the Manila clam by Cigarría and Fernández (2000), whose work in Spain suggested a "refuge size" at shell length 21 mm. Beyond this mortality from common predators such as the shore crab *Carcinus maenas*, the trigger fish *Balistes capricus*, gilt-head bream *Sparus auratus,* and oystercatchers *Haematopus ostralegus* is substantially reduced. Nevertheless, larger clams remain vulnerable to larger predators. In Japan, Manila clams with shells up to at least 45 mm shell length are consumed by the Naticid gastropod mollusk, *Euspira fortune* (Hiromi & Shin'ichi, 2009).

In their native range, laboratory experiments by Takahashi et al. (2016) showed that the Indo-Pacific Portunid crab *Charybdis japonica* consumed *R. philippinarum* across various clam size ranges, which they defined as juvenile (shell length 1–5 mm), young (10–15 mm), and adult (20–25 mm). Smaller crabs (carapace width 19–24 mm) predated only juvenile clams, whereas larger crabs (88–93 mm) favored young and adult clams but were not observed consuming juvenile clams. Foraging theory again suggests that this relative neglect of smaller clams by larger crabs reflects an optimization of net energetic benefit.

Using larger Manila clams, Sun et al. (2017) also reported the relative sizes of crab and clam as material factors in successful predation by another Portunid, *Portunus trituberculatus.* The encounter rate between crab and clam increased with increasing crab and clam sizes. The clam predation rate by larger crabs (100–110 mm carapace width) was greater than that by smaller crabs (80–90 mm), and the predation rate of both crab sizes was higher when preying on smaller (28–31 mm shell length) than on larger (38–41 mm) clams. Handling time increased with clam size. Not every clam encountered was successfully captured, and not every clam successfully captured was consumed.

In the northern Adriatic lagoons, the Manila clam is predated by an incoming non-native blue crab *(Callinectes sapidus)*. In the laboratory, Boschiero et al. (2025) have reported a "crushing" technique used on small clams in contrast to a "dactyl" technique for larger clams, in which, after chipping the valve edge, the movable claw component of a cheliped is inserted between the valves to force them open. Chiesa et al. (2025) reported shell injuries corresponding to the serrations on the crab's chelae and a "massive" loss of Manila clams, which accompanied the blue crab's arrival in Po delta lagoons. The arrival of this crab clearly represents a major economic threat in these Adriatic lagoons.

Sun et al. (2017) have further shown that competition between individual blue crabs for Manila clams can impede predation due to fighting and kleptoparasitism. On encountering another crab, the probability of fighting is greater in higher crab densities, which in turn reduces the probability of clam capture upon encounter. Moreover, when under conditions of short-term starvation, the probability of fighting between crabs increased, which in turn significantly increased the proportion of time spent handling clams and reduced foraging success.

Another voracious predator of Manila clams in NW Europe is the shore crab *C. maenas*. This species is known to respond to both tactile and chemical stimuli in detecting bivalve prey (Elner & Hughes, 1978). Harris (2016) found that whereas over 70% of available 4–5 mm clams are consumed, at 18 mm, consumption was less than 10%. At shell lengths from 10–16 mm, consumption by male shore crabs is significantly greater than by female crabs, which Harris attributes to the greater size of the male chelae. Harris (2016), among others, has also observed that *C. maenas* varies its approach to accessing the soft tissues according to bivalve prey size. Below a clam length of 8 mm, clams once extracted from the sediment were passed straight to the mouthparts, while with larger clams, the valves were first cracked open by the much more powerful chelae or occasionally subjected to pressure to break the ligament.

In Poole Harbour, UK, the abundance of the shore crab *C. maenas* has impacts on clam populations, which vary with sediment type and shore elevation. Higher shore positions provide some protection at low water, when crabs are less active and themselves subject to gull predation. Conversely, the sandier substrates near the mouth of the Harbor appear to provide less protection from crab predation compared to finer substrates upstream. Although the clam is known to be consumed by wading birds such as oystercatchers at low water, the resulting predation pressure on the population appears much less than that from crabs.

Imabayashi et al. (1996) investigated predation on *R. philippinarum* by the intertidal night-feeding polychaete *Halla okudai* (Imajima) from the Seto Sea and Ariake Bay, Japan. Here, the condition index of the polychaete increases with the density of *R. philippinarum,* which is considered its main prey item. This polychaete was observed searching for prey by raising the head 20–30 mms above the sediment, which then rotated such that three sensory tentacles faced towards the prey item. This behavior was taken to indicate a primarily olfactory detection of prey. Handling involves thwarting attempts by *R. philippinarum* to escape by wrapping its anterior round the shell margin and enveloping the clam in a viscous secretion from its body wall, which impedes movement and contains a paralytic toxin. Handling time from around 1 to 6 h correlated significantly with clam length. The worm is able to enter the clam when the shell gap is only 2 mm and subsequently secretes digestive enzymes.

Saito and Imabayashi (1994, 1997) and Saito et al. (2000) calculated the energetics of this predator-prey relationship, comparing small and larger prey clams. In the laboratory, larger *H. okudai* were shown to select larger clams, and the energetically optimum clam prey size was estimated as 10%–20% of worm body weight. Since smaller clams represented less food, the rate of predation was higher, and the time spent searching necessarily increased, being around 28 h for small clams but 16 h for large clams. Manila clams below 10% of worm body weight required a compensatory increase in searching time, which reduced energetic net gain. Predation of clams above 20% of worm body weight required the secretion of greater volumes of viscous material, which also reduced net energy gain.

Various *Octopus* species are known to predate Manila clams. Octopods are known to utilize two distinct methods of entry when presented with bivalves. The first involves the pulling apart of the valves using the muscular suckers on the arms, and the second is to drill a hole in a valve using the toothed radula or abrasive salivary papilla. *Octopus minor* is distributed along the coasts of China, Japan, and the Korean Peninsula. Song et al. (2019) presented relatively large *O. minor* with three bivalve species: *R. philippinarum, Mactra chinensis*, and *Mytilus galloprovincialis,* all of shell length around 33–35 mm and wet weight 10–13 g. Only the pulling method of clam predation was observed. In initial single-prey experiments, *O. minor* was found to predate *R. philippinarum* at a rate of 1.73 ± 0.50 clams per day. When presented with all three bivalve species, *O. minor* preferentially selected *R. philippinarum.*

To investigate this preference in terms of optimal forging theory, the energy content of the three bivalves was compared to estimates of energy consumption required for shell opening. While there was no significant difference in opening time, the pulling power required to open *R. philippinarum* was calculated as 24.84 ± 3.67 N, which was significantly different and fell between that of the other two species, with *Mytilus galloprovincialis* requiring the least force. However, the mean energy content of *R. philippinarum, at* 5.94 ± 0.13 kJ, was significantly higher than that of the other two species, with the result that the energy profitability (net gain) was also significantly higher at 26.78 ± 0.60 kJ/prey•h^{-1} (Song et al., 2019).

While this was consistent with foraging theory, the prey preference may also have been influenced in this case by a greater familiarity with Manila clams, which were abundant at the site in question (see Carlsson et al., 2009).

Ebisawa et al. (2011) investigated predation of Manila clams by *Amphioctopus fangsiao* (synonym *Octopus ocellus*), also a native of the NW Pacific. *Octopus ocellus* specimens of 7.0–52.1 g wet weight pulled open the valves of smaller clams (shell length 0.9–2.9 cm), taking a mean opening time of 3.1 m with energy cost measured as oxygen consumption of 8.0 μg O_2 g^{-1}. For larger clams, *Octopus ocellatus* gains access by initially attempting to pull, then drilling a hole through a valve. Pulling alone was most energetically efficient in terms of net energy gain per unit handling time. The octopus selected the most energetically efficient opening technique according to the clam size.

In Maizuru Bay, Sea of Japan, out of 13 fish species known to be present, 6 fish species from five different families predated Manila clams up to 5 mm shell length. One of these, from the puffer fish family, which opens clams with powerful beak-like front teeth, took Manila clams at least up to 15 mm (Takahashi et al., 2016).

Even where viable wild populations of Manila clams thrive, the prodigious fecundity of this broadcast-spawning species is comparable to, if not entirely matched by, a prodigious depletion, much of which is through predation. While the full extent of this depletion will vary by year and by location, at its most severe, predation may account for the complete extirpation of spat in the months following settlement. On the Sashike Tidal Flats in southern Japan, Tsutsumi et al. (2018) found that a settlement in excess of around 80×10^3 clams m^2, peaking in January 2015, was reduced by April to under 5000 clams m^{-2} of length 2–5 mm. Whilst various predators will have contributed to this decline, on the basis of stomach contents and extensive characteristic shallow excavations in the spring sediment surface, they were able to attribute much of this loss to the ray *Hemitrygon akajei*.

Tsutsumi et al. (2018) also showed the same Sashike sediments to contain virtually no older age group cohorts, suggesting the possibility that predation pressure is capable of entirely suppressing a normal multimodal population demography and raising the prospect that a more age-balanced wild broodstock population supplies larvae and spat from elsewhere in the adjacent Yatsushiro Bay area.

12.5 Siphon cropping

Due to their close proximity to the surface of sediments, the Manila clam's siphon tips are vulnerable to predation. Although *R. philippinarum* can quickly retract its siphons back into the shell in response to movement above, both crustaceans and fish are known to nip off the ends in the sympatric NE Pacific Venerid *Leukoma staminea* (Peterson & Quammen, 1982) and no doubt the Manila clam also. Such siphon nipping is known to provide an important contribution to the diet of small fish and to significantly attenuate the annual growth of the bivalve.

Aside from its direct impact, siphon cropping, which is survivable, can result in a greater risk of whole animal predation. Byers (2005) compared the susceptibility to predation of the Manila clam and the native *Leukoma staminea* in British Columbia. Noting the deeper burrowing habit of the native, they found the latter to be at an advantage in terms of predation avoidance after experiencing siphon snipping by crabs at the sediment surface. This is because before regeneration of the siphons, clams must move nearer the sediment surface to compensate for shorter siphons. In such circumstances, the shallower burrowing Manila clam must occupy a position very close to the surface, resulting in mortality twice that of the native. Nevertheless, these authors observed that Manila clams were present in much higher numbers.

Occasionally, Manila clams can be observed to close their valves faster than the siphons can be withdrawn, resulting in the siphons being left protruding, which, even in the absence of nipping, would result in tissue damage. Although there is little evidence that *R. philippinarum* actively sheds its siphon tips as a defensive strategy (autotomy), this has been observed in the laboratory, and in any event, valve closure with the siphon tips outside suggests that these organs can be sacrificed in aid of survival.

Manila clams are capable of regenerating their siphon tips within about 20 days (Ekaratine & Davenport, 1996). Wang et al. (2021) have elucidated the morphological and transcriptome mechanisms of wound healing and siphon regeneration, including cell mobilization and morphogenesis. In contrast, *Donex hanleyaunus*, which occupies exposed sandy beaches in South America, can regenerate its siphons in 5 days, perhaps suggesting a lower selective pressure for fast regeneration in low-energy intertidal habitats.

12.6 Predation refuge areas

In Arcachon Bay, France, Dang et al. (2010) observed that, after introduction in 1980, *R. Philippinarum* had rapidly colonized intertidal sea-grass meadows (*Zostera noltii*). In Spain, Galvan and Puente (2023) have shown dense *Zostera noltii* meadows to be "nursery" habitats by virtue of the fact that they protect juvenile Manila clams. By creating more structurally complex environments above and below ground, they restrict predator foraging success. However, clam condition index was lower in areas of high seagrass cover, as food availability for static filter feeders is dependent on water movements, which are reduced within seagrass meadows. Roman et al. (2023) have suggested that seagrass beds may also provide Manila clams with a refuge from heatwaves.

Laboratory and field experiments by Takeyama et al. (2020) have demonstrated that Manila clam predation by the naticid snail *Glossaulax didyma* was radically reduced in gravel sediments of particle size 10–20 mm. In the field, clam survival rates in gravel were 88% after 2 months compared to 1% in sand. This result corroborated laboratory experiments indicating a significant protective effect of gravel substrate, which has also been reported in other parts of the world.

In addition to an interaction between sediment and predation rate, there also appears to be an interaction between salinity and predation. Poole Harbour in the UK has had a thriving naturalized Manila clam population and fishery since its introduction in the 1980s (Humphreys et al., 2007). Hydrodynamic and behavioral modeling have demonstrated larval dispersal out of the narrow harbor entrance into the adjacent fully marine Poole Bay (Herbert et al., 2011) (see Chapter 20). However, despite larval supply immediately outside the Harbour mouth, the sublittoral distribution of the Manila in Poole Bay is sporadic beyond areas where freshwater streams run down the adjacent beach. The possibility emerges that salinity may also act to provide an element of refuge from predation, allowing both enhanced survival and an extension of distribution into lower sublittoral elevations where stenohaline epibenthic predators are excluded by diluted waters. The alternative hypothesis of reduced salinity coinciding with reduced competition for resources such as space or food provides an unlikely explanation, not least since in this area, infaunal densities are not high.

References

Anderson, G. J., Miller, M. B., Chew, K. K. 1982 A guide to Manila clam aquaculture in Puget Sound. Technical report Washington Sea grant. 82-4. National Oceanic and Atmospheric Administration.

Boschiero, M., Facca, C., Cavraro, F., Tonolli, M., Malavasi, S., & Franzoi, P. (2025). Feeding strategies of the invasive blue crab (*Callinectes sapidus*) on Manila clam (*Ruditapes philippinarum*): Implications for aquaculture. *Estuarine, Coastal and Shelf Science, 325*, 109486.

Boulding, E. G. (1984). Crab-resistant features of shells of burrowing bivalves: Decreasing vulnerability by increasing handling time. *Journal of Experimental Marine Biology and Ecology, 76*(3), 201–223. https://doi.org/10.1016/0022-0981(84)90189-8.

Bourne, N. (1984). Clam predation by scoter ducks in the Strait of Georgia, British Columbia. Canadian technical report of fisheries and aquatic sciences. 1331, no. 17 Department of Fisheries and Oceans, Fisheries Research Branch, Pacific Biological Station.

Byers, J. E. (2005). Marine reserves enhance abundance but not competitive impacts of a harvested nonindigenous species. *Ecology, 86*(2), 487–500. https://doi.org/10.1890/03-0580.

Caldow, R. W. G., Stillman, R. A., Dit Durell, S. E. A. L. V., West, A. D., McGrorty, S., Goss-Custard, J. D., Wood, P. J., & Humphreys, J. (2007). Benefits to shorebirds from invasion of a non-native shellfish. *Proceedings of the Royal Society B: Biological Sciences, 274*(1616), 1449–1455. https://doi.org/10.1098/rspb.2007.0072.

Carlsson, N. O. L., Sarnelle, O., & Strayer, D. L. (2009). Native predators and exotic prey—An acquired taste? *Frontiers in Ecology and the Environment, 7*(10), 525–532. https://doi.org/10.1890/080093.

Chiesa, S., Petochi, T., Boscolo Brusà, R., Raicevich, S., Cacciatore, F., Franceschini, G., Antonini, C., Vallini, C., Bernarello, V., Oselladore, F., Ciani, M., Di Blasio, L., Campolunghi, M. P., Baldessin, F., Boldrin, L., & Marino, G. (2025). Impacts of the blue crab invasion on Manila clam aquaculture in Po Delta coastal lagoons (Northern Adriatic Sea, Italy). *Estuarine, Coastal and Shelf Science, 312*(2025), 109037.

Cigarría, J., & Fernández, J. M. (2000). Management of Manila clam beds I. Influence of seed size, type of substratum and protection on initial mortality. *Aquaculture, 182*(1–2), 173–182. https://doi.org/10.1016/S0044-8486(99)00257-4.

Dang, C., de Montaudouin, X., Gam, M., Paroissin, C., Bru, N., & Caill-Milly, N. (2010). The Manila clam population in Arcachon Bay (SW France): Can it be kept sustainable? *Journal of Sea Research, 63*(2), 108–118. https://doi.org/10.1016/j.seares.2009.11.003.

Dethier, M. N., Kobelt, J., Yiu, D., Wentzel, L., & Ruesink, J. L. (2019). Context-dependence of abiotic and biotic factors influencing performance of juvenile clams. *Estuarine, Coastal and Shelf Science, 219*, 201–209. https://doi.org/10.1016/j.ecss.2019.02.013.

Dudas, S. E., McGraw, I. J., & Dower, J. F. (2005). Selective crab predation on native and introduced bivalves in British Columbia. *Journal of Experimental Marine Biology and Ecology, 325*(1), 8–17.

Ebisawa, S., Tsuchiya, K., & Segawa, S. (2011). Feeding behavior and oxygen consumption of *Octopus ocellatus* preying on the short-neck clam *Ruditapes philippinarum*. *Journal of Experimental Marine Biology and Ecology, 403*(1–2), 1–8. https://doi.org/10.1016/j.jembe.2011.03.014.

Ekaratne, S. U. K., & Davenport, J. (1996). Autotomy and regeneration of siphonal tips in Manila clam *Tapes philippinarum* (Mollusca: Bivalvia). *Journal of South Asian Natural History, 2*(1), 91–94.

Elner, R. W., & Hughes, R. N. (1978). Energy Maximization in the diet of the shore crab, *Carcinus maenas*. *The Journal of Animal Ecology, 47*(1), 103. https://doi.org/10.2307/3925.

Esteves, E., Pina, T., Chícharo, M. A., & Andrade, J. P. (2000). The distribution of estuarine fish larvae: Nutritional condition and co-occurrence with predators and prey. *Acta Oecologica, 21*(3), 161–173. https://doi.org/10.1016/S1146-609X(00)01078-X.

Galvan, C., & Puente, A. (2023). Exploring the newly emerging effects of native seagrasses on survival and growth of nonnative juvenile clams. *Frontiers in Marine Science, 10*, 1160734. https://doi.org/10.3389/fmars.2023.1160734.

Gosling, E. (2015). *Marine bivalve Molluscs* (second edition). Wiley.

Haddon, M., Wera, R. G., & Packer, H. A. (1987). Depth and density of burial by the bivalve *Paphies ventricosa* as a refuge from predation by the crab *Ovalipes catharus*. *Marine Biology, 94*, 25–30. https://doi.org/10.1007/BF00392896.

Harris, M. R. (2016). *Thesis*. University of Portsmouth.

Herbert, R. J. H., Willis, J., Jones, E., Ross, K., Hübner, R., Humphreys, J., Jensen, A., & Baugh, J. (2011). Invasion in tidal zones on complex coastlines: Modelling larvae of the non-native Manila clam, *Ruditapes philippinarum*, in the UK. *Journal of Biogeography, 39*(3), 585–597.

Hiromi, H., & Shin'ichi, S. (2009). Predatory behaviour of the naticid *Euspira fortunei*: Why does it drill the left shell valve of *Ruditapes philippinarum*? *Journal of Molluscan Studies, 75*(2), 147–151. 2009 https://doi.org/10.1093/mollus/eyp010 .

Houki, S., & Kawamura, T. (2020). Feeding and digestion periodicity of Manila clam *Ruditapes philippinarum* in natural intertidal and subtidal zones estimated from the morphological condition of the crystalline style. *Journal of Molluscan Studies, 86*(4), 361–371. https://doi.org/10.1093/mollus/eyaa027.

Hughes, R. N. (1980). Optimal foraging theory in the marine context. *Oceanography and Marine Biology, 18*, 423–481.

Humphreys, J., Caldow, R. W. G., Mcgrorty, S., West, A. D., & Jensen, A. C. (2007). Population dynamics of naturalised Manila clams *Ruditapes philippinarum* in British coastal waters. *Marine Biology, 151*(6), 2255–2270. https://doi.org/10.1007/s00227-007-0660-x.

Imabayashi, H., Saito, H., Ohmasa, C., & Kawai, K. (1996). Foraging behavior of the polychaete *Halla okudai* in aquariums. *Benthos Research, 1996*(50), 11–17. https://doi.org/10.5179/benthos1990.1996.50_11.

Jensen, A. C., Humphreys, J., Caldow, R. W. G., Grisley, C., & Dyrynda, P. E. J. (2004). Naturalization of the Manila clam (*Tapes philippinarum*), an alien species, and establishment of a clam fishery within Poole Harbour, Dorset. *Journal of the Marine Biological Association of the United Kingdom, 84*(5), 1069–1073. https://doi.org/10.1017/s0025315404010446h.

Johnson, K. B., & Brink, L. A. (1998). Predation on bivalve Veligers by polychaete larvae. *Biological Bulletin, 194*(3), 297–303. https://doi.org/10.2307/1543099.

Kaiser, M. J., Attrill, M. J., Jennings, S., & Thomas, D. (2005). *Marine ecology: Processes, systems, and impacts* (third edition). Oxford University Press.

Lehane, C., & Davenport, J. (2004). Ingestion of bivalve larvae by *Mytilus edulis*: Experimental and field demonstrations of larviphagy in farmed blue mussels. *Marine Biology, 145*(1), 101–107. https://doi.org/10.1007/s00227-003-1290-6.

Peterson, C H, & Quammen, M L (1982). Siphon nipping: Its importance to small fishes and its impact on growth of the bivalve *Protothaca staminea* (Conrad). *Journal of Experimental Marine Biology and Ecology, 63*(3), 249–268.

Quayle, D. B. and Bourne, N. (1972). The clam fisheries of British Columbia. Fisheries Research Board of Canada Bulletin, 179:70.

Richardson, H. (1985). Availability of buried littleneck clams (*Venerupis japonica*) to northwestern crows (*Corvus caurinus*). *Journal of Animal Ecology, 54*(2), 443–457. https://doi.org/10.2307/4490.

Rodrigues, C. L., Nojima, S., & Kikuchi, T. (1987). Mechanics of prey size preference in the gastropod *Neverita didyma* preying on the bivalve *Ruditapes philippinarum*. *Marine Ecology Progress Series, 40*, 87–93. https://doi.org/10.3354/meps040087.

Roman, M., Gilbert, F., Viejo, R. M., Román, S., Troncoso, J. S., Vázquez, E., & Olabarria, C. (2023). Are clam-seagrass interactions affected by heatwaves during emersion? *Marine Environmental Research, 186*, 105906.

Saito, H., & Imabayashi, H. (1994). Food preference of the polychaete *Halla okudai*. *Journal of the Faculty of Applied Biological Science, 33*, 151–157.

Saito, H., & Imabayashi, H. (1997). The effect of prey size on growth of the carnivorous polychaete *Halla okudai* (Lysaretidae). *Benthos Research, 52*, 81–87.

Saito, H., Imabayashi, H., & Kawai, K. (2000). Interaction between searching cost and growth of the bivalve feeder *Halla okudai* under rearing conditions, in relation to prey size. *Fisheries Science, 66*, 908–914.

Short, J., Metaxas, A., & Daigle, R. M. (2013). Predation of larval benthic invertebrates in St George's Bay, Nova Scotia. *Journal of the Marine Biological Association of the United Kingdom, 93*(03), 591–599.

Song, M., Wang, J., & Zheng, X. (2019). Prey preference of the common long-armed octopus *Octopus minor* (Cephalopoda: Octopodidae) on three different species of bivalves. *Journal of Oceanology and Limnology, 37*(5), 1595–1603. https://doi.org/10.1007/s00343-019-8217-7.

Stanley, S. M. (1977). Trends, rates, and patterns of evolution in the Bivalvia. In A. Hallam (Ed.), *Patterns of evolution* (pp. 209–247). Elsevier Science Publishers.

Sun, Y., Wang, F., Su, X., Liu, D., & Dong, S. (2017). Size matters for predator-prey relationships: The case of swimming crab *Portunus trituberculatus* preying on Manila clam *Ruditapes philippinarum*. *Aquaculture, 479*, 638–644.

Takahashi, K., Sawada, H., & Masuda, R. (2016). Reproduction and depletion of the Manila clam in Maizuru Bay, Sea of Japan. *Nippon Suisan Gakkaishi, 82*(5), 699–705. https://doi.org/10.2331/suisan.16-00017.

Takeshita, F., & Maekawa, T. (2020). *Paratectonatica tigrina* (Gastropoda: Naticidae) adjusts its predation tactics depending on the chosen prey and their shell weight relative to its own. *Journal of the Marine Biological Association of the United Kingdom, 100*(6), 921–926. https://doi.org/10.1017/s0025315420000831.

Takeuchi, S., Yamada, F., Shirozu, H., Ohashi, S., & Tamaki, A. (2015). Burrowing ability as a key trait in the establishment of infaunal bivalve populations following competitive release on an extensive intertidal sandflat. *Journal of Experimental Marine Biology and Ecology, 466*, 9–23. https://doi.org/10.1016/j.jembe.2015.01.011.

Takeyama, K., Sassa, S., & Kohno, H. (2020). Development and verification of a new maintenance and management method considering the geoenvironmental dynamics for a habitat of the Manila clam *Ruditapes philippinarum. Global Ecology and Conservation, 24*, e01367. https://doi.org/10.1016/j.gecco.2020.e01367.

Tezuka, N., Takada, Y., Shigeta, T., & Uchida, M. (2021). Identification of potential predators for Asari clam *Ruditapes philippinarum* using time-lapse camera observations. *Japan Agricultural Research Quarterly, 55*(1), 85–96 (2021).

Troost, K., Veldhuizen, R., Stamhuis, E. J., & Wolff, W. J. (2008). Can bivalve Veligers escape feeding currents of adult bivalves? *Journal of Experimental Marine Biology and Ecology, 358*(2), 185–196. https://doi.org/10.1016/j.jembe.2008.02.009.

Tsutsumi, H., Nishioka, T., Kitagawa, N., Fujiyoshi, Y., Kozaki, M., Yamamoto, M., Ichinomiya, M., & Komorita, T. (2018). Seasonal changes in short-neck clam (*Ruditapes philippinarum*) population and its controlling factors on Sashiki tidal flats. *Japanese Journal of Benthology, 73*(1), 1–10. https://doi.org/10.5179/benthos.73.1.

Vermeij, G. J. (1977). The Mesozoic marine revolution: Evidence from snails, predators and grazers. *Paleobiology, 3*(3), 245–258. https://doi.org/10.1017/S0094837300005352.

Wang, Z., Bai, Y., Nie, H., Xu, Q., Yin, Z., Zhang, Y., Yin, X., & Yan, X. (2021). Molecular mechanisms of wound healing and regeneration of siphon in the Manila clam *Ruditapes philippinarum* revealed by transcriptomic analysis. *Genomics., 113*(3), 1011–1025. https://doi.org/10.1016/j.ygeno.2021.02.010.

Yap, W. G. (1977). Population biology of the Japanese little-neck clam, *Tapes philippinarum*, in Kaneohe Bay, Oahu, Hawaiian Islands. *Pacific Science., 31*(3), 223–244.

Yoon, S., Abe, H., & Kishi, M. J. (2013). Responses of Manila clam growth and its food sources to global warming in a subarctic lagoon in Japan. *Progress in Oceanography, 119*, 48–58.

Zwarts, L., & Wanink, J. (1989). Siphon size and burying depth in deposit and suspension feeding benthic bivalves. *Marine Biology, 100*, 227–240.

CHAPTER

Pathogens and parasites 13

13.1 Range of infectious agents

The definitions of pathogen and parasite allow for considerable overlap. While pathogens are by definition disease-causing through infection, they may also derive benefit (typically nutrition) from their hosts, which falls within the definition of parasitism. Moreover, parasites that derive nutrition from their hosts can also cause disease. Which term is applied tends to be a matter of convention, with viruses, bacteria, and some fungi typically referred to as pathogens. Unicellular eukaryotic organisms are often called either, larger metazoans are parasites (Dimijian, 1999).

In many instances, the immune responses of *Ruditapes philippinarum* described in Chapter 9 can combat infectious organisms, at least restricting their impact to chronic low levels with minor implications for fitness. In microbial diseases, defense mechanisms include phagocytosis of pathogens by hemocyte cells along with antibacterial properties of extracellular body fluids (Chapter 9 and Allam & Paillard, 1998; Chu, 1988). However, even low-level infection can reduce the clam's resilience in relation to other challenges, and severe infection can interfere directly with development, growth, reproductive capacity, and also increase mortality (Yang et al., 2021). In such circumstances, populations of clams may be overwhelmed.

Even though Manila clams are usually consumed after cooking, a further implication of Manila clam infection is the possibility of transmission to human consumers. For example, some forms of Hepatitis virus pose a risk to human health (Gao et al., 2015), as does Norovirus, which, when consumed via infected clams, leads to severe gastroenteritis.

Especially, but not only, in intensive cultivation scenarios, epidemic disease and mass mortalities can have considerable consequences for commercial production. Indeed, some commercial farming practices, including high-density culture and the transfer of microbiologically unmonitored stocks, are associated with increased severity and frequency of infections. For example, experiments with subtidal and bottom cage rearing in Korea suggest that the protozoan *Eomarteilia granula*, while not thought to represent a high mortality risk in natural conditions, appears to have caused mass mortalities in the context of intensive high densities (Lee et al., 2025).

The causes of population decline and mass mortality episodes in Manila clam populations are considered further in Chapter 14, but there is strong evidence that infections are often important contributory factors. For instance, heavy infectious loads can involve a range of parasites, resulting in a progressive decline in condition,

The Manila Clam *Ruditapes philippinarum*: Science and Significance of a Globalised Species.
DOI: https://doi.org/10.1016/B978-0-443-36704-5.00013-4
Copyright © 2026 Elsevier Inc. All rights are reserved, including those for text and data mining, AI training, and similar technologies.

as happened in Cheonsu Bay, South Korea. Here, infection rates of up to 93% with a prevalence of the protozoan *Perkinsus olseni* and up to 33% co-infection by the trematodes *Cercaria tapetis* and *Parvatrema duboisa* caused attenuation of both somatic growth and gametogenesis, resulting from inflammation of the gills due to hemocyte infiltration and effective gonad "castration" by the two parasite types, respectively (Yang et al., 2021).

The economic significance of *R. philippinarum* has ensured that considerable research into its pathogens, including disease monitoring and prevention measures, has been conducted across many countries, often stimulated by patterns of increased mortality. For example, Tamaoki et al. (2025) have developed microsatellite markers for the pycnogonid *Nymphonella tapetis,* whose parasitic larval and juvenile stages feed on the body fluids of the host *R. philippinarum*. Moreover, multispecies co-infections can affect disease progression, as reported by Kim et al. (2025), who identified a novel uncharacterized but virulent eukaryotic protist (MCPX) that coinfects Manila clams with *P. olseni*.

Bathige et al. (2025) have provided an extensive overview of the parasitic fauna of Manila clams, ranging from viruses and bacteria to protozoans, including chromistan diatoms and dinoflagellates, to metazoans, including annelids, crustaceans, and pycnogonid sea spiders. Moreover, they have summarized the worldwide epidemiological distribution patterns of these species. Fig. 13.1 lists 73 known parasites of the Manila clam, showing their distribution by country. Other species include the Crustacean decapod *Fabia subquadrata and* the parasitic copepod *Mytilicola orientalis* (Bonham et al., 2018). Although found within Manila clams, it cannot be assumed that all these species are significantly detrimental. Nevertheless, many are known to be pathogenic to bivalves, including *R. philippinarum*, among the most significant being the protozoan *P. olseni* and the bacterium *Vibrio tapetis* (*V. tapetis*) (Jiang et al., 2020). These are both known to be extensively distributed across Europe and East Asia and are considered in turn below.

13.2 Brown ring disease

The causative agent of brown ring disease (BRD) is the gram-negative motile bacterium V*ibrio tapetis* (Borrego et al., 1996; Paillard & Maes, 1990). Although this pathogen is detected in various shellfish species, it appears that the Manila clam is especially susceptible, and BRD has been associated with mass mortality episodes on the European Atlantic coast and in Korea (Park et al., 2006) (See Chapter 14).

This bacterium colonizes the periostracum at the edge of the mantle, resulting in the deposit of melanized conchiolin within the extrapallial region. Such deposits are a bivalve defense response not confined to *V. tapetis* infection and may also be associated with "Brown Mussel Disease" as reported in Arcachon Bay, France. BRD initially shows as a brown discolored ring around the outer margins of the inner periostracum (Paillard & Maes, 1990, 1995a, 1995b) and progresses through seven stages as originally described by Paillard and Maes (1994). Fig. 13.2A shows the

No.	Virus	Country	No.	Bacteria	Country	No.	Protozoa	Country	No.	Arthropoda	Country
1	Hepatitis A virus		1	*Aliivibrio finisterrensis*		1	*Ancistrum crissum*		1	*Fabia subquadrata*	
2	Hepatitis E virus		2	*Rickettsia* sp.		2	*Ancistrum edajimanum*		2	*Myticola-like*	
3	Herpes-like virus		3	*Rickettsia* sp. /*Chlamydia* sp.		3	Coccidia-like Apicomplexan		3	*Mytilicola orientalis*	
4	Marine Birnavirus		4	Rickettsiales-like		4	*Cryptosporidium* sp.		4	*Nymphonella tapetis*	
5	Norovirus		5	*Vibrio alginolyticus*		5	*Eomarteilia granula*		5	*Pinnixa faba*	
6	OsHv-1		6	*Vibrio antiquaries*		6	Gregarine-like Apicomplexan		6	*Pinnixa littoralis*	
7	Oyster herpesvirus (OsHV)		7	*Vibrio artabrorum*		7	*Haplosporidium nelsoni*		7	*Pseudomyicola ostreae*	
8	Polyomavirus like		8	*Vibrio atlanticus*		8	*Ichthyosporea* sp.				
9	Virus-like particles (VLPs)		9	*Vibrio cholerae*		9	*Marteila refringens*			**Annelida**	
	Metazoa		10	*Vibrio cortegadensis*		10	*Marteila tapetis*		1	*Polydora ciliata*	
1	*Cercaria elegans*		11	*Vibrio diabolicus*		11	*Marteilia* sp.				
2	*Cercaria harengulae*		12	*Vibrio fluvialis*		12	*Marteilioides* sp.				
3	*Cercaria pectinate*		13	*Vibrio gallaecicus*		13	*Mikrocytos*-like				
4	*Cercaria tapes*		14	*Vibrio harveyi*		14	*Minchinia tapetis*				
5	*Cercaria tapetis*		15	*Vibrio logei*		15	*Nematopsis* sp.				
6	*Cercaria tapidis*		16	*Vibrio mediterranei*		16	*Perkinsus chesapeaki*				
7	Cestode		17	*Vibrio parahaemolyticus*		17	*Perkinsus honshuensis*				
8	*Derogenes varicus*		18	*Vibrio* sp.		18	*Perkinsus olseni*				
9	Graffillidae-like Turbellaria		19	*Vibrio splendidus*		19	*Sphenophyra*-like				
10	*Parvatrema duboisi*		20	*Vibrio tapetis*		20	*Thraustochytriidae* sp.				
11	*Proctoeces orientalis*		21	*Vibrio tubiashii*		21	*Trichodina* sp.				
12	*Proctoeces* sp.		22	*Vibrio vulnificus*		22	*Urosporidium* sp.				

FIGURE 13.1

Worldwide distribution of 73 organisms found within *Ruditapes philippinarum*. Color-coded by country. Not all these organisms are known to be pathogenic, but those that are include the protozoan *Perkinsus spp.* and the bacterium *Vibrio spp.*, which are considered further below. Worldwide distribution of some of these species may be more extensive than shown in this figure. *Vibrio* and *Perkinsus* are also reported in North American bivalves, including the Manila clam (Allam et al., 2001).

From Bathige et al. (2025).

stages of progression of the brown discoloration of the shell interior, and Fig. 13.2B shows progression from the extra-pallial compartment and into the soft tissues before death (Paillard, 2004). The internal shell appearance at Stage 7 BRD infection is shown in Fig. 13.3. The brown discoloration adjacent to the shell crack suggests infection through this fault.

Recovery can be achieved by a "nacrezation," process involving the deposit of new shell material over the brown deposit (Paillard, 1992). However, when the *V. tapetis* bacterium overwhelms the defenses of the extrapallial fluids, it enters the soft tissues and hemolymph, where it proliferates to an extent that causes severe damage.

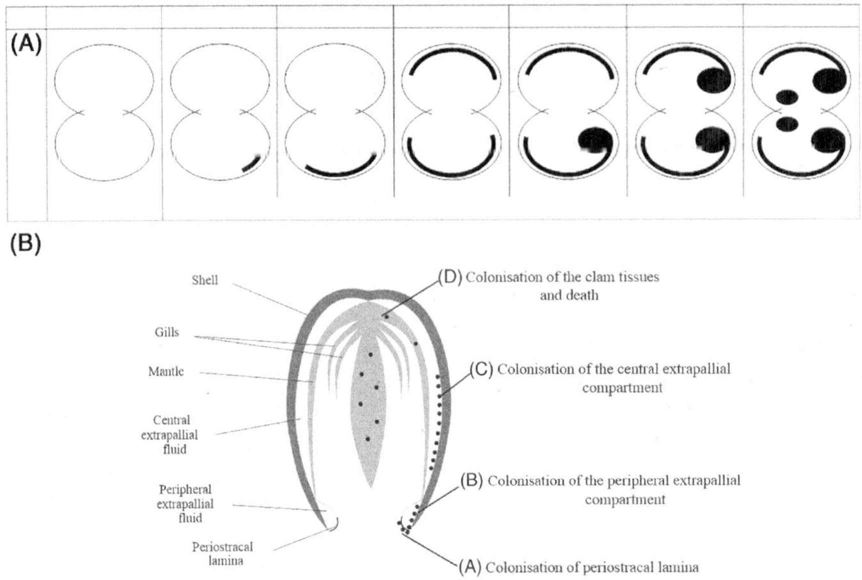

FIGURE 13.2

The progression of brown ring disease in *Ruditapes philippinarum.* (A) The seven stage classification scale of Brown Ring disease (BRD) symptoms established by Paillard and Maes (1994). Symptoms can be observed with the naked eye from Stage 2. (B) Progressive spread of BRD from the shell into the soft tissues, leading to death. After Paillard (2004).

From Flyte-Sainte Marie (2008) with permission and thanks.

The disease imposes significant physiological and energetic stresses, which have been modeled by Flyte-Sainte Marie (2008). As infection progresses, the condition index declines, and mortalities are usually reported in the winter season, suggesting that warmer temperatures may enhance the clams' defense responses (Park et al., 2006). Conversely, disease prevalence is higher in conditions of greater environmental stress, such as low salinity around S = 20, and also appears to vary with sediment type (Reid et al., 2003). The presence of *V. tapetis* in sediments may be one mechanism of transmission, although Mortensen et al., 2007, among others, have suggested the most likely transmission route is by direct contact with infected clams or their feces. Therefore, microbiological monitoring of potential seed clam infection before laying is an important means of disease control.

Paillard et al. (2014) estimated a prevalence in a natural environment in France to be 10%–20% of Manila clams. Allam et al. (2001) have shown that resistance to *V. tapetis* varies with population. North American Manila clam populations in Puget Sound appear more resistant than those in the Bay of Brest, France, and the disease is not widely reported in Canada.

Possible treatments for *Vibrio* infection in hatcheries include the exposure of larval Manila clams to broad-spectrum antimicrobial peptides (Zhang et al., 2025).

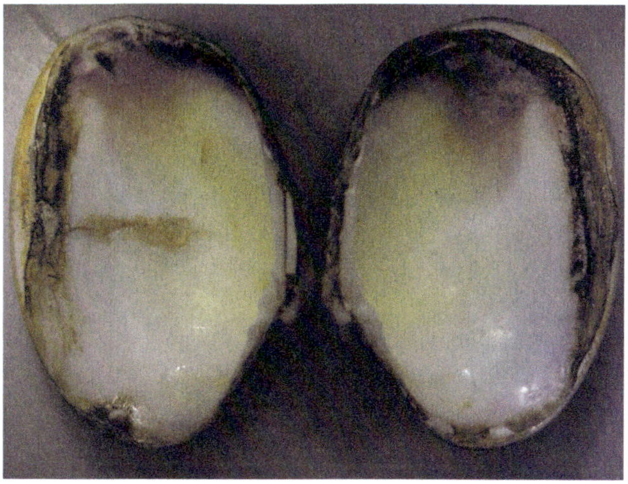

FIGURE 13.3

Symptoms of Stage 7 Brown Ring disease on Manila clam valve interiors. The discoloration adjacent to the left valve fracture suggests an alternative entry point for infection.

From Christine Paillard, with permission and thanks.

13.3 Perkinsosis

Caused by the protist species *Perkinsus spp.*, the first reported occurrence of perkinsosis occurred in oysters in the Gulf of Mexico (Mackin et al., 1950). Subsequently, the genus has been recognized as a pathogen of many Veneroid bivalves, including *R. philippinarum*. In *R. philippinarum, P. olseni* was first reported in European populations in 1988 In Korea, *Perkinsus* infections in Gomso Bay were initially reported by Choi and Park (1997) and confirmed as *P. olseni* using molecular markers, morphology, and sporulation patterns (Park et al., 2005). It is now widespread on the west coast of South Korea (Park & Choi, 2001), in Japan (Choi et al., 2002), and in China (Wu et al., 2011). Subsequently, *P. olseni* has been recognized as a pathogen of many Veneroida bivalves, including *R. philippinarum* in East Asia and Europe.

Perkinsosis has been catalogued by the World Health Organization for animal health as a notifiable disease, and international measures have been established to control it. In the United States, the risk of importation of *Perkinsus* infection with Manila clams imported from East Asia has been recognized and mitigated (Elston et al., 2004), and the United States Pacific coasts have been declared free of *P. olseni* (2025). The pathogen is also thought to be absent from Canadian coasts.

Infection is transmitted via free-swimming zoospores. Heavy infection prevents gonad maturation and damages the gill and mantle tissues, reducing the filtration activity. A key symptom of *Perkinsus* infection is visible white cysts on the gills,

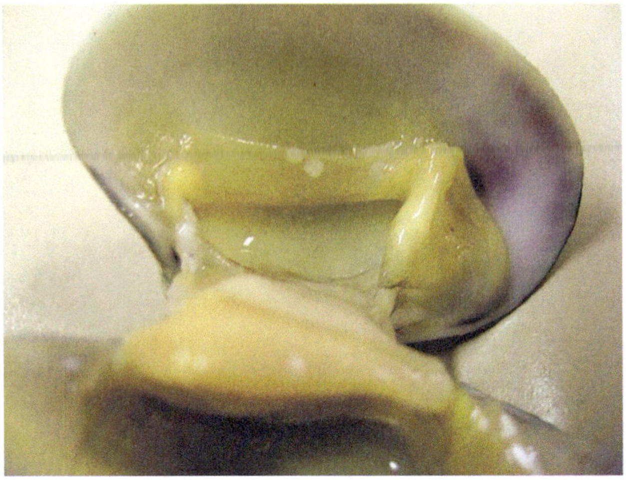

FIGURE 13.4

Symptoms of *Perkinsus* infection. Pale, round cysts on soft tissues are characteristic of *Perkinsus olseni* infection.

From Bruno Chollet and Isabelle Arzul, with permission and thanks.

mantle, or digestive gland (Fig. 13.4), and clams may be found gaping on the sediment surface when infected, presumably due to resulting oxygen deficiency as a consequence of decreased gas exchange (Nam et al., 2018). In Japan, co-infections of *P. olseni* and *Perkinsus honshuensis* have been shown to be especially lethal to smaller Manila clams of length 3–15 mm (Waki et al., 2012). In Arcachon Bay, France, *P. olseni* infection was found to be very low in a low-salinity area (Binias et al., 2014).

P. olseni has been implicated in many mass mortality events in East Asia and Europe to an extent that threatens commercial Manila clam production. For example, in the Venice Lagoon, Italy, a severe mass mortality event was associated with severe *P. olseni* infection (Pretto et al., 2014). Moreover, the prevalence of *Perkinsus* disease increases with water temperature, raising concerns about the implications of climate change, including population declines and mass mortality risks (Rowley et al., 2014).

There are no practical treatments for *Perkinsus* infection, and attention is therefore focused on quarantine measures along with the potential of selective breeding for greater resistance (Smits et al., 2020).

References

Allam, B., Ashton-Alcox, K. A., & Ford, S. E. (2001). Haemocyte parameters associated with resistance to brown ring disease in *Ruditapes spp.* clams. *Developmental and Comparative Immunology, 25*(5–6), 365–375. https://doi.org/10.1016/S0145-305X(00)00072-0.

Allam, B., & Paillard, C. (1998). Defense factors in clam extrapallial fluids. *Diseases of Aquatic Organisms, 33*(2), 123–128. https://doi.org/10.3354/dao033123.

Bathige, S. D. N. K., Kim, S.-H., Jeon, H.-B., Kim, H. J., & Park, K.-I. (2025). Parasitic fauna of the Manila clam (*Ruditapes philippinarum*): Insights, threats, and control measures – A review. *Estuarine, Coastal and Shelf Science, 319*, 109263. https://doi.org/10.1016/j.ecss.2025.109263.

Binias, C., Do, V. T., Jude-Lemeilleur, F., Plus, M., Froidefond, J.-M., & Montaudouin, X. (2014). Environmental factors contributing to the development of brown muscle disease and *Perkinsosis* in Manila clams (*Ruditapes philippinarum*) and trematodiasis in cockles (*Cerastoderma edule*) of Arcachon Bay. *Marine Ecology, 35*(s1), 67–77. https://doi.org/10.1111/maec.12087.

Bonham, V., Humphreys, J., Yang, N., Haiying, O., & Caiping, Y. 2018 *Ruditapes philippinarum* (Japanese carpet shell) CABI https://doi.org/10.1079/cabicompendium.61697.

Borrego, J. J., Castro, D., Luque, A., Paillard, C., Maes, P., Garcia, M. T., & Ventosa, A. (1996). *Vibrio tapetis sp.* nov., the causative agent of the brown ring disease affecting cultured clams. *International Journal of Systematic Bacteriology, 46*(2), 480–484. https://doi.org/10.1099/00207713-46-2-480.

Choi, K. S., & Park, K.-I. (1997). Report on the occurrence of *Perkinsus sp.* in the Manila clams, *Ruditapes philippinarum*, in Korea. *Journal of Aquaculture, 10*, 227–237.

Choi, K. S., Park, K. I., Lee, K. W., & Matsuoka, K. (2002). Infection intensity, prevalence, and histopathology of *Perkinsus sp.* in the Manila clam, *Ruditapes philippinarum*, in Isahaya Bay, Japan. *Journal of Shellfish Research, 21*(1), 119–125.

Chu, F. L. E. (1988). Humoral defense factors in marine bivalves. *American Fisheries Society, 18*, 178–188.

Dimijian, G. G. (1999). Pathogens and parasites: Insights from evolutionary biology. *Baylor University Medical Center Proceedings, 12*(3), 175–187. https://doi.org/10.1080/08998280.1999.11930169.

Elston, R. A., Dungan, C. F., Meyers, T. R., & Reece, K. S. (2004). *Perkinsus sp.* infection risk for Manila clams, *Venerupis philippinarum* (A. Adams and Reeve, 1850) on the Pacific coast of north and central America. *Journal of Shellfish Research, 23*(1), 101–105.

Flyte-Sainte Marie, J. G. (2008). *Ecophysiology of brown ring disease in the Manila clam* Ruditapes philippinarum*, experimental and modelling approaches*. University of Amsterdam.

Gao, S., Li, D., Zha, E., Zhou, T., Wang, S., & Yue, X. (2015). Surveillance of Hepatitis E virus contamination in shellfish in China. *International Journal of Environmental Research and Public Health, 12*(2), 2026–2036. https://doi.org/10.3390/ijerph120202026.

Jiang, K., Nie, H., Li, D., & Yan, X. (2020). New insights into the Manila clam and PAMPs interaction based on RNA-seq analysis of clam through in vitro challenges with LPS, PGN, and poly(I:C). *BMC Genomics, 21*(1). https://doi.org/10.1186/s12864-020-06914-2.

Kim, S.-H., Zhao, B., Bathige, S. D. N. K., & Park, K.-I. (2025). Comparative study of structural features and parasite–parasite interactions between MCPX, a novel pathogen identified from the Manila clam *Ruditapes philippinarum*, and *Perkinsus olseni*. *Estuarine, Coastal and Shelf Science, 312*, 109062. https://doi.org/10.1016/j.ecss.2024.109062.

Lee, H.-M., Kang, H.-S., Kajino, N., Hong, H.-K., Cho, Y.-G., & Choi, K.-S. (2025). First report of mass mortality event caused by Eomarteilia granula in Manila clam *Ruditapes philippinarum* reared in subtidal and bottom cages on the south coast of Korea. *Estuarine, Coastal and Shelf Science, 323*, 109408. https://doi.org/10.1016/j.ecss.2025.109408.

Mackin, J. G., Owen, H. M., & Collier, A. (1950). Preliminary note on the occurrence of a new protistan parasite, *Dermocystidium marinum* n. *sp.*, in *Crassostrea virginica* (Gemlin). *Science, 111*, 328–329.

Mortensen, S., Arzul, I., Miossec, L., Paillard, C., Feist, S., Stentiford, G., ... Gregory, A. (2007). Molluscs and crustaceans, 5. 3. 7 brown ring disease (infections with *Vibrio tapetis*). In R. Raynard, T. Wahli, I. Vatsos, & S. Mortensen (Eds.), *Review of disease interactions and pathogen exchange between farmed and wild finfish and shellfish in Europe* (pp. 341–347). Oslo: VESO on behalf of DIPNET.

Nam, K.-W., Jeung, H.-D., Song, J.-H., Park, K.-H., Choi, K.-S., & Park, K.-I. (2018). High parasite burden increases the surfacing and mortality of the Manila clam (*Ruditapes philippinarum*) in intertidal sandy mudflats on the west coast of Korea during hot summer. *Parasites & Vectors, 11*(1). https://doi.org/10.1186/s13071-018-2620-3.

Paillard, C. (1992). *Etiology and characterisation of brown ring disease in breeding Ruditapes philippinarum clams. (Étiologie et caractérisation de la maladie de l'anneau brun chez la palourde d'élevage Ruditapes philippinarum)*. Université de Brest.

Paillard, C. (2004). A short-review of brown ring disease, a vibriosis affecting clams, *Ruditapes philippinarum* and *Ruditapes decussatus*. *Aquatic Living Resources, 17*(4), 467–475. https://doi.org/10.1051/alr:2004053.

Paillard, C., Jean, F., Ford, S. E., Powell, E. N., Klinck, J. M., Hofmann, E. E., & Flye-Sainte-Marie, J. (2014). A theoretical individual-based model of brown ring disease in Manila clams, *Venerupis philippinarum*. *Journal of Sea Research, 91*, 15–34. https://doi.org/10.1016/j.seares.2014.03.005.

Paillard, C., & Maes, P. (1994). Brown ring disease in the Manila clam *Ruditapes philippinarum*: Establishment of a classification system. *Diseases of Aquatic Organisms, 19*(2), 137–146. https://doi.org/10.3354/dao019137.

Paillard, C., & Maes, P. (1995a). The brown ring disease in the Manila clam *Ruditapes philippinarum*, I: Ultrastructural alterations of the periostracal lamina. *Journal of Invertebrate Pathology, 65*, 91–100. https://doi.org/10.1006/jipa.1995.1015.

Paillard, C., & Maes, P. (1995b). The brown ring disease in the Manila clam *Ruditapes philippinarum*, II: Microscopic study of the brown ring syndrome. *Journal of Invertebrate Pathology, 65*, 101–110. https://doi.org/10.1006/jipa.1995.1016.

Paillard, C., & Maes, P. (1990). The brown ring disease in *Tapes philippinarum*: Adherence of the *Vibrio* P1 strain to the periostracum. In A. Figueras (Ed.), *Abstracts, fourth international colloquium on pathology in marine aquaculture, 17-21 Sept. 1990* (pp. 20–21). Instituto de Investigaciones Marinas - CSIC, Vigo.

Park, K.-I., & Choi, K. S. (2001). Spatial distribution and infection intensity of the protozoan parasite *Perkinsus sp.* in the Manila clam *Ruditapes philippinarum* in Korea. *Aquaculture, 203*, 9–22.

Park, K. I., Paillard, C., Le Chevalier, P., & Choi, K. S. (2006). Report on the occurrence of brown ring disease (BRD) in Manila clam, *Ruditapes philippinarum*, on the west coast of Korea. *Aquaculture, 255*(1-4), 610–613. https://doi.org/10.1016/j.aquaculture.2005.12.011.

Park, K. I., Park, J. K., Lee, J., & Choi, K. S. (2005). Use of molecular markers for species identification of Korean *Perkinsus sp.* isolated from Manila clams *Ruditapes philippinarum*. *Diseases of Aquatic Organisms, 66*(3), 255–263. https://doi.org/10.3354/dao066255.

Pretto, T., Zambon, M., Civettini, M., Caburlotto, G., Boffo, L., Rossetti, E., & Arcangeli, G. (2014). Massive mortality in Manila clams (*Ruditapes philippinarum*) farmed in the lagoon of Venice, caused by *Perkinsus olseni*. *Bulletin of the European Association of Fish Pathologists, 34*(2), 43–53.

Reid, H. I., Soudant, P., Lambert, C., Paillard, C., & Birkbeck, T. H. (2003). Salinity effects on immune parameters of *Ruditapes philippinarum* challenged with *Vibrio tapetis*. *Diseases of Aquatic Organisms, 56*(3), 249–258. https://doi.org/10.3354/dao056249.

Rowley, A. F., Cross, M. E., Culloty, S. C., Lynch, S. A., Mackenzie, C. L., Morgan, E., O'Riordan, R. M., Robins, P. E., Smith, A. L., Thrupp, T. J., Vogan, C. L., Wootton, E. C., & Malham, S. K. (2014). The potential impact of climate change on the infectious diseases of commercially important shellfish populations in the Irish Sea - A review. *ICES Journal of Marine Science, 71*(4), 741–759. https://doi.org/10.1093/icesjms/fst234.

Smits, M., Enez, F., Ferraresso, S., Rovere, G. D., Vetois, E., Auvray, J.-F., Genestout, L., Mahla, R., Arcangeli, G., Paillard, C., Haffray, P., & Bargelloni, L. (2020). Potential for genetic improvement of resistance to *Perkinsus olseni* in the Manila Clam, *Ruditapes philippinarum*, using DNA parentage assignment and mass spawning. *Frontiers in Veterinary Science, 7*. https://doi.org/10.3389/fvets.2020.579840.

Tamaoki, M., Yamada, K., Tomiyama, T., & Miyazaki,, K. (2025). Development and characterization of novel microsatellite markers in the Manila clam parasitic arthropods, *Nymphonella tapetis* (*Pycnogonida, Ascorhynchidae*), using next-generation sequencing. *Estuarine, Coastal and Shelf Science, 318*(286), 109230. https://doi.org/10.1016/j.ecss.2025.109230.

O.I.E., 2019. Manual de las pruebas diagnostico para los animales acuaticos. *Perkinsus olseni*. Retrieved 13.10.2025 from: https://www.oie.int/es/que-hacemos/normas/codigos-y-manuales/acceso-en-linea-al-manual-acuatico/?id=169&L=1&htmfile=chapitre_perkinsus_olseni.htm.

Waki, T., Shimokawa, J., Watanabe, S., Yoshinaga, T., & Ogawa, K. (2012). Experimental challenges of wild Manila clams with *Perkinsus* species isolated from naturally infected wild Manila clams. *Journal of Invertebrate Pathology, 111*(1), 50–55. https://doi.org/10.1016/j.jip.2012.05.009.

Wu, S., Wang, C., Lin, X., Wang, Z., Li, X., Liu, J., Deng, J., & Qiu, S. (2011). Infection prevalence and phylogenetic analysis of *Perkinsus olseni* in *Ruditapes philippinarum* from East China. *Diseases of Aquatic Organisms, 96*(1), 55–60. https://doi.org/10.3354/dao02353.

Yang, H.-S., Cho, Y.-G., Shin, J.-S., Park, H.-S., & Choi, K.-S. (2021). Pathological survey of the Manila clam *Ruditapes philippinarum* from the Hwangdo tidal flat in Cheonsu Bay on the west coast of Korea. *Ocean Polar Research, 43*, 365–370. https://doi.org/10.4217/OPR.2021.43.4.365.

Zhang, Y., Liao, K., Zhu, Y., Liu, Y., Huang, H., Zhou, H., Chen, D., Ma, B., Ge, H., Zhang, W., & Xu, J. (2025). Effects of microencapsulated antimicrobial peptides on the growth, immunity, infection resistance, gut microbiota, and gene transcriptome in Manila clam (*Ruditapes philippinarum*) larvae. *Fish & Shellfish Immunology, 165*. https://doi.org/10.1016/j.fsi.2025.110542.

USDA. (2025) Retrieved September 02, 2025, from https://www.aphis.usda.gov/sites/default/files/p-olseni-country-freedom-2024-final-report.pdf.

CHAPTER

Population dynamics, declines and crashes

14

14.1 Populations and stocks

A population is a group of individual organisms of the same species occupying a particular space at a particular time. This concept is underpinned by the notion that the individuals in a population can at least potentially interbreed, although in practice opportunities for interbreeding are often more constrained than the population as a whole, and limited to spatially proximal subsets. In many cases, the idea of a population corresponds approximately to the concept of a "stock" in fisheries management, and the economic importance of the Manila clam has therefore ensured that the attributes of its populations have received research attention.

In practice, population size is normally estimated by the more achievable number of individuals per unit area, rather than from a comprehensive and typically implausible count. Population size varies in response to four primary population parameters: natality, mortality, immigration, and emigration. How these change over time is the subject of population dynamics. For benthic species, natality (birth rate) is usually replaced by the number of larvae settling in an area (recruitment), with the consequence that the focus of population dynamics is, in practice, on post-metamorphosis settlers, thereby to some extent obscuring the difference between natality and immigration and between populations and metapopulations.

The concept of a metapopulation requires a group of spatially separate populations that can interact because of individuals moving between them (Hanski & Gilpin, 1991). It originated in the 1960s in the context of habitat fragmentation and the viability over time of ostensibly isolated populations. For benthic animals such as the Manila clam, whose dispersal is achieved through pelagic larvae, the idea of mutually supportive but spatially discrete metapopulations is theoretically highly relevant. Larval dispersal studies such as those described in Chapters 10 and 20 suggest metapopulation relations, although in practice such metapopulations can be difficult to identify. Nevertheless, Toba et al. (2007) were able to suggest metapopulation relations across ostensibly local populations in Tokyo Bay by finding larval cohorts during periods when spawning adults were not locally present.

In any event, for the Manila clam among many other benthic species, patterns of recruitment and mortality over time are the main preoccupations of population dynamics. Which in turn requires the analysis of the age structures within populations.

The Manila Clam *Ruditapes philippinarum*: Science and Significance of a Globalised Species.
DOI: https://doi.org/10.1016/B978-0-443-36704-5.00025-0
Copyright © 2026 Elsevier Inc. All rights are reserved, including those for text and data mining, AI training, and similar technologies.

14.2 Length at age

More pronounced concentric rings on the shell are assumed to reflect the cessation or deceleration of growth in the winter season and can therefore be used to estimate age for the purposes of locating individuals into year classes (see Chapter 5). Measurements of these ring dimensions along the long axis can be used to plot mean length at age and therefore growth rates. Von Bertalanffy' growth functions can then be fitted to describe the growth curve mathematically. The growth function assumes a maximum achievable theoretical length (asymptote), and the rate at which this is achieved is represented by a growth coefficient. Fig. 14.1 shows von Bertalanffy growth functions derived from various geographical populations. The solid line is derived from the population in Poole Harbour, UK indicating an asymptotic length of around 40 mm, which in this case corresponded to the minimum fishery landing size at that time.

Growth ring length at age data can be checked against length-frequency distributions as described below. Together, these approaches can provide reliable information relevant to several significant population parameters. For example, patterns of natural and fishery mortality rates with age and, when combined with other data such as mass/length ratios, significant ecological estimates such as total secondary production. Using such approaches, Humphreys et al. (2007) determined that the UK Poole Harbour fishery was sustainable, taking around 75% of Manila clams above minimum landing size, which, when all sizes were considered, represented only 20% of annual production. Moreover, using such information in conjunction with predation modeling, Caldow et al. (2007) were able to achieve quantitative estimates of the benefit to protected shorebirds of this non-native species as a food source, as is discussed in Chapter 21.

14.3 Length–frequency distributions

Fig. 14.2 shows size-frequency distributions for a Japanese Manila clam population over time. With the help of software-based statistical methods (e.g., Macdonald & Green, 1988), each year class (cohort) can be identified from the multimodal nature of the curves and the histograms from which they are derived. This Japanese population consists of more or less five cohorts, with the oldest (C1) reaching a shell length of between 40 and 50 mm. Although the growth rate varies with the fertility of particular estuaries, this growth over 5 years is broadly consistent with that found in many parts of the world. In exploited populations, maximum length is attenuated by fishing, which, in effectively regulated fisheries, may remove most individuals above the minimum landing size, but not below.

Through repeated sampling, each cohort is tracked over time in terms of both abundance and shell length. Such histograms rarely follow an entirely logical progression. For example, the most persistently abundant cohort C3 appears numerically greater in December 2005 than it was 2 months earlier. Moreover, the September 2005 histogram is an anomaly in that there is no clear multimodal distribution beyond C5. Although

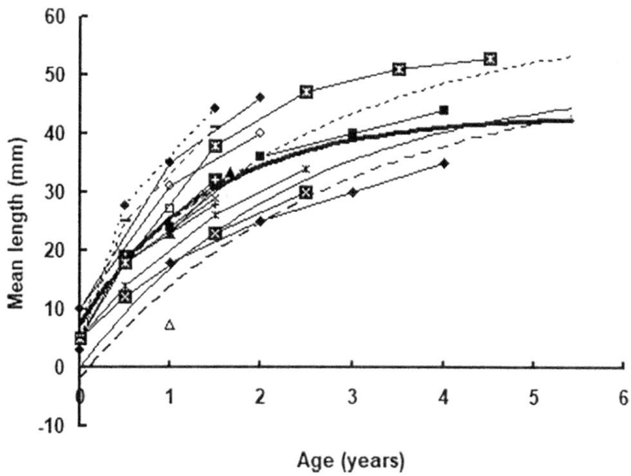

FIGURE 14.1

Length at age graphs for *Ruditapes philippinarum* from various published reports as follows: Ponurovski and Selin (1988) Vostok Bay (long dash); Goshima et al. (1996) Saroma Lagoon, Hokkaido, Japan (thinsolid); Melia et al. (2004) Sacca Di Goro and Arcachon, Le Ferret (solid line and filled dots); Arcachon, LesJacquets (solid line and open dots); Nosho and Chew (1972) Hood Canal, Washington, USA (solid lineand filled squares); Ikematsu (1957) Ariake Bay, Japan (solid line and open squares); Ohba (1959) Ariake Bay, Japan (solid line and filled diamonds); Ikematsu (1957) Ariake Bay, Japan (solid line and filled triangles); Matsukawa-Ura lagoon (solidline and open triangles); Toba et al., 1992 North Puget Sound (solidline and crosses); Clam Bay, Central Puget Sound (solid line andasterisks); Filucy Bay, Central Puget Sound (solid line and plus signs); Marrowstone Island, Central Puget Sound (solid line and whitecrosses); Kopachuck State Park, South Puget Sound (solid line andwhite plus signs); Hartstene Island #1, South Puget Sound (solid lineand white asterisks); Hartstene Island #2, South Puget Sound (solid lineand minus signs), Miller et al. (1978) Puget Sound (dashed line and filled dots). The thick solid line is from Humphreys et al. (2007), showing a von Bertalanffy growth function fitted to all Poole Harbour, UK Manila clams of different ages sampled in 13 months, with the thin short-dashed line showing a von Bertalanffy function fitted only to clams less than or equal to 19 months old.

From Humphreys et al. (2007).

theoretically, clam behaviors such as vertical or horizontal migrations may contribute, more likely explanations for such anomalies include random sampling error resulting from estimating population characteristics as densities derived from sub-samples.

Size-frequency distributions provide insights into various aspects of the demographics of a population, especially when (as in Fig. 14.2) both small (retained by a 1 mm mesh sieve) and larger (retained by a 5 mm mesh) clams have been sampled to give a good representation of the population as a whole. The data reveals a wild

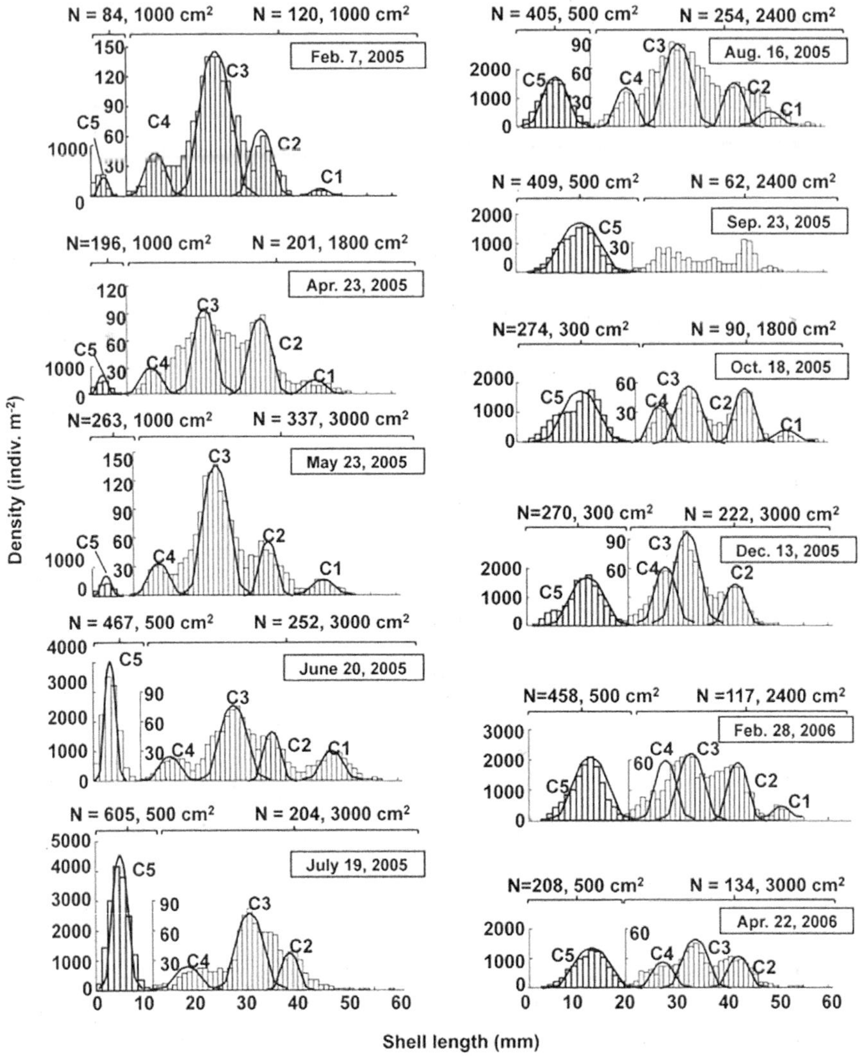

FIGURE 14.2

Size–frequency distribution of a *Ruditapes philippinarum* population in Hichirippu Lagoon, Hokkaido, Japan. The left and right scales on each histogram correspond to samples obtained using sieve mesh sizes of 1 and 5 mm respectively (C = cohorts, n – number of individuals in the sample).

From Komorita et al. (2014).

indigenous population with low-level recruitment from February to May 2005, which greatly increased in June and July, suggesting a main spring spawning period leading to a peak density of C5 settlers. This cohort can be seen to grow significantly in length until about September, when it enters the cold season and cessation of growth.

Despite the good range of sizes captured in this study, the use of the 1 mm mesh means that smaller C5 settlers will have been missed after settlement. In any event, the surviving C5 cohort is shown to suffer a steep decline in abundance until August 2005; however, those that remain experienced a much lower subsequent attrition.

Decline in the abundance of the smallest and most vulnerable juvenile clams may, to a large extent, be due to high susceptibility to predation. However, predation is not the only possible explanation. Kurihara et al. (2025) determined the densities of juvenile (0–2 mm shell length), subadult (5–15 mm), and adult (20–30 mm) Manila clams on an intertidal flat in western Japan. For juvenile and subadult clams, they found a significant negative correlation between wind speed and related wave height fluctuations, with the subsequent abundance of these smaller-sized groups. Since wave heights will have increased the physical disturbance of the sediment, explanations for the decline in abundance likely includes mortality of these smaller individuals due to sediment instability or their dispersal beyond the sampling area.

When settlers survive such disturbance, secondary settlement may occur. Toba et al. (2011) found that the mobilization of juvenile clams could be caused by wave-generated oscillatory turbulence and inshore advection currents. They reported Manila clams moving several hundred meters because of these effects and concluded that physical conditions could account for changing patterns of juvenile abundance (Chapter 10). Therefore, although the density of larval settlers will be a key factor in population dynamics, a range of post-recruitment processes is also important. These include seasonal hydrodynamic conditions which can modify abundance and mortality due to a range of effects from mechanical damage to the impact of turbidity on gill filtration function (Fujimoto et al., 1984), and high vulnerability to predation, either while exposed in the boundary layer or expedited by limited ability for repeated reburial (Chapter 6).

14.4 Secondary production

Ecological approaches seeking to understand the trophic contribution of a population within its ecosystem require the estimation of secondary production, i.e., the annual quantity of organic animal matter contributed to the ecosystem. Since the shell is largely inorganic material, it is the organic soft body only that contributes to the food chain. This organic fraction of the animal is expressed as ash-free dry weight (AFDW). By correlating shell length with the AFDW of the soft tissues (i.e., excluding the inorganic shell), a sequence of size-frequency distributions can provide a basis for estimates of secondary production. This can be achieved by the increment summation method originated by Crisp (1984), which involves summing the monthly AFDW growth increments of all cohorts across the full year. Using this general approach,

Humphreys et al. (2007) calculated that the annual production of a wild population in Poole Harbour, UK, in the absence of fishing would be 8594.96 mg AFDM m^{-2}. Here, the fishery was estimated to take 20% of annual production by removing 75% of clams on or above the minimum landing size. These estimates, while they capture the character of a viable Manila clam population and sustainable fishery, are smaller than equivalent values further south and in many indigenous areas.

An alternative approach, as used by Komorita et al. (2014), which tracks the movement of carbon through the food chain, estimated the secondary production of the Hokkaido Manila clam population as 130g C m^{-2} $year^{-1}$. From a fisheries standpoint, production is measured as whole animal wet weight, expressed in tonnes (t) of landings per year. The metric ton (t = 1000 kg) is the international standard for trade.

14.5 Fishery decline and collapse

Bivalve populations tend to exhibit high natural volatility due to the vulnerability of annual recruitment success and the susceptibility of mature individuals to irregular but natural variability of conditions in estuaries and coastal seas. Abundances may change gradually over time or suffer mass mortality episodes, but the fecundity of bivalves often means that within a few years, populations recover. Therefore, it should not be a surprise to find periodic declines in Manila clam populations (Burdon et al., 2014; Parada & Molares, 2008; Parada et al., 2012).

However, upon natural variations are superimposed the impacts of anthropogenic disturbances, and although it may be difficult to distinguish between the two, there is good reason to believe that human activity has become a major source of mortality. Consequently both chronic declines and acute mass mortalities of Manila clam populations across the world have been observed, with serious economic and social consequences.

In Japan, there have been two major periods of Manila clam fishery decline. The first was from the 1980s to 1990s, and the second from the mid-2000s. In the 1980s, Japan reported annual production from the Manila clam fishery of up to 160,000t. However, in the 1990s, this estimate fell to 40,000t and reportedly to 4000–5000t by the 2020s. This massive collapse of stocks resulted in harvesting activity ceasing altogether in some areas (Toba, 2017, 2025). In Ise Bay, for example, landings fell to 400t by 2014 and to 1t or less after 2016 (Shiba et al., 2021). In Korea, over the three decades from the 1990s, West Coast Manila clam landings fell from 60,000t to 30,000–40,000t (KOSIS, 2024). China has also seen declines in Manila clam production (Liu et al., 2023; Wu et al., 2011).

Similar declines are reported in Europe. In Galicia, Spain, in 2023, a 95% mortality event was recorded, and in Italy's Venice Lagoon a 75% decrease in annual production has occurred in recent years (Milan et al., 2023): The latter being associated with *Perkinsus* infection (Pretto et al., 2014), (see Chapter 13), and thermal stress linked to heatwaves (Bertolini & Pastres, 2021).

Table 14.1 summarizes causes of decline. Many or all these factors have likely played a part in the decline of Manila clam stocks, but it is clear from the literature that each estuary and bay will have a different combination of responsible circumstances. Moreover, artificially high population densities as a result of aquacultural "enhancement" with seed clams will also almost certainly sometimes play a role through expedited disease transmission and resulting epidemics, feeding competition within the clam population (Liu et al., 2023), and the clam's own contribution to accumulating organic matter.

When environmental monitoring data is compared with known tolerances, mortality may be found to occur even when each individual environmental variable remains within the known tolerance limits. Much evidence now suggests that synergies of challenging variables can result in accumulated aggregate stress, which collectively achieve lethality and account for mortality episodes.

Tsutsumi (2006) has reviewed the evidence relating to a Japanese Manila clam population collapse in an area of Ariake Bay, where annual landings fell from 50,000t in 1980 to around 400t in 1997. A decline was also seen in the macro-benthic community generally. They attributed the decline in the intertidal flats to the reduction of fluvial sediment supply. However, in subtidal areas, a combination of three factors was thought causative: a marked increase in extensive autumn phytoplankton blooms (toxic "red tides"); large areas of hypoxic bottom waters in early summer; large (3500 hectare area) land reclamation projects since 1986; and heavy metal pollution. The absence of stock recovery when harvesting ceased suggested that fishing pressure was not a causal factor in the Manila clam population collapse. Where a general decline in benthic infauna coincides with Manila clam stock declines, classic severe pollution mechanisms may be the cause, sometimes exacerbated by higher temperatures consequent on climate change.

In China, Liu et al. (2023) have investigated temperature-related environmental influences on the survival of Manila clams in Laizhou Bay. Here, clam aquaculture is threatened by high summer mortality, which varies by year. Their analysis revealed that high mortalities were associated with three interactive variables: elevated temperature, hypoxia, and hydrogen sulphide concentration. These were found to cause physiological responses which resulted initially in falls in clam glycogen reserves followed by condition declines (measured as condition index), resulting in a stock reduction of 80%. This mortality occurred in a year of unusually high summer water temperatures (mean 26°C and maximum 31°C). No large-scale mortalities were observed in a year with less severe summer temperatures.

It is of interest that while such extrinsic factors will be familiar in many coastal waters, their combined effects only caused a mass mortality event when combined with high summer temperatures: a phenomenon likely to increase in both frequency and severity under climate change scenarios. These results further support the hypothesis that mass mortality events can result from sublethal levels of multiple individual factors, resulting in an overall lethal degree of physiological and energetic stress.

Hypoxic conditions with high hydrogen sulphide levels are typically caused by pollution such as effluents high in organic particulates or inorganic nutrients.

Table 14.1 Anthropogenic factors considered to contribute to the decline of Manila clam populations.

Category	Causal factor	Direct effect	Secondary effect (examples)
Civil engineering	Seawall construction	Changes in current patterns and velocity	Sediment erosion
	River improvements	Sediment supply	Granular fraction changes, such as increased mud in sediment
	Land reclamation	Habitat loss	Reduced tidal flats and shallows
Water quality degradation	Inorganic pollution	Eutrophication	Hypoxia; Toxic blooms; algal mats
	Pharmaceutical pollutants	Physiological and endocrine disruption	Reproductive interference
Water regulation improvements	Enhanced effluent processing	Reduced organic particulates	Reduced non-living food particulates
	Reduced nutrient release	Oligotrophication	Reduced algal food availability
Disease/predator globalization	*Perkinsus* spp *Vibrio* spp	Perkinsus disease Brown Ring Disease (BRD)	Physiological stress and reproductive impairment
	Moonsnail/ Blue crab arrivals etc	Excessive predation	Population decline, or decimation
Fishing effort	Improved mechanization	Overfishing	Reduced density-dependent fitness

(continued on next page)

Aquaculture practices	Hatchery seed laying	High population density	Intraspecific competition and epidemics
	Genetic homogenization	Reduced fitness in the population	Increased susceptibility to challenges
Climate change	Higher mean temperatures	Increased metabolic rate and energetic demand	Over-reliance on food abundance.
	Increased precipitation	Increases in hyposaline conditions	Sublethal physiological stress
	Heat wave episodes	Prolonged intense thermal challenge	Sublethal or lethal thermal stress
	Severity and regularity of storm events	Increased wave energy impacts	Acute sediment destabilization
	Sea level rise	Hydrodynamic changes and estuarine squeeze	Large-scale distribution change and/or reduction

From Toba et al. (2016).

The latter can lead to plankton blooms followed by high oxygen demand through aerobic decomposition of excess particulate matter, including expired phytoplankton. The resulting anoxic conditions favor sulphate-reducing bacteria and the production of toxic hydrogen sulphide. The solution to this is to reduce pollution. Ironically, however, measures to reduce pollution can themselves be complicit in Manila clam decline and mortality events, as studies in Tokyo Bay and elsewhere have elucidated.

In SW Japan, including Tokyo Bay, Manila clams have two spawning peaks per year, in spring and autumn. Various studies have shown that it is the autumn spawning that primarily contributes to stock formation (Ikematsu, 1957; Toba et al., 2025; Tsutsumi et al., 2018). On the Banzu tidal flat within Tokyo Bay, the autumn settler cohort has higher densities and lower post-recruitment mortality than the spring recruitment cohort. However, Toba et al. (2025) also found this important autumn cohort in particular to exhibit a size-dependent mortality involving a combination of environmental factors, including temperature, salinity, and predation, acting synergistically on sexually mature clams (over 20 mm shell length).

Nutrient levels in Japanese estuaries and bays have significantly decreased since the 1990s due to the implementation of advanced wastewater schemes (Kubo et al., 2018). Pollution reduction measures have had a number of beneficial effects, such as reducing the frequency of toxic "red tides" and hypoxic water masses (Ando et al., 2021). However, they can also shift prevailing inorganic nutrient conditions from eutrophic to oligotrophic, with consequently reduced primary production. From in situ growth experiments in the range of conditions found along a nutrient gradient (eutrophic to oligotrophic) in Osaka Bay, Fujiwara et al. (2020) determined that the condition index of the Manila clams was strongly correlated with primary production, which was in turn significantly correlated with nutrient levels.

Noting a 90% drop in Manila clam catch between 2013 and 2017 in Isle and Mikawa Bays, Kamohara et al. (2021) examined the relationship between Manila clam catches and growth (including gonad weight) with concentrations of phytoplankton (as Chlorophyll a), total nitrogen, and total phosphorus. From 1998 to 2006, when these inorganic nutrient levels remained high, clam growth (and gonad weight) remained stable, whereas between 2007 and 2017, gonad weight was lower. They suggested that the higher nutrient concentrations, especially between June and September of each year, were needed to maintain phytoplankton levels and clam fishery resources.

Drawing together such evidence with their own work based on long-term observations at Banzu tidal flat, Toba et al. (2025) have concluded that the Manila clam decline is attributable to mortality associated with spawning by mature clams in a late summer environment that combines high water temperature with oligotrophic nutrient conditions.

It has been established that *R. philippinarum* clams are highly effective in terms of growth and reproduction, and direct much of their available energy to these activities. Moreover, it is also known that the response of metabolic rate to temperature is uncontrolled (Chapter 8). Toba et al. (2025) postulate that the decline in Tokyo Bay Manila clam stocks is associated with mortality due to compromised energy equilibrium in hot autumn temperatures due to the unregulated energy demand of spawning under oligotrophic conditions and consequent relative food shortage. Should this combination of factors not trigger mortality events directly, they would decrease resistance to disease, physiological shock such as salinity variation, and reduce burrowing ability. The latter increasing susceptibility to predation by increased numbers of the black seabream (*Acanthopagrus schlegelii*) in the bay, also consequent on warming waters.

These explanations for stock declines provide more general insights into the global spread of the Manila clam and its particular success in enclosed eutrophic estuaries, bays, and lagoons. They also exemplify how complex and sometimes synergistic interactions of anthropogenic factors, from pollution to climate change, can impact natural coastal systems, including clam populations. This is considered further in Chapter 22 about the particular success of the Manila clam in anthropogenically altered environments such as eutrophic estuaries and lagoons. In any event, as Japan improves its water quality, it is thought unlikely in the future to achieve the levels of aquaculture production of the past (Satoshi Watanabe, personal communication).

References

Ando, H., Maki, H., Kashiwagi, N., & Ishii, Y. (2021). Long-term change in the status of water pollution in Tokyo Bay: Recent trend of increasing bottom-water dissolved oxygen concentrations. *Journal of Oceanography, 77*(6), 843–858. https://doi.org/10.1007/s10872-021-00612-7.

Bertolini, C., & Pastres, R. (2021). Tolerance landscapes can be used to predict species-specific responses to climate change beyond the marine heatwave concept: Using tolerance landscape models for an ecologically meaningful classification of extreme climate events. *Estuarine, Coastal and Shelf Science, 252*, 107284. https://doi.org/10.1016/j.ecss.2021.107284.

Burdon, D., Callaway, R., Elliott, M., Smith, T., & Wither, A. (2014). Mass mortalities in bivalve populations: A review of the edible cockle Cerastoderma edule (L.). *Estuarine, Coastal and Shelf Science., 150*, 271–280. https://doi.org/10.1016/j.ecss.2014.04.011.

Caldow, R. W. G., Stillman, R. A., Durell, S. E. A. V., West, A. D., McGrorty, S., Goss-Custard, J. D., Wood, P. J., & Humphreys, J. (2007). Benefits to shorebirds from invasion of a non-native shellfish. *Proceedings of the Royal Society B: Biological Sciences, 274*(1616), 1449–1455. https://doi.org/10.1098/rspb.2007.0072.

Crisp, D. J. (1984). Energy flow measurements. In N. A. Holme, & A. D. McIntyre (Eds.), *Methods for the study of marine benthos* (second edition). Blackwell.

Fujiwara, T., Higuchi, K., & Fujii, T. (2020). Minimum nutrient concentrations for sustaining marine ecosystem productivity: Measurement during in situ growth experiments of bivalve and sessile organisms. *J, 43*, 175–182. ournal of Japan Society on Water Environment https://doi.org/10.2965/jswe.43.175 .

Goshima, S., Ide, N., Fujiyoshi, Y., Noda, T., & Nakao, S. (1996). Reproductive cycle and shell growth of transplanted Manila Clam *Ruditapes philippinarum* in Saroma Lagoon. *Nippon Suisan Gakkaishi, 62*(2), 195–200. https://doi.org/10.2331/suisan.62.195.

Hanski, I., & Gilpin, M. (1991). Metapopulation dynamics: Brief history and conceptual domain. *Biological Journal of the Linnean Society, 42*(1-2), 3–16. https://doi.org/10.1111/j.1095-8312.1991.tb00548.x.

Humphreys, J., Caldow, R. W. G., Mcgrorty, S., West, A. D., & Jensen, A. C. (2007). Population dynamics of naturalised Manila clams *Ruditapes philippinarum* in British coastal waters. *Marine Biology, 151*(6), 2255–2270. https://doi.org/10.1007/s00227-007-0660-x.

Ikematsu, W. (1957). Ecological studies on the clam, Tapes japónica (Reeve)—II. On the setting season and the growth in the early young stage. *Nippon Suisan Gakkaishi, 22*(12), 736–741. https://doi.org/10.2331/suisan.22.736.

Kamohara, S., Shiba, S., Tsurushima, D., & Suzuki, T. (2021). Relationship of Manila clam (*Ruditapes philippinarum*) growth with aged changes of total nitrogen and total phosphorus in Mikawa Bay, Japan. *Bulletin of the Japanese Society of Fisheries Oceanography, 85*, 69–78.

Fujimoto, T., Shitani, M., & Oda, S. (1984). Factors necessary for persistence of fishery ground of asari clam. Bulletin of Fukuoka Fisheries and Marine Technology Research Center, 1982, 34–105.

KOSIS 2024 Korean Statistical Information Service. Retrived from http://kosis.kr. (Accessed 13.10.2025).

Komorita, T., Kajihara, R., Tsutsumi, H., Shibanuma, S., Yamada, T., & Montani, S. (2014). Food sources for *Ruditapes philippinarum* in a coastal lagoon determined by mass balance and stable isotope approaches. *PLoS One, 9*(1), e86732. https://doi.org/10.1371/journal.pone.0086732.

Kubo, A., Hashihama, F., Kanda, J., Horimoto-Miyazaki, N., & Ishimaru, T. (2018). Long-term variability of nutrient and dissolved organic matter concentrations in Tokyo Bay between 1989 and 2015. *Limnology and Oceanography, 64*(1), S209–S222. https://doi.org/10.1002/lno.10796.

Kurihara, T., Nakano, T., Kuroki, Y., Uchikawa, J., Ikushima, N., & Nasu, H. (2025). Year-to-year density fluctuations in clams *Ruditapes philippinarum* linked to wind-induced disturbance during juvenile to subadult stages. *Marine Ecology Progress Series, 758*, 43–60. https://doi.org/10.3354/meps14818.

Liu, Y., Ren, J. S., Wang, X., Wu, W., & Zhang, J. (2023). Environmental influence on summer survival of Manila Clam *Ruditapes philippinarum*: A case study in an aquaculture bay. *Marine Environmental Research, 192*, 106242. https://doi.org/10.1016/j.marenvres.2023.106242.

Macdonald, P. D. M., Green, P. E. J. (1988) User's Guide to Program MIX: An interactive program for fitting mixtures of distributions. Guenther Printing Hamilton, ON. Release 2.3. Ichthus Data Systems.

Melia, P., De Leo, G. A., & Gatto, M. (2004). Density and temperature dependence of vital rates in the Manila clam *Tapes philippinarum*: A stochastic demographic model. *Marine Ecology Progress Series, 272*, 153–164.

Milan, M., Bernardini, I., Bertolini, C., Rovere, G. D., Manuzzi, A., Pastres, R., Peruzza, L., Smits, M., Fabrello, J., Breggion, C., Sambo, A., Boffo, L., Gallocchio, L., Carrer, C., Sorrentino, F., Bettiol, C., Lodi, G. C., Semenzin, E., Varagnolo, M., Matozzo, V., Bargelloni, L., & Patarnello, T. (2023). Multidisciplinary long-term survey of Manila clam grown in farming sites subjected to different environmental conditions. *Science of The Total Environment, 863*, 160796. https://doi.org/10.1016/j.scitotenv.2022.160796.

Miller, M. B., Chew, K., Jones, C. R., Goodwin, L., & Magoon, C. D. (1978). *Manila clam seeding as an approach to clam population enhancement*. Washington Sea Publisher, University of Washington.

Nosho, T. Y., & Chew, K. K. (1972). The setting and growth of the Manila clam, Venerupis japonica (Deshayes), in Hood Canal, Washington. *Proceedings of the National Shellfisheries Association, 62*, 50–58.

Parada, J., & Molares, J. (2008). Natural mortality of the cockle Cerastoderma edule (L.) from the Ria of Arousa (NW Spain) intertidal zone. *Journal of Marine Biology and Oceanography, 43*(3). https://doi.org/10.4067/S0718-19572008000300009.

Parada, J. M., Molares, J., & Otero, X. (2012). Multispecies mortality patterns of commercial bivalves in relation to estuarine salinity fluctuation. *Estuaries and Coasts, 35*(1), 132–142. https://doi.org/10.1007/s12237-011-9426-2.

Ponurovskii, S. K., & Selin, N. I. (1988). Distribution, population structure, and growth of the bivalve mollusk *Ruditapes philippinarum* in Vostok Bay, Sea of Japan. *Soviet Journal of Marine Biology, 1*, 11–15.

Pretto, T., Zambon, M., Civettini, M., Caburlotto, G., Boffo, L., Rossetti, E., & Arcangeli, G. (2014). Massive mortality in Manila clams (*Ruditapes philippinarum*) farmed in the Lagoon of Venice, caused by *Perkinsus olseni*. *Bulletin of the European Association of Fish Pathologists, 34*(2), 43–53.

Shiba, S., Yoshida, T., Tsurushima, D., Kamohara, S., Tanaka, Y., & Suzuki, T. (2021). A comparative examination of individual growth models, as the method to assess the food environment in relation to the resource level of Manila clam *Ruditapes philippinarum*. *Journal of Advanced Marine Science and Technology Society, 26*(2), 1–12.

Toba, M., Kobayashi, Y., Kakino, J., Yamakawa, H., Ishii, R., & Okamoto, R. (2016). Stocks and fisheries of asari in Japan. *Bulletin of the Japan Fisheries Research and Education Agency., 42*(9-21), 2016.

Toba, M., Yamakawa, H., Kobayashi, Y., Sugiura, Y., Honma, K., & Yamada, H. (2007). Observations on the maintenance mechanisms of metapopulations, with special reference to the early reproductive process of the Manila clam *Ruditapes philippinarum* (Adams & Reeve) in Tokyo Bay. *Journal of Shellfish Research., 26*, 121–130.

Toba, D. R., Thompson, D. S., Chew, K. K., Anderson, G. J., & Miller, M. B. (1992). *Guide to Manila clam culture in Washington*. Washington Sea Publisher, University of Washington.

Toba, M. (2017). Revisiting recent decades of conflicting discussions on the decrease of Asari clam *Ruditapes philippinarum* in Japan: A review. *Nippon Suisan Gakkaishi, 83*(6), 914–941. https://doi.org/10.2331/suisan.17-00007.

Toba, M., Ito, M., & Kobayashi, Y. (2011). Bedload transport of newly-settled juveniles of the Manila clam *Ruditapes philippinarum* observed in situ at Banzu tidal flat, Tokyo Bay. *Journal of Shellfish Research, 30*(3), 777–789. https://doi.org/10.2983/035.030.0318.

Toba, M., Kobayashi, Y., Hayashi, T., & Kagami, Y. (2025). Possible processes responsible for the decline in the stock of Manila clam based on long-term observations in Banzu tidal flat. *Estuarine, Coastal and Shelf Science, 318*, 109232. https://doi.org/10.1016/j.ecss.2025.109232.

Tsutsumi, H. (2006). Critical events in the Ariake Bay ecosystem: Clam population collapse, red tides, and hypoxic bottom water. *Plankton and Benthos Research, 1*(1), 3–25. https://doi.org/10.3800/pbr.1.3.

Tsutsumi, H., Nishioka, T., Kitagawa, N., Fujiyoshi, Y., Kozaki, M., Yamamoto, M., Ichinomiya, M., & Komorita, T. (2018). Seasonal changes in short-neck clam (*Ruditapes philippinarum*) population and its controlling factors on Sashiki Tidal Flats. *Japanese Journal of Benthology, 73*(1), 1–10. https://doi.org/10.5179/benthos.73.1.

Wu, S. Q., Wang, C. X., Lin, X. M., Wang, Z. X., Li, X. F., Liu, J., Deng, J. H., & Qiu, S. Y. (2011). Infection prevalence and phylogenetic analysis of *Perkinsus olseni* in *Ruditapes philippinarum* from East China. *Diseases of Aquatic Organisms, 96*(1), 55–60. https://doi.org/10.3354/dao02353.

CHAPTER

Aspects of the ecological niche 15

15.1 Introduction

The concept of the ecological niche remains fundamental in ecology and is especially relevant in ecological studies where the factors determining distribution and abundance are of interest. The modern formalized concept originates with Hutchinson (1957), who distinguished between the fundamental niche, which represents the full range of environmental conditions under which a species can exist, and the realized niche: the variables within which the species actually does exist. These variables include both physicochemical (Chapter 11) and biological factors such as trophic characteristics, competition, and predation (Chapter 12). It can be the case that while the former are highly influential in determining the fundamental niche, the difference between the fundamental and realized niche of a species may be determined by biological factors that prevent the species from occupying the whole of its fundamental niche. Therefore, while dimensions of the fundamental niche can be elucidated in the laboratory, the realized niche must usually be informed by field research.

On the other hand, using distribution data alone can also jeopardize attempts to discern the niche, especially with species like the Manila clam, which combine broadcast spawning and a pelagic larval stage with a capacity to thrive in volatile habitats such as estuaries. For example, in the field, it is possible to find Manila clams in habitats incompatible even with experimentally determined components of its fundamental niche. Alternatively, a species may be absent from habitats where its presence would be predicted. This is because Hutchinsonian niches are descriptions of the conditions within which a species could exist indefinitely, which in turn requires the fundamental niche dimensions of a habitat to be compatible with the entire life cycle (Pulliam, 2000).

Since each reproductive stage of *Ruditapes philippinarum* has different thermal tolerances (Chapter 10), pelagic larvae can settle and grow at temperatures lower than those required for gonad maturation and gametogenesis. Consequently, the broadcast spawning habit and high fecundity of *R. philippinarum* (and many other bivalves), combined with coastal advection, may result in successful and repeated settlement and growth in habitats beyond the scope of the full life-cycle fundamental niche, resulting in populations that are entirely dependent for recruitment on a habitat elsewhere. Such populations cannot be seen as spatially distinct components of proper

The Manila Clam *Ruditapes philippinarum*: Science and Significance of a Globalised Species.
DOI: https://doi.org/10.1016/B978-0-443-36704-5.00023-7
Copyright © 2026 Elsevier Inc. All rights are reserved, including those for text and data mining, AI training, and similar technologies.

meta-populations, since they cannot contribute to recruitment anywhere. In niche theory, this phenomenon is referred to as a "source-sink" effect.

A second cause of presence outside the niche is the temporal instability of habitats. This requires a change from habitat conditions which fall within the niche to those which do not, sometimes leaving a residual but ultimately unsustainable population to be found in fieldwork. In estuarine and coastal habitats, variables are spatially and temporally dynamic. For example, salinity in estuaries occupied by *R. philippinarum* can vary over quite small distances and in response to river input, rainfall and tidal cycles. Sediment composition can also change through both erosion and deposition in response to flow rates and storm effects.

In these circumstances, a population may temporarily survive outside its true niche variables (time-delayed local extinction). Such phenomena have led Pulliam (2000) to suggest that dispersal may be as important as competition in determining the relationship between niche and distribution. While at a larger scale, distribution may relate to the limits of dispersal, even on smaller scales, benthic species may exhibit complex spatial patterns in uniform environments. Hanski (1989), in the context of metapopulation theory, regards local "extinction" (extirpation) as a recurrent rather than a unique event due to demographic stochasticity and/or environmental instability. In any event, the niche cannot always be reliably determined from habitat variables extant in places where the species can be found. However, places where longitudinal studies have revealed a persistent population exhibiting full reproductive cycles and multimodal demographic structure, as exemplified in Chapter 14, can confidently be concluded as located somewhere within the niche dimensions.

A further difficulty in seeking to establish niche boundaries is the complex effects of multiple interacting variables. Even in single-factor laboratory experiments, causal relationships may be difficult to extract from correlations, and small differences in methods can compromise the comparability of different studies. For example, exact tolerance determinations on clams are sometimes conducted in experimental chambers removed from any sediment matrix, making results, although both physiologically interesting and of direct relevance for commercial husbandry practices and transportation, incompatible with those implied by distribution studies.

Therefore, in practice, the full ecological niche, although theoretically fundamental to ecology, is difficult to elucidate, not least for estuarine and coastal species. On the other hand, many ecological studies investigate single "niche dimensions" such as the "trophic niche." If applied comparatively to sympatric species, such studies can also elucidate "niche partitioning" between species. The concept of niche partitioning addresses the processes by which species utilize resources within a community in such a way that facilitates interspecific coexistence in otherwise more directly competitive situations.

Niche partitioning is a process in which competition is a key driver. Patterns of partitioning may result from initially slight differences in the range of optimal conditions, tolerances, behavior, etc of two species, which enable each to compete most effectively in different dimensions of overlapping ecological niches. Moreover, competitively driven niche partitioning may, through natural selection, lead to character displacement involving more permanent shifts in physiological or anatomical fitness patterns, such as physico-chemical tolerance ranges in related species.

For example, spatial niche partitioning describes the utilization of distinct spaces by two species whose fundamental niche might allow both to occupy identical local distributions. In this sense, the concept of resource partitioning can lie behind the difference between the fundamental and realized niches of sympatric species. Similarly, niche partitioning may be trophic, involving food specialization within more extensive possible food sources. However, where the Manila clam has been anthropogenically introduced as a non-native species, the process of niche partitioning is unlikely to have progressed in the time available, although the selective mechanisms may have commenced. Moreover, for reasons discussed in Chapter 20, the intensity of competition in estuarine soft sediments may, with some exceptions, be quite weak as a driver for niche partitioning or as an explanation for distribution.

The species with which *R. philippinarum* has most frequently been directly compared in experimental and field contexts is the European sympatric venerid *Ruditapes decussatus*. Such comparisons have been motivated primarily as investigations into the invasive potential of the non-native with respect to European ecosystems. Below, the reported temperature and salinity tolerances, along with food preferences, are compared, after which a field-based niche-analysis study of the two in a Spanish estuary is considered.

15.2 Species trait comparisons

Even if the habitats occupied by the Manila clam are sometimes relatively unsaturated in terms of species diversity (Chapter 20), it does not necessarily follow that its arrival is not subject to competition from native species. Insights into the abilities of the Manila clam to naturalize, or become invasive, can be provided by comparing the traits of the species with those of native sympatrics. In this respect, comparisons of the results from different single-species experiments are of limited use due to inevitable differences in uncontrolled variables.

Measured tolerances of species to environmental variables such as temperature and salinity are not properly comparable unless clams from the same locality are subjected to identical treatments in a single experimental setup. Otherwise, conditions, water composition, endogenous cycles, pathogen load, acclimation, and adaptation, etc., are among the extraneous factors that can confound any notion of establishing reliable relative tolerance limits and interspecific comparative validity.

However, in Europe, several direct comparative studies have been conducted with *R. decussatus*. Here, results in relation to food utilization, growth, and salinity are considered with a view to elucidating both competitive advantage and resource partitioning.

Using ^{14}C-labeled food sources, Sorokin and Giovanardi (1995) demonstrated that in the Venice lagoon, *R. philippinarum* can consume, digest, and assimilate a wide variety of foods, including cyanobacteria, planktonic bacteria, green algae, diatoms, and microzooplankton such as rotifers (Chapter 7). They also found that the food "preferences" (measured as carbon assimilation rate) of *R. philippinarum* and *R. decussatus* are distinctly different. Using ^{14}C to label the diatom *Nitzschia* and the cyanobacterium *Coelosphaerium,* Sorokin and Giovanardi, by direct comparison, demonstrated that, while both *R. philippinarum* and *R. decussatus* can assimilate diatoms and bacteria, they differ in their preferences, with *R. philippinarum* favouring the diatom and *R. decussatus* the cyanobacterium. This could be taken to suggest some degree of trophic partitioning. However, their results also imply a competitive advantage in the non-native, by virtue of much greater rates of carbon assimilation from these food sources, which were between three and six times that of *R. decussatus*. High assimilation efficiency has implications for growth and reproduction.

By analyzing acetate peel replicas of shells, Moura et al. (2017) compared the performance of *R. philippinarum* and *R. decussatus* using samples from the Tagus estuary, Portugal. From von Bertalanffy growth functions and growth performance indices, they concluded that *R. philippinarum* reached 40 mm length just before the age of 2, noting that Tagus Manila clam growth performance was much higher than in any populations of *R. decussatus* in Portugal.

Several comparative studies have indicated some reproductive advantage for *R. philippinarum* in European Atlantic coast habitats. Based on direct examination of female gonads, Beninger and Lucas (1984) estimated the reproductive activity of the two clam species in a common habitat at a hatchery pond in Barfleur, France. They found a significantly more prolonged season of reproductive activity in *R. philippinarum* as indicated by a longer period during which oocytes were present. No oocytes were found in the gonad of *R. philippinarum* in the four months from mid-November to mid-March, compared to 7 months without oocytes in *R. decussatus* (mid-September to mid-April).

Based on gonad histology, Delgado and Perez-Camacho (2007) reported similar relative findings for the two species when exposing the clams to the temperature conditions of the coastal waters of Galicia, Spain. They suggested that a higher gonadal development rate in *R. philippinarum* represented an advantage over the native species. Likewise, Lareuelle et al. (1994) at two sites in Brittany, France, found *R. philippinarum* to exhibit a greater speed of maturation, a more extended breeding season, and a greater number of spawning events than the native species.

Other reproductive differences between the two species relate to spawning patterns. Whereas *R. philippinarum* accumulates oocytes before their release, *R. decussatus* gametes are liberated continuously. In the early phases of gametogenesis, the gonad of *R. philippinarum* is heterogeneous, simultaneously exhibiting up to three reproductive conditions, while in *R. decussatus,* gonadal development is more uniform. Another difference concerns the reabsorption of gametes in stress conditions, thereby retaining energy. This is common in *R. philippinarum* but rare in *R. decussatus* (Vasquez et al., 2021).

Comparisons with *R. decussatus* suggest that *R. philippinarum* is more effective in spreading and then maintaining spatially separate spawning locations. Larval behavior and dispersal are effective in supporting both self-recruiting populations and the persistence of spatially separate meta-populations. In comparison to *R. decussatus*, the connectivity between locations is greater. Self-recruitment grounds, which are less likely to receive larval inputs from other spawning grounds, are more susceptible to decline. In contrast, well-connected grounds can compensate for a deficit with larval transport from elsewhere (Bidegain & Juanes, 2013).

As temperature increases, clams may become subject to oxidative damage due to an increase in mitochondrial respiration. This can damage lipids, proteins, and DNA. Various antioxidant responses can mitigate this (Chapter 9). Using several biochemical markers, Anacleto et al. (2014) demonstrated temperature to have a stronger influence on metabolism and oxidative status in *R. decussatus*, indicating that *R. philippinarum*, rather than "up-regulating" energetically expensive cellular responses to warming, has evolved a less demanding metabolic "arrest strategy" based on isolation of the tissues to guarantee longer survival. Consequently, progressive temperature increases in a constantly immersed situation have a stronger deleterious impact on the native *R. decussatus*.

Unlike for salinity stress, valve closure will provide only very transient direct relief from high temperatures. Nevertheless, indirectly, by diminishing gas exchange and restricting metabolic rate, this behavioral response may have a role to play in reducing temperature stress, particularly in *R. philippinarum*. In Anacleto et al.'s (2014) study, as water temperatures progressed from 28°C to 34°C, *R. philippinarum* kept its valves completely closed and demonstrated a decreasing metabolic rate. Conversely, *R. decussatus* kept its valves open and exhibited an increase in metabolic rate. Therefore, based on direct laboratory comparisons, *R. philippinarum* shows a markedly greater propensity to close valves in response to acute warming episodes than the European *R. decussatus*. These authors subjected constantly immersed Portuguese specimens of both species to progressively increasing temperatures. Mortality in samples *of R. philippinarum* and *R. decussatus* commenced at 40°C and 36°C. respectively. However, Macho et al. (2016) reported no mortality in either species up to 32°C, but 34% and 24% mortality, respectively, after 2 days of tidal exposure to 36°C.

The avoidance of such oxidative stresses may be achieved by the behavioral avoidance response of seeking refuge deeper within the sediment. In this respect, *R. decussatus,* with its longer siphons and deeper burrowing habit, may have an advantage (Macho et al., 2016), but both species are known to burrow deeper in response to heat-wave episodes (Roman et al., 2023).

Macho et al. (2016), working with clams from Galicia, Spain, have compared the burrowing depths of the European native *R. decussatus* with the introduced *R. philippinarum* using both reported field data and experimental verification. While *R. philippinarum* is a relatively shallow burrowing species (typically around 4 cm with a maximum of 7–10 cm), *R. decussatus* has a deeper burrowing habit (around 8 cm with a maximum of 15–20 cm) and correspondingly longer siphon extension capability. For example, at an extreme of 40°C at the sediment surface, *R. decussatus* at 6–8 cm might experience a possibly sublethal 30°C–31°C, while much nearer the surface *R. philippinarum* a lethal 34°C–36°C. Spikes in burrowing activity in both species suggest that, in contrast to *R, philippinarum, R. decussatus* burrows progressively deeper as temperatures rise from 21°C to 36°C (Macho et al., 2016).

All this suggests that the heat response strategies differ in the two species, *R. philippinarum* relying more on valve closure, thereby depressing metabolically induced oxidative stress, while *R. decussatus* puts a greater emphasis on deeper burial. It is not clear which of these two strategies is more effective in the wild, although in the laboratory, *R. philippinarum* appears more resistant to temperature extremes.

In their comparative laboratory study of gonad development Delgado and Perez-Camacho (2007) subjected *R. philippinarum* and *R. decussatus* to temperatures of 14°C and 18°C, as representative of the spring and summer temperature conditions of the Galician Rias in Spain. Whereas both species were well adapted to Galician conditions, these authors concluded that *R. philippinarum's* responses are relatively advantageous. Similarly, Vasquez et al. (2021) have also provided evidence suggesting that at warmer temperatures, reproductive success is greater in *R. philippinarum* than in *R. decussatus*.

On the basis of biochemical indicators of cellular damage, Carregosa et al. (2014) reported both *R. philippinarum* and *R. decussatus* to have an optimum salinity around 28 PSU. A similar optimum does not, however, imply equal competitive advantage in the wild, where tolerances of rates of change and extremes may be critical. Using artificial seawater with adult clams from the Ria de Aveiro, Portugal, Carregosa et al. (2014) demonstrated that *R. decussatus* can tolerate a wider range of salinities than *R. philippinarum*.

Dominguez et al. (2021) subjected three Galician Venerid clam species to experimental conditions resembling the salinity variations of a natural tidal cycle over 6 days, while testing their cumulative scope for growth. In December, where the (submerged) clams experienced approximately steady salinities of 30 PSU, over 12 h, both mean clearance rate and scope for growth (SFG, see Chapter 8) of *R. philippinarum* were markedly higher than those of *R. decussatus*, by a factor of approximately 3. Although individuals of both species exhibited a wide range of SFG measures, the mean December basal (day zero) SFG of *R. decussatus* was 8.2 $Jh^{-1}g^{-1}$ compared to 24.1 $Jh^{-1}g^{-1}$ for *R. philippinarum.* Their data suggest that in winter and early spring, at near optimum salinity conditions for both species, the growth potential of the Manila clam greatly outstrips that of the native species.

Two other commercially important European species, *Venerupis corrugata* and *Cerastoderma edule,* are relatively intolerant of hyposaline conditions, suggesting that if, as predicted, extreme rainfall events increase, these may be replaced by the *Ruditapes* species (Blanco et al., 2022), with *R, decussatus* the most tolerant. Overall, *R. philippinarum* in NW European estuaries occupies a niche dimension between that of *Venerupis corrugata* (lower tolerance to dilution) and *R. decussatus* (higher tolerance to dilution). Within this range, in protected eutrophic estuaries, where benthic microalgae add to an abundant food supply, and many potential but stenohaline predators are excluded, the traits of *R. philippinarum* in terms of food assimilation, growth rate, and reproductive potential are advantageous.

15.3 Distribution and niche dimensions

Field-based correlations between distribution and environmental variables provide insights into niche dimensions. Such studies include field-validated habitat suitability assessments to predict the aquaculture yield in different areas and niche modeling to assess the Manila clam's potential as a non-native competitor. Although it is almost inevitable that occasional and extreme values of some environmental variables will not be picked up by field sampling, such approaches do provide valuable information contributing to the elucidation of niche dimensions.

In the northern Adriatic, Vincenzi et al. (2006) have investigated suitable sites for Manila clam farming in the Sacca di Goro lagoon by deriving a suitability index (SI) for each of a range of exogenous variables as reported in the clam rearing publications of Barillari et al. (1990) and Paesanti & Pellizzato (2000). According to these reports optimal growth conditions include approximate ranges as follows: salinity from 25% to 35%; water current (hydrodynamism) 0.3–1.0 ms^{-1} (thought to represent a balance between food circulation and sediment stability); depth (bathymetry) of less than 2 m; sediment sand percentage of around 60–80 (the residual being largely silt and clay); Chlorophyll *a* concentrations from 2 to 11 mg l^{-1}, and oxygen saturations from 40% to 100%. Fig. 15.1 shows Vincenzi et al.'s (2006) SIs for these variables, along with the predicted relationship between overall habitat suitability and aquaculture yield. The yield predictions of their model were substantiated by landings in 6 of 10 of the sites studied, but their representation of optimal conditions remains of interest, not least as they are broadly consistent with other studies in terms of sediment, salinity, current velocity, and depth as significant niche factors.

In an estuary of two confluent rivers flowing into the Bay of Santander, Spain, Bidegain et al. (2015) examined the relationships between the distributions of *R. philippinarum* and *R. decussatus* and known ranges of physico-chemical variables. These variables were salinity, water depth below mean sea level, current velocity, and sediment composition with particular regard to percentages of sand and organic matter. Table 15.1 shows the range and means for each variable in relation to the study area generally (which extended the full length of the estuary), and the known distributions of the two species as revealed by sampling. Neither species was present in the more oceanic conditions at the downstream end of the estuary, and both species occupied the lower half of the shore, indicating their specialized estuarine niche dimensions.

For *R. philippinarum*, the percentages of sand and organic matter in the sediment were both positively associated with high habitat suitability. Conversely, higher values of depth, current velocity, and salinity were negatively associated with habitat suitability. Salinity was described as the key variable in determining the niche specialization of *R. philippinarum*, although depth and current speed also played a role in narrowing the overall niche dimension.

However, the mechanisms by which depth, current velocity, and salinity operate to limit distribution are not revealed by such modeling studies. These variables may

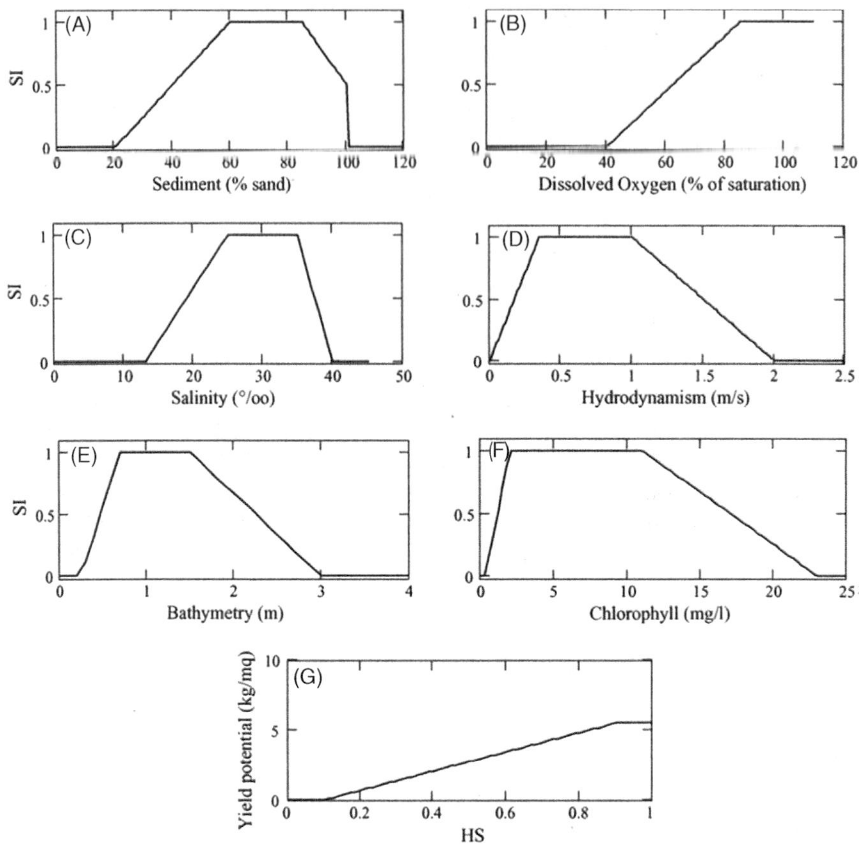

FIGURE 15.1

Habitat suitability parameters. Parameter-specific suitability function relationships between habitat suitability model variables in the form of indices normalized between zero (i.e., habitat not suitable for clam rearing) and 1 (habitat best suitable for clam rearing with respect to that specific parameter). (A) sediments, (B) dissolved oxygen, (C) salinity, (D) hydrodynamism, (E) bathymetry, (F) chlorophyll a (G) relationship between habitat suitability and clam yield potential.

From Vincenzi et al. (2006).

act directly or indirectly and may interact. For example, current velocity may link with depth. Moreover, depth limits may relate to trophic factors such as marine predator accessibility, microbenthic algal growth, or combinations of the two. In the same estuary, Bidegain and Juanes (2013) concluded predation to exert a significant effect on both clam species, but greater on *R. philippinarum,* which they linked to its shallower burrowing habit.

Table 15.1 The values of environmental variables considered in ecological niche factor analysis.

		Maximum	Minimum	Mean	SD
A. Whole study area					
	Depth (m)	16.30	0.00	4.70	3.91
	Current velocity ms^{-1}	1.10	0.01	0.25	0.08
	Sand (%)	99.60	11.45	64.00	23.60
	Salinity	34.16	0.00	31.80	23.60
	Organic matter (%)	10.70	1.20	3.57	2.40
B. *Ruditapes philippinarum* locations	Depth (m)	1.50	0.05	1.00	0.43
	Current velocity ms^{-1}	0.35	0.01	0.20	0,10
	Sand (%)	98.30	11.45	65.20	30.80
	Salinity	34.10	26.00	31.40	2.35
	Organic matter (%)	10.68	1.12	4.78	3.16

(continued on next page)

C. *Ruditapes decussatus* locations					
	Depth (m)	1.50	0.60	1.30	0.35
	Current velocity ms^{-1}	0.40	0.05	0.29	0.09
	Sand (%)	98.60	11.70	67.60	31.50
	Salinity	34.09	26.00	31.50	2.45
	Organic matter (%)	10.68	1.12	4.96	3.27

For each variable, maximum, minimum, mean value, and standard deviation are presented for the whole study area and for the presence locations of Ruditapes decussatus *and* Ruditapes philippinarum. *From Bidegain et al. (2015).*

In any event, from comparing the two venerid species distributions, it is clear that the physico-chemical ranges within which they are both present show considerable overlap. In fact, Bidegain et al. (2015) report that in 71.1% of the study area, the two clam species coexisted, and in relation to presence observations, significant differences between the distributions were only found for current velocity: *R. philippinarum* favouring lower velocity locations.

These patterns of coexistence and predominance analyzed by Bidegain et al. (2015) suggested that despite overlapping niche dimensions, interspecific competition between the two species did not have strong significance in determining the realized niche of either.

15.4 The realized niche as refuge

Based on evidence from experimental and field-based studies along with physiological energetics research, considered in this Chapter and in Chapter 8, it is possible to achieve some general qualitative conclusions on the commonly exhibited realized niche dimensions within which self-recruiting populations of multimodal age-size cohorts can persist.

The Manila clam is specialized and well adapted for optimal success on the lower elevations of temperate intertidal shores and sometimes shallow subtidal locations in the middle and lower reaches of protected transitional areas, where freshwater influence on waters of largely marine origin is usually significant but moderate. Hydrodynamically, these shores are protected from storm events and long-fetch wave action through their location within coastal estuaries, lagoons, and bays. Tidal flows and wind-driven water movements are necessarily low energy, and while sufficient to resuspend benthic microalgae and replenish phytoplankton above inhalant siphons, are generally insufficient to significantly destabilize sediments.

These habitats are typically micro or mesotidal regions whose often unsorted benthic substrates, while including mud and pebbles, must necessarily have a significant sand fraction. In such areas, *R. philippinarum* can become the predominant benthic organism on the basis of versatility of food sources, high assimilation efficiency, and consequently high physiological commitment to growth and reproduction. These characteristics make the species particularly effective in the sort of anthropogenically eutrophic environments commonly found in protected coastal waters, provided dissolved oxygen levels remain above about 40%. It is in such eutrophic habitats that the Manila clam is most effective and has occasionally been described as invasive (but see Chapter 22).

The species is well adapted to survive temporary severe and seasonal variations in the conditions of salinity and temperature to which its intertidal coastal habitats are subject. However, occasional extremes of stochastic volatility, as might be caused by severe weather such as rare storm events capable of destabilizing otherwise stable sediments, or causing prolonged episodes of very low salinity or temperature, are associated with significant mortality. When such stresses do not destabilize or erode sediments sufficient to expose the clam, its behavioral and physiological adaptations can offer temporary survival; however, the shallow burrowing habit, commitment to a fixed position implied by shell tumidity (Chapter 5), and consequent limited ability for repeated reburial are problematic, representing a survival limit.

Whereas the upper shore limit of the Manila clam can be understood in terms of physical tolerances, factors affecting the limits of lower elevation are biologically determined. Although in the laboratory the clam demonstrates the physiological capacity to survive and grow well in conditions of constant immersion in full seawater, its subtidal distribution is usually, but not always, highly attenuated. In this respect, the example of Jiaozhou Bay, China (Chapter 17), is interesting since subtidal Manila clams extend up to around 10 km from the shore. In reflecting on the circumstances that enable an often exclusively intertidal clam to achieve such an extensive sub-tidal distribution, it is informative to consider the conditions within the Bay.

Jiaozhou Bay is a very large natural inlet of the Yellow Sea, to which it is connected with a relatively narrow entrance of 3.1 km width. It is surrounded by urban centres, including the major city of Qingdao with a population of over 8 million. The Bay's area is around 370 km^2 with dimensions of approximately 32 × 27 km, most parts of which are shallower than 5 m. Twenty-six rivers discharge in the vicinity, including ten which have been characterized as canals of industrial, domestic, and agricultural effluent discharging directly into the Bay (Dai et al., 2007). The Bay therefore has eutrophic characteristics, is rich in phytoplankton, and is subject to algal blooms, which have been reported to correlate with major nutrient events triggered by high rainfall. Even in coastal waters outside the Bay, salinity (reported during August algal bloom monitoring) appears reduced, between 27.9 and 29.1. Sometimes, in severe winters, the Bay freezes (Yuan & Yu, 2011).

In most cases, explanations for the lower limit of distribution involve increased intensity of predation by exclusively subtidal, essentially marine predators, which need both the immersion and higher salinities of high water tides to access Manila clams. Some such predators are known to be able to decimate the shallow burrowing Manila clam. This predator hypothesis receives corroboration from various evidential directions. Firstly, from studies which report enhanced survival among the roots of seagrasses and amongst pebbles, both of which are thought to provide refuges by inhibiting predator effectiveness (Chapter 12). Secondly, from observations that subtidal Manila clams are restricted to areas where ostensibly fully marine waters are locally diluted. This phenomenon is known to occur outside Poole Harbour in the UK, where, in the protected but otherwise marine environment of Poole Bay, Manila clams can be found in locations where small streams can be seen running down the beach at low water.

Such reduced salinity areas may be conceptualized as refuge areas insofar as they offer some level of protection from marine predation. The idea of refuges has been widely applied in ecology, not least in relation to protection from predation offered by estuarine environments as juvenile fish "nurseries." Such refuge areas are thought also to protect small fish from the higher energy hydrodynamics of open shores, while also providing zones of high productivity and abundant food. Cases such as Jiaozhou

Bay and Poole Bay suggest a similar role not only for Manila clam juveniles but also for adults.

Essentially, it is the combination of challenging (e.g., salinity and temperature) and beneficial (e.g., low energy) conditions within estuaries that makes them effective refuges for those species that can tolerate the former sufficiently to exploit the latter through physiological advantages. The same can be said of intertidal low-energy sediment shores. However, the case of the Manila clam suggests wider benefits in modern industrialized coastal environments.

The classic Manila clam sites, such as Venice lagoon and the Tagus estuary in Europe, and Tokyo and Jiaozhou bays in east Asia, have been characterized by eutrophic conditions in which nutrient loads provide for abundant phytoplankton and microbenthic algal production. In such places, low-intensity hydrodynamics and moderately reduced salinities combine with eutrophic conditions to generate high fishery and aquaculture yields and the greatest chance of "invasive" concerns. Moreover, the attribution of reduced nutrient inputs, sufficient to convert eutrophic to oligotrophic waters, as causes for manila clam population crashes and fishery failure (Chapter 14), is also indicative of benefits from anthropogenically altered conditions.

While similar patterns no doubt apply to other estuarine taxa, particular Manila clam traits such as range of food sources, assimilation efficiency, energetic "scope for growth," pollution tolerance (Chapter 11), and their consequences for growth rate and reproductive commitment can be seen as beneficial not just in estuarine lagoons and bays, but especially in such environments when anthropogenically altered. As an alternative to a stenohaline, deeper burrowing and less tumid predator-resistant condition, the Manila clam has evolved a niche which, while physiologically demanding, also provides refuge from the hazards of fully marine environments and the mechanical and biological challenges they represent. However, the traits that specialize the species for such environments can also be seen as "preadaptations" for life in anthropogenically altered environments and a mutualistic relationship with *Homo sapiens*. This proposition is further developed in Chapter 25.

References

Anacleto, P., Maulvault, A. L., Lopes, V. M., Repolho, T., Diniz, M., Nunes, M. L., Marques, A., & Rosa, R. (2014). Ecophysiology of native and alien-invasive clams in an ocean warming context. *Comparative Biochemistry and Physiology Part A: Molecular & Integrative Physiology, 175*(1), 28–37. https://doi.org/10.1016/j.cbpa.2014.05.003.

Barillari, A., Boldrin, A., Pellizzato, M., & Turchetto, M. (1990). Condizioni ambientali nell'allevamento di *Tapes philippinarum*. In Tapes philippinarum *biologia e sperimentazione* (pp. 183–195). E.S.A.V, Regione Veneto.

Beninger, P. G., & Lucas, A. (1984). Seasonal variations in condition, reproductive activity, and gross biochemical composition of two species of adult clam reared in a common habitat: *Tapes decussatus* L. (Jeffreys) and *Tapes philippinarum* (Adams & Reeve). *Journal of Experimental Marine Biology and Ecology, 79*(1), 19–37. https://doi.org/10.1016/0022-0981(84)90028-5.

Bidegain, G., Bárcena, J. F., García, A., & Juanes, J. A. (2015). Predicting coexistence and predominance patterns between the introduced Manila clam (*Ruditapes philippinarum*) and the European native clam (*Ruditapes decussatus*). *Estuarine, Coastal and Shelf Science, 152*, 162–172. https://doi.org/10.1016/j.ecss.2014.11.018.

Bidegain, G., & Juanes, J. A. (2013). Does expansion of Manila clam *Ruditapes philippinarum* cause competitive displacement of the European native clam *Ruditapes decussatus*? *Journal of Experimental Marine Biology and Ecology, 445*, 44–52.

Blanco, S., Morán, P., Diz, A. P., Olabarria, C., & Vázquez, E. (2022). Effects of short-term hyposalinity stress on four commercially important bivalves: A proteomic perspective. *Environmental Research, 215*(Pt 3), 114371. https://doi.org/10.1016/j.envres.2022.114371.

Carregosa, V., Figueira, E., Gil, A. M., Pereira, S., Pinto, J., Soares, A. M. V. M., & Freitas, R. (2014). Tolerance of *Venerupis philippinarum* to salinity: Osmotic and metabolic aspects. *Comparative Biochemistry and Physiology Part A: Molecular & Integrative Physiology, 171*, 36–43. https://doi.org/10.1016/j.cbpa.2014.02.009.

Dai, J., Song, J., Li, X., Yuan, H., Li, N., & Zheng, G. (2007). Environmental changes reflected by sedimentary geochemistry in recent hundred years of Jiaozhou Bay, North China. *Environmental Pollution, 145*(3), 656–667. https://doi.org/10.1016/j.envpol.2006.10.005.

Delgado, M., & Perez-Camacho, A. (2007). Comparative study of gonadal development of *Ruditapes philippinarum* (Adams and Reeve) and *Ruditapes decussatus* (L.) (Mollusca: Bivalvia): Influence of temperature. *Scientia Marina, 71*(3), 471–484.

Dominguez, R., Olabarria, C., Woodin, S. A., Wethey, D. S., Peteiro, L. G., Macho, G., & Vázquez, E. (2021). Contrasting responsiveness of four ecologically and economically important bivalves to simulated heat waves. *Marine Environmental Research, 164*, 105229.

Hanski, I., & Gilpin, M. (1991). Metapopulation dynamics: Brief history and conceptual domain. *Biological Journal of the Linnean Society, 42*, 3–16.

Hutchinson, G. E. (1957). Concluding remarks. *Cold Spring Harbour Symposium Quantitative Biology, 22*, 415–427. https://doi.org/10.1101/SQB.1957.022.01.039.

Lareuelle, F., Guillou, J., & Paulet, Y. M. (1994). Reproductive pattern of the clams *Ruditapes decussatus* and *R. philippinarum* on intertidal flats in Brittany. *J, 74*, 351–366 Journal of the Marine Biological Association of the United Kingdom.

Macho, G., Woodin, S. A., Wethey, D. S., & Vázquez, E. (2016). Impacts of sublethal and lethal high temperatures on clams exploited in European fisheries. *Journal of Shellfish Research, 35*(2), 405–419. https://doi.org/10.2983/035.035.0215.

Moura, P., Garaulet, L. ía, Vasconcelos, P., Chainho, P., Costa, J. é, & Gaspar, M. (2017). Age and growth of a highly successful invasive species: The Manila clam *Ruditapes philippinarum* (Adams & Reeve, 1850) in the Tagus Estuary (Portugal). *Aquatic Invasions, 12*(2), 133–146. https://doi.org/10.3391/ai.2017.12.2.02.

Paesanti, F., & Pellizzato, M. (2000). *Tapes Philippinarum. Veneto Agricoltura*, 73.

Pulliam, H. R. (2000). On the relationship between niche and distribution. *Ecology Letters, 3*(4), 349–361. https://doi.org/10.1046/j.1461-0248.2000.00143.x.

Roman, M., Gilbert, F., Viejo, R. M., Román, S., Troncoso, J. S., Vázquez, E., & Olabarria, C. (2023). Are clam-seagrass interactions affected by heatwaves during emersion? *Marine Environmental Research, 186*, 105906.

Sorokin, Y. I., & Giovanardi, O. (1995). Trophic characteristics of the manila clam (*Tapes philippinarum* Adams and Reeve). *ICES Journal of Marine Science, 52*(5), 853–862. https://doi.org/10.1006/jmsc.1995.0082.

Vasquez, E., Woodin, S. A., Wethey, D. S., Peteiro, L. G., & Olabarria, C. (2021). Reproduction under stress: Acute effect of low salinities and heat waves on reproductive cycle of four ecologically and commercially important bivalves. *Frontiers in Marine Science, 8*, 685282. https://doi.org/10.3389/fmars.2021.685282.

Vincenzi, S., Caramori, G., Rossi, R., & Leo, G. A. D. e (2006). A GIS-based habitat suitability model for commercial yield estimation of *Tapes philippinarum* in a Mediterranean coastal lagoon (Sacca di Goro, Italy). *Ecological Modelling, 193*(1-2), 90–104. https://doi.org/10.1016/j.ecolmodel.2005.07.039.

Yuan, Y., & Yu, Z. (2011). Report of HAB case studies on coastal waters of Qingdao and Dalian in China, Qingdao. *Institute of Oceanology, Chinese Academy of Sciences*. https://cearac.nowpap.org/hab-web/wp/content/uploads/2023/11/HAB_Case_Study_China_2010.pdf.

PART

Aquaculture and fisheries

III

CHAPTER

Worldwide production and socio-economics 16

16.1 Global production and economic significance

The human exploitation of Manila clams takes many forms. Recreational hand-gathering and small-scale commercial fishing of wild populations can be contrasted with industrial-scale aquaculture operations involving advanced hatchery technology, large polyvalent fishing vessels, and often sophisticated mechanized harvesting. Moreover, clear distinctions between such categories are confused by many overlapping elements: large-scale industrial aquaculture enterprises may rely on hand gathering of their harvests, whereas small businesses may develop and use efficient mechanized harvesters. Approaches to aquaculture vary from the relocation of collected newly settled wild "spat" clams to favored areas for fattening, to industrial scale seed supply by extensive hatcheries with 100,000l capacity fully enclosed stainless steel larval culture vessels. What such approaches have in common, however, is that they are coastal fishery enterprises with socio-economic links within coastal communities.

Although the sea accounts for almost half the world's biological production, it provides only about 17% of human protein intake. It is argued that overstretched land-based resources in the context of a growing population (predicted as 9.8 billion by 2050) make this relatively small contribution untenably low. On the other hand, it is also argued that overfishing, pollution, and other unsustainable practices can make growth in seafood production unacceptable (Costello et al., 2020; EC, 2017). Can more seafood be extracted without compromising the marine environment upon which many ecological assets and services, including seafood production itself, depend? Of the three plausible possibilities: wild fisheries, finfish aquaculture, and bivalve aquaculture, the latter provides the most environmentally sustainable forms of production and therefore a necessary part of growing future supply (see Chapter 18).

Many countries collect Manila clam production data, which they submit to various regional and global organizations, including the UN FAO. The reliability of the data is variable for several reasons, none of which are unique to Manila clam reports. In England, for example, licensed fishers submit catch returns to their regional fisheries authorities, while wholesalers report their receipts of clams to an agency of central government. Each type of report is conducted under a different statutory instrument, and the data rarely match. In China, molluscan aquaculture production in the 1990's

The Manila Clam *Ruditapes philippinarum*. Science and Significance of a Globalised Species
DOI: https://doi.org/10.1016/B978-0-443-36704-5.00020-1
Copyright © 2026 Elsevier Inc. All rights are reserved, including those for text and data mining, AI training, and similar technologies.

was estimated by multiplying an average yield per unit area by the total culture area, which is thought to have resulted in a significant overestimate of production (Guo et al., 1999). Other causes of error include confusion between, or conflation of, similar species and reporting underestimations (sometimes intentional).

In addition, understanding of the meaning in practice of categories such as fishery or aquaculture production also varies amongst those submitting and compiling statistics, not least as these approaches lie on a spectrum which includes somewhat intermediate methods such as the "enhancement" of originally natural fisheries by the laying of hatchery seed amongst otherwise wild populations on subtidal offshore sea beds. In some cases, fishers point out that the proportions of their catch deriving from seed clams, wild clams, or the feral progeny of seed clams are unknown (Wang Ti Lun, personal communication).

Various forms of Manila clam fishing also occur in the informal economic sector: unregulated, often governed by custom and sometimes for protein subsistence. A tradition of collection for family consumption is widespread, particularly in East Asian countries (Fig. 16.1). Illegal Manila clam fishing is also common, both from land and in boats (see Chapter 19). These forms of fishing go unrecorded.

While such difficulties are widespread in reported fishery statistics, aggregate reports from national, regional, and global institutions provide the best available evidence of production, value, growth over time, and the relative contributions of different species. Table 16.1 shows UN FAO statistics on worldwide Manila clam aquaculture production by tonnage. The ascent of Manila clam production has resulted in establishing the species as pre-eminent amongst bivalves. While ostensibly higher production derives from "cupped oyster" species, this category includes several commercially important oyster species whose individual contributions are not reported (Gosling, 2015). The nearest species-specific data relates to the Pacific oyster *Magallana gigas* (the now accepted synonym of *Crassostrea gigas*), and some of the production of this species may be cryptically contained within the "cupped oyster" category. This would therefore increase the *M. gigas* production statistic, but whether this would raise it above *Ruditapes philippinarum* is highly unlikely (Humphreys, 2026; Humphreys et al., 2025).

Therefore, the statistics show the Manila clam to be the world's economically preeminent cultured mollusk. By 2022, the global value of the Manila clam was estimated at USD 8602 million, and indirect added value greatly increases this estimate (FAO, 2025), see Fig. 16.1. Among the conspicuous aspects of UN FAO statistics is the general growth rate of all aquaculture production. It has been estimated that by 2030, 65% of all seafood will be supplied by aquaculture (World Bank, 2013). Data also suggest that by 2015, accelerating bivalve production already made up around 15% of total marine production (wild and cultured), with 89% of this from aquaculture (Wijsmann et al., 2019). Of this, oysters, clams, scallops, and mussels together contribute 80% (e.g., FAO, 2014).

Significant growth in Manila clam production has occurred in the past half century since the 1980s. While this coincides with much of the second phase of the worldwide spread of the species (Chapter 3), the continued dominance of China in Manila

Table 16.1 Worldwide aquaculture of animals in marine areas in order of 2023 production (annual average live weight × 1000 tonnes).

Species	1 1980s	2 1990s	3 2000s	4 2010s	5 2020	6 2021	7 2022	8 2023	9 Share[a]
Cupped oysters[b] *Crassostrea spp.*								6706	18.2%
Whiteleg shrimp *Penaeus vannamei*								6520	17.7%
Other mollusks[c]								6326	17.2%
Manila clam *Ruditapes philippinarum*	112	1006	2499	3862	4265	4342	4426	4449	12.1%
Atlantic salmon, Salmo salar								2713	7.4%
Pacific cupped oyster *Magallana gigas*								531	1.4%

From FAO (2025).

[a] *Share of total marine aquaculture production by weight in 2023.*

[b] *Due to difficulties reliably distinguishing individual cupped oyster species, this category includes all such species, excluding the Pacific oyster Crassostrea gigas (= Magallana gigas).*

[c] *This category includes less commonly cultured mollusks such as abalone and conchs but excludes clams, oysters and other major cultured groups.*

FIGURE 16.1

"Recreational" Manila clam gathering for family consumption in Japan and South Korea. (A) Yokohama, Japan, (B) Jeju Island, South Korea.

Photos by he author with permission from hand gatherers.

clam production suggests that this growth is underpinned by general expansion in response to demand, along with technological developments in hatchery and nursery production. The latter of which stimulated much of the Phase 2 globalization of the species.

Since 2010, the Manila clam has superseded each of the two oyster categories in terms of total global value and today approximates the total value of all oyster species (Table 16.1). Of all named global aquaculture species, only the crustacean Whiteleg shrimp and Atlantic salmon surpass it in terms of tonnage. Of the major East Asian, European, and North American centers of Manila clam production, China is by far the biggest producer. In Europe, 92% of farmed Manila clams are reportedly produced in Italy (Eurostat, 2023; FAO, 2022).

FIGURE 16.1, cont'd.

16.2 Socio-economics and supply chains

The economic benefit of bivalve fishing to coastal communities has been apparent since prehistoric times (Chapter 1). In many parts of the world, clams remain freely available through common property rights (e.g., Appleby, 2005), which, combined with the low costs of entry, makes for an accessible resource of value either for subsistence or as a commodity. Total production metrics, while useful in the policy context, are also indicative of the socio-economic significance of the Manila clam, and benefits from shellfish production are increasingly recognized as an explicit component of marine planning processes. Nevertheless, while socio-economic benefits are often known to be large, generalizations are difficult to achieve and vary according to social, cultural, and political contexts (McCay & Jenks,1998).

In the Sacco di Goro lagoon on Italy's northern Adriatic coast, Manila clam farming makes an important contribution to the local economy. Shellfish farming supports about 1600 jobs directly, corresponding to around 60% of the age 14–65 population, and many other jobs in related industries. The indirect benefit of clam farming in terms of such related industries raises the question of how direct benefits alone, when represented economically by value at first sale, greatly underestimate the full value of these enterprises. To economic transactions directly in the value chain as the harvest passes from laying to production and purification (depuration), must be added transactions involving the supply of equipment and services by which producers procure and maintain their operations. Once packaged, Manila clams often go through further stages, such as wholesale distribution to retail outlets, all of which add further to the total economic benefit.

Fig. 16.2 illustrates these value and supply chains as determined for Manila clam aquaculture in the Sacco di Goro lagoon. As the clams progress through the horizontal stages of the value chain, value accumulates. All these stages may be conducted by a single organization, or the clams may be sold on, for example, for depuration, before they are packaged for bulk wholesale transport or, in some cases, sold on in smaller quantities for retail in local shops, wet markets, or restaurants. Fig. 16.2 also shows many of the indirect business-to-business transactions needed to support clam production. In this Italian case, these range from boat manufacture, maintenance, and operation; protective clothing and fishing gear; maintenance of clam grounds; construction and operation of depuration facilities, and packaging. Further induced added value in a community is generated by personal household spending by employees of an enterprise and its suppliers.

Full economic significance is therefore best represented by the aggregation of "value added" at each stage of the value chain, combined with indirect and induced financial transactions. Indirect costs of production can be estimated using standard industrial "multipliers" to achieve comprehensive estimates expressed as gross output and gross value added. In fact, Fig. 16.2 was originally constructed as part of an assessment of carbon sustainability (see Chapter 18), and no estimations of gross output can be found specifically for the Manila clam industry. However, some level of approximation may be provided by an analysis of the economics of the Pacific oyster (*Magallana gigas*), which shares many supply and value chain features with the Manila clam. Humphreys et al (2014) estimated the gross output of the Pacific oyster industry in England to be more than five times the value at first sale. Even allowing for the fact that the Manila clam may be only one of several target species requiring common resources (polyvalent boats, etc.), the idea of gross output in contrast to value at first sale gives more realistic insights into the true value of Manila clam in local and national economies.

16.3 Small-scale fisheries, access and women

Definitions of small-scale fisheries vary according to economic and social context. The concept is therefore often thought of as a relative term in contrast to what might be

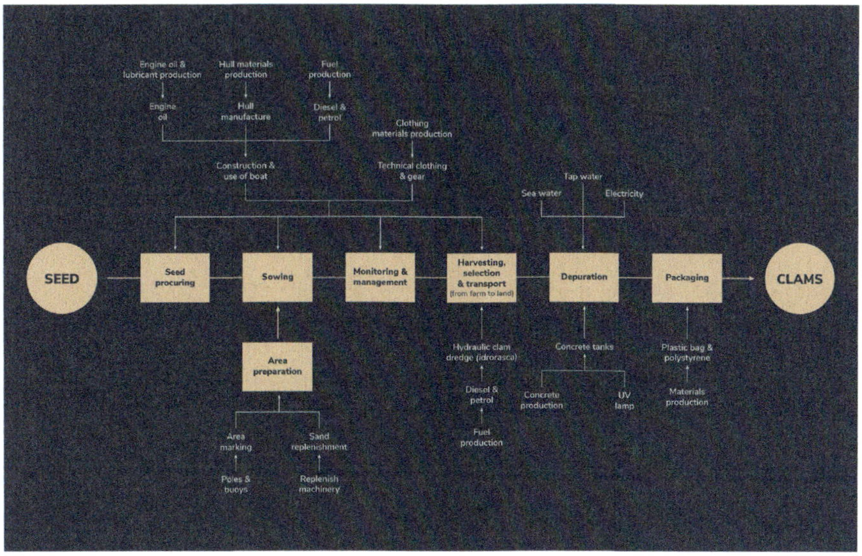

FIGURE 16.2

Supply chain model for Manila clam aquaculture in the Italian Adriatic Sacca di Goro Lagoon.

Models of this type can provide frameworks both for product lifecycle assessments of environmental impacts and for assessment of the full economic contribution of a fishery. Here, seed clams are collected from natural populations and moved to favored plots for on-growing. The horizontal chain shows the main stages from seed collection to packaging for sale. Each stage involves direct expenditure by the producer. Vertical supply chains show the indirect value of production. The value of each transaction contributes to economic estimates of "gross output" and "gross value added," which, in contrast to simple "value at first sale" statistics, represent the full economic contribution of a fishery, much of which benefits local coastal communities. This model was drawn up to estimate the environmental impacts of Manila clam production, so it omits transactions into wholesale operations and retailers, both of which add further to total economic output.

Modified from Turolla et al. (2020).

termed industrial-scale fishing. Nevertheless, in any economic context, a key defining attribute is the cost of entry, and therefore the extent to which a fishery is accessible to new independent participants. Platteau (1989) argued for definitions which exclude larger-scale fisheries as, in developing economies at least, "inaccessible except to a new class of capitalists arising from outside of fishing communities."

This question of accessibility to small-scale fisheries is an important consideration in socio-economics, not least in the context of the relative poverty sometimes experienced in the coastal communities of both developed and less developed countries (Bene et al., 2007). The existence of large wholesale enterprises and hatcheries in many parts of the world should not distract from the fact that the [illegible] for Manila

FIGURE 16.3

Commercial Manila clam industry involvement by women. (A) Bare hand picking, Arcachon Bay, France. (B) Dredge catch sorting in Poole Harbour, UK. (C) Rake gathering in Galicia, Spain. (D) Retail trading in Busan, Korea.

Images (A) and (D) by the author. Courtesy (A) Maud Vial; (B) Mike Bailey, Natalie Bailey; (C) With permission from Mulleres Salgadas (https://mulleressalgadas.es/la-asociacion/).

clam fisheries is frequently small-scale family and community enterprises. Even where boat-based fishing is involved, the inshore nature of Manila clam distribution, normally on protected coasts such as estuaries, bays, and lagoons, means that small boats are sufficient. Moreover, such areas are typically intertidal, also allowing shore-based clam collecting at low water. Whether boat or shore-based, Manila clam fisheries typically involve relatively low entry costs, which fall on a spectrum from small, motorized boats and dredges to simple hand rakes. In some cases, such as Arcachon Bay in France, even hand rakes are prohibited (Chapter 19), and Manila clam fishers are limited to their gloved but otherwise bare hands.

FIGURE 16.3, cont'd.

The inshore nature, combined with the economic accessibility of Manila clam fisheries, often leads to a close association between coastal communities and their clam fisheries and grounds. This association can be reflected in the approaches adopted for fishery management in which communities achieve significant influence through "co-management" schemes such as that exemplified by the Skokomish First Nation tribe in Washington, United States, which are discussed in Chapter 19.

A notable feature of Manila clam fisheries is the involvement of women. Women's contribution to and involvement in fishing is regarded as insufficiently recognized and under-researched (Kleiber et al., 2014a;Zhao et al., 2014). Consequently, women's catch may be neglected in fishery statistics (Kleiber et al., 2014b). It has been argued that women's work in fishing is often overlooked because it does not fit into a conventional mindset of "women on land, men at sea." Moreover, sometimes women's intertidal fishing is not counted as fishing (Frangoudes et al., 2019), even though its

FIGURE 16.3, cont'd.

contribution both economically and in the context of individual families and coastal communities may be substantial.

Gustavsson (2020), noting an increasing recognition and interest in the significant role of women in fishing families, communities and industries, has examined women's changing roles to develop a systematic understanding of "feminization" in fishing, highlighting the wide range of fishing-related practices from undertaking direct fishing activities, to establishing formal business partnerships and comanagement involvement (Frangoudes et al., 2008). Women's contributions to Manila clam fishing and in all aspects of the value chain including marketing becomes quickly apparent in field investigations into Manila clam fisheries. Although patterns differ according to varying cultural norms, often small-scale fisheries involve nuclear or extended family teams with women acting in a range of roles from fishers to professional partners in family businesses to sole small business proprietors (Fig. 16.3).

Changes in women's roles combined with feminist movements are leading to new forms of activism. In the UK, where women make up an estimated 15% of the fisheries and aquaculture sector, a *Women in Fisheries* network has been established to address some of the reported barriers to further progress (Women in Fisheries, 2025). In Galicia, Spain, *La Asociación de Mulleres Salgadas* (literally "salty women" meaning women of the sea) was established in 2016 as a regional feminist association aimed at properly recognizing the value of women's work in the fishing industry across all trades. Over 2000 individuals, associations, and cooperatives are affiliated. Their work includes proposing the creation of an "Equality Observatory" for the Galician fishing sector to research and establish measures for improving the situation, with special attention to the under-representation of women in management and representative positions. Fig. 16.3C shows a member raking for Manila clams.

References

Appleby, T. (2005). The public right to fish: Is it fit for purpose? *Journal of Water Law, 16*(6), 201–205.

Bene, C., MacFadyen, G., & Allison, E. H. (2007). *FAO fisheries technical paper no 481*. Rome: FAO.

Costello, C., Cao, L., Gelcich, S., Cisneros-Mata, M. Á., Free, C. M., Froehlich, H. E., Golden, C. D., Ishimura, G., Maier, J., Macadam-Somer, I., Mangin, T., Melnychuk, M. C., Miyahara, M., Moor, C. L., Naylor, R., Nøstbakken, L., Ojea, E., O'Reilly, E., Parma, A. M., Plantinga, A. J., Thilsted, S. H., & Lubchenco, J. (2020). The future of food from the sea. *Nature, 588*(7836), 95–100. https://doi.org/10.1038/s41586-020-2616-y.

EC. (2017). *Food from the Oceans. How can more food and biomass be obtained from the oceans in such a way that does not deprive future generations of their benefits? Scientific opinion No 3/2017*. Brussels: European Commission.

EUROSTAT, 2023. Production from aquaculture excluding hatcheries and nurseries (from 2008 onwards). Retrieved June 10, 2025, from https://ec.europa.eu/eurostat/databrowser/view/fish_aq2a__custom_10352407/default/bar?lang=en.

FAO. (2014). *The state of world fisheries and aquaculture opportunities and challenges. Technical Report*. Rome: FAO.

FAO. (2022). *The state of world fisheries and aquaculture (2022). Towards blue transformation*. Rome: FAO https://doi.org/10.4060/cc0461en.

FAO. (2025). *FAO yearbook of fishery and aquaculture statistics*. Rome: FAO.

Frangoudes, K., Gerrard, S., & Kleiber, D. (2019). Situated transformations of women and gender relations in small-scale fisheries and communities in a globalized world. *Maritime Studies, 18*(3), 241–248. https://doi.org/10.1007/s40152-019-00159-w.

Frangoudes, K., Marugan-Pintos, B., & Pascual-Fernandez, J. J. (2008). From open access to co-governance and conservation: The case of women shellfish collectors in Galicia (Spain). *Marine Policy, 32*, 223–232. http://ec.europa.eu/research/sam/pdf/sam_food-from-oceans_report.pdf.

Gosling, E. (2015). *Marine bivalve Molluscs* (second edition). John Wiley & Sons.

Guo, X., Ford, S. E., & Zhang, F. (1999). Molluscan aquaculture in China. *Journal of Shellfish Research, 18*(1), 19–31.

Gustavsson, M. (2020). Women's changing productive practices, gender relations and identities in fishing through a critical feminisation perspective. *Journal of Rural Studies, 78*, 36–46. https://doi.org/10.1016/j.jrurstud.2020.06.006.

Humphreys, J. (2026). Bivalves with benefits: The ascendance of the Manila Clam, *Ruditapes philippinarum. Bulletin of Estuarine and Coastal Sciences, 125*, 15–19.

Humphreys, J., Choi, K.-W., & Watanabe, S. (2025). *Ruditapes philippinarum*: The $7 billion mollusc, 2025. In J. Humphreys, K-W. Choi, & S. Watanabe (Eds.). *The Manila clam:* Ruditapes philippinarum: 325. Special Issue Estuarine Coastal and Shelf Science.

Humphreys, J., Herbert, R. J. H., Roberts, C., & Fletcher, S. (2014). A reappraisal of the history and economics of the Pacific oyster in Britain. *Aquaculture, 428–429*(2014), 117–124.

Kleiber, D., Harris, L. M., & Vincent, A. C. J. (2014a). Improving fisheries estimates by including women's catch in the Central Philippines. *Canadian Journal of Fisheries and Aquatic Sciences, 71*(5), 656–664. https://doi.org/10.1139/cjfas-2013-0177.

Kleiber, D., Harris, L. M., & Vincent, A. C. J. (2014b). Gender and small-scale fisheries: A case for counting women and beyond. *Fish and Fisheries, 16*(4). https://doi.org/10.1111/faf.12075.

McCay, B. J., & Jenks, W. P. (1998). The importance of shellfisheries to coastal communities. In C. L. MacKenzie, V. G. Burrell, A. Rosenfeld, & W. L. Hobart (Eds.). In *The history, present condition, and future of the molluscan fisheries of North and Central America and Europe: 2* (pp. 131–145). Pacific Coast and supplemental topics. NOAA Technical Report NMFS 128.

Platteau, J.-P. (1989). The dynamics of fisheries development in developing countries: A general overview. *Development and Change, 20*(4), 565–597. https://doi.org/10.1111/j.1467-7660.1989.tb00358.x.

Turolla, E., Castaldelli, G., Fano, E. A., & Tamburini, E. (2020). Life cycle assessment (LCA) proves that Manila clam farming (*Ruditapes Philippinarum*) is a fully sustainable aquaculture practice and a carbon sink. *Sustainability, 12*(13), 5252. https://doi.org/10.3390/su12135252.

Wijsmann, J. W. M., Troost, K., Fang, J., & Roncarati, A. (2019). Global production of marine bivalves. Trends and challenges. In A. C. Smaal, J. Ferreira, J. Grant, J. Petersen, & Ø Strand (Eds.), *Goods and services of marine bivalves*. Springer https://doi.org/10.1007/978-3-319-96776-9_2.

Women in Fisheries. (2025). *Women in fisheries voices from the industry. Policy brief* https://www.women-fisheries.com/news/policy-brief.

World Bank. (2013). Technical Report 83177-GLB. Washington D.C: World Bank.

Zhao, M., Tyzack, M., Anderson, R., & Onoakpovike, E. (2014). In J. Urquhart, T. Acott, D. Symes, & M. Zhao (Eds.). In *Social issues in sustainable fisheries management: 9* (pp. 233–254). MARE Publication Series. https://doi.org/10.1007/978-94-007-7911-2_13.

CHAPTER

From hatchery and seabed to market

17

17.1 Introduction

In areas where the Manila clam is indigenous or naturalized, wild fisheries exist using a wide range of capture methods similar to those employed for farmed clams. The most basic form of Manila clam culture involves the collection of wild clams for relocation to areas suitable and convenient for further growth and harvesting. Relaying can achieve improved harvest through the preparation of the ground, manipulation of clam densities, and protection from predation. The earliest occurrence of this approach is possibly ancient, but it is thought to have been used in Ariake Bay, Japan, by 1830, as a method for local communities to develop their shellfishery (Shimoyame et al., 2015). The approach continues to be used in China and other indigenous areas, but also with naturalized non-native wild populations such as in Poole Harbour, UK.

Larval collection devices, which encourage the settlement of wild spat, can be used to procure clams for translocation; however, various factors have motivated the development of industrial-scale Manila clam hatchery and nursery techniques for "seed" production. These include severe population declines in indigenous areas (Chapter 14) and economic growth opportunities presented by an expanding and global market.

In modern Manila clam aquaculture, the successful production of seed clams from broodstock is a complex and energy-intensive process demanding considerable skill and experience. Hatchery and nursery techniques require the scrupulous avoidance of contamination, careful regulation of temperatures, the production of large quantities of phytoplankton of various species, and the maintenance of an appropriate relationship between larval and juvenile clam densities and the concentration of microalgal food supplied. Various countries and treaty organizations have produced technical guides to all aspects of culture and farming (Jones et al., 1993; Helm et al., 2004; Seafish, 2005), most of which building on the earlier development of hatchery techniques for species such as the Pacific oyster *(Magallana gigas)*, whose global spread and exploitation preceded similar growth of the Manila clam. Here, an outline of the main stages will be provided with attention to variations in practice.

The Manila Clam *Ruditapes philippinarum*: Science and Significance of a Globalised Species.
DOI: https://doi.org/10.1016/B978-0-443-36704-5.00029-8
Copyright © 2026 Elsevier Inc. All rights are reserved, including those for text and data mining, AI training, and similar technologies.

17.2 Broodstock and spawning

Hatcheries and nurseries supply seed clams for laying, either to enhance natural settlement or as a crop in more formally established clam farms. The regular laying of seed clams is an essential component of the farming cycle, not least where natural spawning and recruitment are economically inadequate and cannot meet demand.

Hatchery techniques involve inducing broodstock to spawn in controlled conditions. The selection of broodstock may sometimes be based on desirable features such as shell color. Broodstock conditioning typically involves the optimization of temperature and food supply for ripening the gonads, at which point spawning can be induced by thermal manipulation. Broodstock collected in the winter period may need 6–9 weeks of conditioning. Where the species is non-native, large broodstock clams may be induced to spawn several times with conditioning periods between, while in the Korean west coast hatchery at Gwang Seung, for each spawning in summer, broodstock are collected from the shore and transferred in baskets to large indoor tanks for a single cycle, after which they are discarded. Once commenced, spawning episodes may last for up to some hours.

The optimization of fertilization and larval growth requires controlled water temperature and food supply. According to the scale of operation, larval rearing may be achieved in large tanks or smaller continuous flow systems. Large rearing tanks become contaminated with larval waste; therefore, every few days, water is discarded via filters, which remove the larvae for release into another tank of fresh, plankton, and aerated seawater. Fig. 17.1 shows larvae being introduced into one such aerated larval tank in which abundant algal food has been cultured in preparation. In France, large, steel, fully enclosed larval bins are also used.

Alternatively, smaller continuous flow systems provide for the constant renewal of phytoplankton-loaded temperature-regulated water, typically in several small, inverted cone-shaped "upwelling" tanks as developed by the Seasalter Shellfish hatchery in Kent, UK. In continuous flow systems, water is introduced from below and escapes via a filter that retains larvae, but which must be monitored to avoid clogging.

The stages of larval development are described in Chapter 10. After about 14 days, the larvae have progressed to the pediveliger stage (180–210 μm length). Metamorphosis and settling (setting) can be anticipated on the basis of microscopic and behavioural observations. Typically, a golden brown shell of length 180–220 μm can be observed alternating between swimming and crawling on the bottom. Eventually, no longer able to swim, it can move using its foot and once finding a suitable location attaches itself with byssus threads. A wide variety of approaches for setting clams includes shallow-sided "water tables" lined with screens, which may be stacked and with continuous water flow downward. A substrate such as ground shell can be put on the screens to provide greater surface area for pediveligers to crawl, but this complicates subsequent counting and separation while making no difference to growth rates.

FIGURE 17.1

Korean Manila clam larvae being introduced into large tanks prepared with abundant phytoplankton culture and continuous aeration.

Courtesy Gomso Bay hatchery, South Korea (Photo credit: The author).

As metamorphosed and settled clams grow, they need grading (sorting by size) and cleaning until such time as they are transferred to the nursary. Manila clam nurseries are generally outdoor areas whose conditions may be less strictly regulated than hatchery conditions. The simplest approaches involve locating the clams in bags or trays suspended in otherwise naturally protected areas; however, more often, nurseries have clam seed containers or troughs through which phytoplankton-enhanced seawater flows. Fig. 17.2 shows Hawaiian Manila clam seed on a "raceway" at the Taylor Shellfish hatchery. Once large enough they are supplied to field-based growing-on enterprises.

FIGURE 17.2

Hawaiian hatchery-produced blue variant Manila clam seed in "raceways" before bagging and sale.

Courtesy Greg Jacob, Taylor Shellfish (Photo credit: The author).

17.3 Triploidy

The nuclei of somatic Manila clam cells contain 38 chromosomes in 19 chromosome pairs. This is the diploid condition in which one of each pair of "homologous" chromosomes comes from each parent. For reproduction, the gonads must produce gametes with half the normal number of chromosomes (the haploid condition) so that on fertilization, the normal diploid condition is restored. Genetic manipulation of larvae in hatcheries can achieve an extra set of chromosomes such that there are three, rather than two, chromosomes in each homologous set (the triploid condition)

(Piferrer et al., 2009). Two advantages can derive from such measures, both relating to the resulting interference with gamete formation and therefore reproduction.

Firstly, attenuation of the ability to reproduce prevents the clams from investing energy in this activity, thereby leaving more available for growth. On average, triploid clams grow faster and have a better meat-to-shell ratio (condition), which in turn increases their commercial quality and value. In France, a government agency (IFREMER) supplies stocks of triploid seed clams to aquaculture enterprises.

A second potential benefit relates to concerns that when introduced as a non-native, the Manila clam may compete with indigenous species or otherwise affect native functional ecology. For this purpose, triploidy is used as a biocontainment strategy preventing farm-to-wild introgression. However, in this respect, triploidy has proved unreliable in terms of the effectiveness of the genetic manipulation processes which can revert after release. For example, Lining et al. (1996) reported that 2–3 years after release, a high proportion (37%) of triploid Manila clams could nevertheless be induced to spawn, compared to 75% of normal diploids.

Triploidy can be induced by several techniques ranging from chemical inhibitors of normal gametogenesis, such as cytochalasin B (Utting & Childs, 1994), heat shock, or the crossing of normal diploid clams with tetraploid clams, which can be produced using similar techniques. High larval losses can result from such processes but are compensated by the improved condition of mature clams. In the United States, this loss is estimated to cost USD 0.25 million annually. Conversely, yield value increase from triploidy, if adopted universally on the US west coast, has been estimated at $5million (USDA, 2012).

17.4 Algal culture and species

A major part of hatchery production is the provision of phytoplanktonic food. This high-cost activity involves the maintenance of pure pathogen-free starter cultures fed with special growing medium, from which larger culture vessels and tanks are inoculated. As an alternative to batch culture in tanks, large-scale production can also be achieved in continuous flow systems in which the rate of removal for use is matched by an input of new culture medium so as to maintain a steady algal concentration. Either way, optimal conditions of hygiene, lighting, temperature, and nutrients must be maintained to maximize growth. Algal cells produced in these ways are usually fed directly to developing larvae, although for some purposes, they can be centrifuged to form a paste for sale or redilution and use as necessary in the production cycle.

For optimal clam growth, the algal food must be highly nutritious, appropriately sized for larval consumption, nontoxic, and with easily digestible cell walls. Of the many microalgal marine species, only a few are used in Manila clam aquaculture. They include diatoms and naked "athecate" flagellates. Commonly used species include *Tetraselmis suecica, Isochrysis galbana, Skeletonema costatum, Pavlova lutheri, Chaetoceros gracilis,* and *Chaetoceros calcitrans*. These species are high in protein, micronutrients, and fatty acids, and when consumed, promote strong growth, reproduction, and immunity (Kumar et al., 2023).

No single microalga possesses all the necessary nutrient requirements for optimal larval growth; therefore, mixed diets of species are most effective. Smaller microalgae are more suitable for early-stage clam larvae. An initial diet may consist of a mixture of smaller flagellates such as *Nannochloropsis oculata* and *Isochrysis galbana,* and small diatoms such as *Chaetoceros calcitrans*. Larger species such as *Chaetoceros gracilis* may be added as the larvae reach 150 μm.

In Japan, an insufficient microalgal supply for the outdoor growth of clam seed greater than 1 mm during colder periods has led to research into new forms of Manila clam diet. Approaches range from the cultivation of novel algal species such as *Eutreptiella eupharyngea*, which is capable of good growth outdoors at low temperatures (Yamasaki et al, 2019), to diet enhancement with manufactured microcapsules of organic matter, to supplements including dissolved organic matter. Uchida et al. (2010) demonstrated significantly enhanced growth in Manila clam soft tissue when a diet of the diatom *Chaetoceros calcitrans* was supplemented with dissolved glucose. Subsequent experiments with the polysaccharide alginate, obtained from the cell walls of brown macroalgae, in soluble hydrolysate form, showed that shell length and total weight can be significantly promoted when added to a diet of *Nannochloropsis sp.* (Yamasaki et al., 2018). Yamasaki, 2021 has argued that in the future, greater efficiency and reliable Manila clam seed production may be achieved by novel combinations of algal species, microencapsulated particulate organic matter, and dissolved supplements.

17.5 Seed laying and predator protection

The size at which hatchery clams are laid on the seabed is a balance between the greater vulnerability of smaller clams and the cost of clam seed, which increases with size. Mortality is also a function of environmental conditions, including water quality. Each producer buys seed at a size they consider optimal, often between 2 and 5 mm shell length. Mass production approaches developed in China can differ markedly from those adopted elsewhere.

While stony substrates impede predation, in primarily sandy sediments predation can effectively rule out successful clam farming. On the other hand, sandy substrates are conducive to more efficient mechanized methods of harvest. An approach to predator protection on sandy seabeds involves laying netting over the seed clams. In the UK, 25 × 2 m lengths of netting of mesh size 5 × 5 mm have been recommended (Seafish, 2005); however, in Poole Harbour, early experimental introductions of seed clams were protected by long strips of narrower, approximately 1 m wide netting. This approach proved effective and has persisted in Europe and North America, having the added advantage that clam beds arranged in long, relatively narrow strips also proved conducive to the development of mechanized forms of laying, maintaining, and harvesting (see Fig. 17.4).

At the Chuckanut Shellfish Farm in Samish Bay, Washington, predator protection is achieved by reusable 6 or 10 mm mesh polypropylene netting, laid in strips and pinned down at 50 m intervals by steel staples. Seed clams of shell length 3.2–3.5 mm are dropped onto the seabed after the predator netting has been laid. During an incoming tide, the seed clams settle through the netting, where they remain protected until harvest 2–3 years later. The solution to net fouling by the algae *Ulva, Fucus,* and *Sargassum* involves periodic *in situ* cleaning with a tractor-towed rotating sweeper brush (Suhrbier et al., 2017). If the predator netting is removed by storms, predation can be 100% within 24 h.

In China, the mass production of Manila clams may be achieved without predator protection. In the Zhangzhou City region, Fujian Province, Great areas of previously tidal flats have been reclaimed by the creation of large culture ponds. Each such pond may be several hectares in area, and extensive areas of adjacent ponds may extend kilometres inland from the new shoreline. These ponds can be filled via creeks and sluices at high water and emptied when required at low tide. Once filled, the ponds are fertilized to trigger phytoplankton production and sometimes inoculated with microalgal food (Nie, 1991). Manila clams seed is sown from small wooden boats (Fig. 17.3). These ponds may also be used for the culture of pufferfish and shrimps. Here, the controlled environment alleviates the need for Manila clam protection. Once grown, these Manila clams are hand-gathered from the drained ponds.

Another approach to mass production in China involves the laying of seed clams on offshore sub-tidal seabeds, such as in Hong Dao Harbour, Qingdao. Although not protected, survival is sufficient that seed clams laid in spring may be harvested in September when large polyvalent boats shown in Fig. 17.5B can each dredge 4–5 tons of Manila clams per day.

17.6 Suspended clam culture

To compensate for a decline in Manila clam catches in Japan (Chapter 14), various authors have investigated the use of suspended culture to supplement conventional production. This practice involves placing clams in net cages or containers suspended along long lines, each with a sediment substrate such as sand, pumice, anthracite coal, or oyster shells. Clams treated this way in the Hokkaido fishing port of Suttsu by Sakuri et al. (2021) were found to have less than 90% mortality and reasonable growth: 28 to 40 mm in one year when suspended near a channel with relatively high flow fluctuations. In the Ria de Arosa, Spain, Camacho and Cuna (1985) demonstrated the potential of "raft culture" in which Manila clams were suspended in oyster-growing cages, apparently without sediment. Although bottom-cultured control clams on the lower intertidal appeared to grow better, they were more susceptible to mortality due to storm-deposited sediment.

In Korea, Lee et al. (2020) observing better growth in subtidal Manila clams due to extended feeding time, experimented with cages suspended at 2 m depth. Suspended clams performed markedly better in terms of both shell and somatic growth than

FIGURE 17.3

One of many adjacent culture ponds in Zhangzhou, Fujian Province, China. These ponds bring seed clams to market size in 8 months. They are constructed from reclaimed tidal flats and are aerated and fertilized for microalgal growth.

Courtesy Zhangzhou Ocean and Fisheries Bureau, China (Photo credit: The author).

those of control clams placed in intertidal bottom cages. In addition, suspended clams spawned earlier, suffered less *Perkinsus olseni* infection (Chapter 13), and thereby avoided a high mortality episode correlated with a mid-summer increase in infection.

17.7 Gathering, fishing, and harvesting

There is considerable overlap between the methods used for Manila clam capture fishing from wild populations and those used in aquaculture harvesting. These methods range from manual "picking" of individual clams from the sediment using just gloved hands (as is required in Arcachon Bay, France: Fig. 16.3A) to the use of mechanized gear such as dredges and harvesters. The methods used for gathering Manila clams in particular localities depend on factors including the tidal regime, the character of the seabed, the availability of labor, and the stage of development of the fishery. Where the habitat is of conservation interest, this may also influence the method adopted. Below, gathering techniques are considered according to the tidal conditions in which they are employed.

A common approach worldwide is to gather clams at low water. That is to say, from exposed intertidal sediments. The common technique for achieving this is to disturb the substrate with a handheld implement such as a fork or, more commonly, a rake (Fig. 16.3C). Although this method is generally used for recreational gathering, it is not exclusive to recreational or subsistence fishing. In fact, this method is the commonest approach worldwide, including in mass or industrialized production scenarios. In Washington, United States, at the States' largest commercial shellfish operation, despite the availability of sophisticated mechanized harvesting equipment, 75% of its Manila clams are supplied through hand gathering, not least by independent fishers working smaller areas of seabed, but also where clam populations occupy stony habitat, making mechanized sorting of clams from substrate ineffective.

Often, gangs of pickers get casual employment either from independent "gang-masters" who sell on to established shellfish companies, or directly from such shellfish companies. Where availability of such clam pickers is high, the motivation for mechanising gathering is reduced, such that, for example, on Korea's largest Manila clam production area at Gomso Flat, gathering is done by hand with pickers taken out by tractor-towed trailers. Conversely, where sediment character, the cost of labor, and the stage of development of the fishery favour mechanized harvesting, many technical innovations have been developed.

Among these mechanized systems are those developed by the Chuckanut Shellfish company in Samish Bay (Puget Sound area, Washington, United States), who supply to Taylor Shellfish in nearby Skelton. In this case, self-propelled tractor vehicles with vibrating griddles at the front rake the clams up, shake out residual sediment, and by conveyor deposit them into baskets at the back (Fig. 17.4).

In subtidal locations or coasts where the low water period is too short, clam fishing and harvesting are achieved using fishing boats. Such places include the sort of lagoon-like habitats with prolonged high water stands often favoured by Manila clams. A peculiarity of Poole Harbour in southern England is an extended "double" high-water period which leaves tidal flats covered for most of the day (Humphreys, 2005a). Here, extensive tidal flats fall within a Marine Protected Area, which, combined with the sediment type, has necessitated the development of a shallow penetrating "pump scoop dredge" in which a pressure water jet of maximum 15 Horse Power washes out sediment such that the dredge retains only clams. This technology has developed at two scales: Smaller batch dredges for the wild cockle and Manila clam fishery and larger for clams harvested with Pacific oysters in designated parts of the harbor. The larger equipment brings clams up continuously on a conveyor.

FIGURE 17.4

Low-water mechanized harvesting at Samish Bay, Washington, USA. Note the strip laying arrangement, which facilitates mechanized approaches to laying predator protection netting, depositing seed clams, net cleaning and harvesting (Image courtesy of Bill Dewey).

Courtesy Chuckanut Shellfish, Washington State, United States (Photo credit: Bill Dewey).

In non-native locations, the arrival of the Manila clam is typically followed by gradual improvement in dredge gear as a result of fisher innovations. In the Venice Lagoon, Italy, for example, Pravoni et al. (2003) described an early improvised approach similar in principle to the pump-scoop dredge used in Poole Harbour. "Rusca" dredge gear consisted of a 60 cm wide iron cage on two runners, with a net attached behind. The rusca dredge is deployed over the side of small moving clam boats with a 25 hp outboard positioned in front of the dredge, whose propeller causes suspension of clams and sediment; the former being captured by the net. Today, harvesting techniques in Italy include dredges with integrated water injection, as shown in Fig. 17.5A, which also shows a typical clam boat of size similar to those used in various European locations. In contrast, Fig. 17.5B shows the fleet of polyvalent vessels equipped with large dredges as used in offshore subtidal fisheries in Hong Dao, China. (Humphreys, 2025b)

Field experiments on the effectiveness of different mechanical fishing gears have been conducted for a new emerging wild Manila clam fishery in the Thames estuary, UK. Here, the regional inshore fisheries authority tested three types, including a simple "dry" dredge, similar to local oyster dredges, a "water injection" dredge which fluidizes the sediment in front of the blade (similar in principle to Fig. 17.5A), and

FIGURE 17.5

Contrasting Manila clam dredge vessels. (A) Italian clam boat with hydraulic dredge in the Sacca degli Scardovari, the largest lagoon of the Po delta region. (B) Chinese clam boats in Hongdao Fisheries Harbour. Clam dredges can be seen on either side of the superstructure on the left and centre boats. In contrast to the much smaller European boats, these large polyvalent vessels fish subtidal Manila clams over 100,000 hectares of seabed up to 10 km offshore. Photo credits: (A) Courtesy of Matteo Ciani and Stefania Chiesa of ISPRA, Sustainable Aquaculture Unit. (B) The author Thanks to Xiujun Sun, Yellow Sea Fisheries Research Institute, Qingdao.

a suction dredge sometimes used for cockle fishing in the area. The dry dredge was found to be unsuitable for the sediment type, and the suction dredge was found to damage too high a proportion of the catch. The water injection type was therefore selected for the developing fishery (Stuart, 2025).

The effects of mechanical clam dredging on the physical and biological properties of the sediment are described in Chapter 18.

17.8 Postharvest processes

In many cases, the transportation of harvested clams to fish markets is periodic, with the result that the catch must be held for days until the next transportation opportunity. This may require the clams to be suspended in sacks or racks or located in purpose-built holding facilities such as tanks. Either way, during the holding period, the catch is typically submerged in ambient water from the adjacent bay or estuary. Mortality during this period can result from fluctuations in water conditions. In Korea, Bae et al. (2021), noting the increased frequency of climate-linked events such as high rainfall and temperature episodes, examined holding mortalities and stress behaviours relating to temperature and salinity, and combinations of the two, in aerated conditions. They reported that cumulative mortality increases with time, higher temperatures, and lower salinities. Findings broadly compatible with laboratory studies (Chapter 11). After 8 days at 24°C and 18 PSU, although they found no significant reduction in condition index (CI), mortalities exceeded 50%.

The disposal of effluents into estuarine and coastal waters is a longstanding and worldwide issue. Sewage and agricultural effluents, in particular, include bacteria and viruses in suspension or attached to particulate matter, which can be collected and accumulated by filter-feeding bivalves. Such contamination represents a risk to human health upon consumption. This is a particular problem in bivalves traditionally consumed raw, such as oysters. Most often, Manila clams are cooked before consumption, which, if of sufficient temperature and duration, will destroy bacteria such as *Escherichia coli* and inactivate viruses such as Hepatitis A. Nevertheless, raw consumption or inadequate cooking remains a risk, and the removal of such pathogens remains a necessary precaution for clams harvested from contaminated waters.

If relocated into clean coastal waters, the pathogen load within clams will be cleared out naturally, and this approach remains a practical, although expensive, solution. A widely used alternative is a process of depuration (purification) involving closed systems in which seawater is cycled repeatedly over the shellfish and an ultraviolet sterilising unit. Spencer (2002) provided a review of the operating conditions of UV depuration along with alternative approaches to purification.

Once receiving any necessary purification, many bivalve species are distributed live. Such products are perishable with a shorter shelf life than many other food types. Therefore, transportation creates challenges in relation to ensuring low mortality and maintenance of condition. For this purpose, low temperatures, which work to reduce metabolic rate and oxygen consumption, are critical. Different species of bivalves vary in their tolerances to distribution conditions. Huang et al. (2025) investigated the effects of packaging on quality changes in *Ruditapes philippinarum.* They reported that lower temperatures worked to inhibit microbial activity and lipid oxidation, resulting in higher survival rates. Through modeling, they were able to recommend optimal packaging and container arrangements.

In the United States, it is recommended that live shellfish should be stored and transported at 7.3°C or less (NSPCC, 2023). Therefore, as elsewhere, "cold-chain logistics" typically involves refrigerated trucks transporting the product in bulk between fishers and wholesalers. Smaller-scale fisheries may use only ice packs and insulated containers, which are often also used for smaller deliveries to consumer-facing businesses such as retailers and restaurants.

Bi et al. (2023) determined the optimal conditions for *R. philippinarum* subjected to 3 days of transport, comparing temperatures of 4°C and 15°C in both submerged ("watery") and anhydrous conditions in which the clams, although kept moist, were exposed to air. Watery conditions resulted in lower glycogen content, CI, and survival rate at 4°C (93% survival) than at 15°C (99% survival). In contrast, in anhydrous conditions, survival at 4°C was 97%, and 63% at 15°C, with glycogen content and CI maintained higher when transported at 4°C. They concluded that anhydrous transportation at 4°C was optimal for the maintenance of quality and condition. Such results indicate that Manila clams are generally quite tolerant of live transport and have the added advantage that anhydrous transport, since it involves less weight of water, is more cost-effective.

References

Bae, H., Im, J., Joo, S., Cho, B., & Kim, T. (2021). The effects of temperature and salinity stressors on the survival, condition, and valve closure of the Manila Clam, *Venerupis philippinarum* in a holding facility. *Journal of Marine Science and Engineering, 9*(7), 754. https://doi.org/10.3390/jmse9070754.

Bi, S., Xue, C., Xu, L., Wen, Y., Wang, L., Li, Z., & Liu, H. (2023). Physiological responses of clam (*Ruditapes philippinarum*) to transport modes with different temperatures. *Journal of Ocean University of China, 22*(2), 517–526. https://doi.org/10.1007/s11802-023-5244-y.

Camacho, A. P., & Cuna, M. (1985). *Report C.M1985/F: 43/Ref. K.* Spain: International Council for the Exploration of the Sea.

Helm, M. M., Bourne, N., & Lovatelli, A. (2004). (comp./ed.) Hatchery culture of bivalves. A practical manual. In *FAO Fisheries Technical Paper. No: 471* (p. 177p). Rome: FAO.

Huang, Y., Xie, X., Younas, S., Liu, C., & Wang, X. (2025). Quality changes in live *Ruditapes philippinarum* during "Last Mile" cold chain breakage: Effect of packaging. *Foods, 14*(6), 1011. https://doi.org/10.3390/foods14061011.

Humphreys, J. (2005a). Salinity and tides in Poole Harbour. In J. Humphreys, & V. May (Eds.), *Ecology of Poole Harbour*. Elsevier.

Humphreys, J. (2025b). Bivalves with Benefits: The ascendance of the Manila clam, *Ruditapes philippinarum. Bulletin of Estuarine and Coastal Sciences, 125*, 15–19.

Jones, G. G., Sanford, C. L., & Jones, B. L. (1993). *Manila clams: Hatchery and Nursery Methods. Innovative Aquaculture Products Ltd.* Canada: Skerry Bay Lasqueti Island B.C.

Kumar, N. A, Sridhar, S., Jayappriyan, K. R., & Raja, R. (2023). Applications of microalgae in aquaculture feed. In E. Jacob-Lopes, M. I. Quieroz, M. M. Maroneze, & L. Q. Zepka (Eds.), *Handbook of food and feed from microalgae* (pp. 421–433). Academic Press.

Lee, H.-M., Kim, H.-J., Park, K.-I., & Choi, K.-S. (2020). Enhanced growth, gonad maturation, and low-level parasite infection in juvenile Manila clam *Ruditapes philippinarum* cultured in subtidal cages on the south coast of Korea. *Aquaculture, 526*, 735410. https://doi.org/10.1016/j.aquaculture.2020.735410.

Lining, S. D., Millican, P. F., & Laing, I. (1996). The breeding potential and biochemical composition of triploid Manila clams, *Tapes philippinarum* Adams and Reeve. *Aquaculture Research, 27*(8), 573–580. https://doi.org/10.1111/j.1365-2109.1996.tb01289.x.

Nie, Q.-Z. (1991). The culture of marine bivalve molluscs in China. In W. Menzel (Ed.), *Estuarine and marine bivalve mollusc culture* (pp. 261–276). CRC Press.

NSSP. (2023). Guide for the control of molluscan shellfish 2023 revision. *Interstate shellfish sanitation conference*. U.S. Food and Drug Administration.

Piferrer, F., Beaumont, A., Falguière, J.-C., Flajšhans, M., Haffray, P., & Colombo, L. (2009). Polyploid fish and shellfish: Production, biology and applications to aquaculture for performance improvement and genetic containment. *Aquaculture, 293*(3-4), 125–156. https://doi.org/10.1016/j.aquaculture.2009.04.036.

Pravoni, F., Libralato, S., Raicevich, S., Granzotto, A., Pastres, R., & Giovanardi, O. (2003). Mechanical clam dredging in Venice lagoon: Ecosystem effects evaluated with a trophic mass-balance model. *Marine Biology, 143*, 393–403.

Sakurai, I., Mori, T., & Hisatomi, Y. (2021). Experimental study on the possibility of a suspended culture of the Japanese littleneck clam *Ruditapes philippinarum* in Suttsu fishing port, Hokkaido. *Fisheries Science, 87*, 239–251. https://doi.org/10.1007/s12562-021-01494-w.

Seafish. (2005). *Clam cultivation*. Hull, UK: Sea Fish Authority.

Shimoyama, S. (2015). Historical occurrence of the short-neck clam, *Ruditapes philippinarum* (Adams & Reeve, 1850), on the sandy flats of Ariake Bay, Kyushu, western Japan. *Plankton Benthos Research, 10*(4), 202–214.

Spencer, B. E. (2002). *Molluscan shellfish farming oxford*. Fishing News Books/Blackwell Publishing.

Stuart, K. 2025. Manila clam handbook. Kent & Essex Inshore Fisheries and Conservation Authority, Retrieved from www.kentandessex-ifca.gov.uk.

Suhrbier, A.D., Cheney, D.P., Cordell, J.R., Dewey, W.F., Davis, J.P., & Ferreira, J.G. (2017). Innovative farming methods for production and harvest of Manila clams in Washington State, USA. World Aquaculture, December 2017 issue 49–57.

Uchida, M., Kanematsu, M., & Miyoshi, T. (2010). Growth promotion of the juvenile clam, *Ruditapes philippinarum*, on sugars supplemented to the rearing water. *Aquaculture, 302*, 243–247.

USDA, 2012. Production of all-female triploid Manila clam seed for US west coast growers. US Department of Agriculture. Retrieved from https://portal.nifa.usda.gov/web/crisprojectpages/1022846-production-of-all-female-triploid-manila-clam-seed-for-us-west-coast-growers.html#:⟨:text=If%20we%20are%20not%20successful, with%20eggs%20from%202N%20females. (Accessed 13.10 2023).

Utting, S., & Childs, A. R. (1994). Genetic manipulation of the Manila clam (*Tapes philippinarum*) using cytochalasin B to induce triploidy. *Aquaculture, 120*(3-4), 271–282.

Yamasaki, Y. (2021). New diets with potential for enhancement of juvenile bivalve seed production and culture techniques. *Bulletin of the Fisheries Research and Education Agency, 50*, 149–158.

Yamasaki, Y., Ishii, K., Hikihara, R., Ishimaru, M., Sato, F., Taga, S., Kishioka, M., Matsunaga, S., Shikata, T., Abe, M., Kato, S., Tanaka, R., & Murase, N. (2019). Usefulness of the euglenophyte Eutreptiella eupharyngea as a new diet alga for clam culture. *Algal Research., 40*, 101493.

Yamasaki, Y., Ishii, K., Taga, S., & Kishioka, M. (2018). Enhancement of dietary effect of *Nannochloropsis sp.* on juvenile *Ruditapes philippinarum* clams by alginate hydrolysates. *Aquaculture Reports, 9*, 31–36.

CHAPTER

Environmental performance 18

18.1 Low trophic aquaculture and sustainability

Manila clam farming falls within the category of "Low Trophic Aquaculture" (LTA), which refers to the farming of primary producers (such as macroalgae) and primary consumers (such as bivalves) that occupy the basal regions of food chains. In contrast to "high trophic aquaculture" of species such as the Atlantic salmon, a predator that is two or more trophic steps away from primary producers, LTA has a higher efficiency of natural resource use and lower environmental impact.

Moreover, LTA organisms consume suspended particles or inorganic nutrients already circulating in the natural environment. Involving the consumption of pelagic or benthic microalgae, Manila clam farming does not require the provision of feed, a significant difference from widespread Atlantic salmon and whiteleg shrimp farming. Also, LTA of bivalves in particular can be recognized as a form of mitigation of climate change due to sequestration of carbon in shells, thereby removing it from the anthropogenically amplified short carbon cycle responsible for global warming. Nutrients sequestered by microalgal food sources are also removed from the water column, thereby reducing eutrophication risk.

Because of such characteristics, LTA, with comparatively low greenhouse gas emissions per unit weight of food produced, is widely considered a desirable solution to the growing need for sustainable protein (Tett et al., 2025). It is further argued that a thriving future for aquaculture could be based on a shift toward LTA (Pounds & Adamson, 2023), which would also contribute to a decarbonization of food production as part of a necessary transformation toward a sustainable "blue economy" and potential food security in a "net-zero carbon world" (Krause et al., 2022).

This is not to say that innovations could not further reduce environmental impacts: such as reducing the use of plastic-based gear, which "undermine the sustainability narrative" (Miguel, 2025), and increasing the use of sustainable energy sources in hatcheries (Martini et al., 2023).

18.2 Carbon impact of Manila clam farming

Product life cycle assessment involves scrutiny of all aspects and stages of production in terms of environmental performance. In contrast to wild Manila clam fisheries,

DOI: https://doi.org/10.1016/B978-0-443-36704-5.00026-2
Copyright © 2026 Elsevier Inc. All rights are reserved, including those for text and data mining, AI training, and similar technologies.

the commercial cultivation of *Ruditapes philippinarum* mostly involves three distinct processes: seed production in industrial hatchery and nursery operations; seed laying and on-growing; subsequent harvesting, packaging, and transport (Chapter 17). In some localities, depending on water quality and consumption traditions, clam purification (depuration) may be necessary after harvesting. In addition to the direct impacts of these processes, each stage requires preparatory activity and supply networks, which are shown in Chapter 16 as Fig. 16.2. All of these represent demand on the environment. Based on such an analysis, Turolla et al. (2020) have determined an inventory associated with the production of Manila clams harvested from the Italian Adriatic Sacca di Goro Lagoon. Their inventory covers all materials and fuels used and all emissions produced.

These authors conclude that Manila clam farming has a lower environmental impact as a consequence of being grown on the seabed, compared with shellfish that are not. They calculate that shell formation (i.e., excluding soft tissue carbon metabolism, which is part of the short carbon cycle) removes enough carbon for the aquaculture operation as a whole to act as a net carbon sink. Even allowing for CO_2 released during farming and distribution operations (fossil fuel consumption, etc.), carbon calcified as shell $CaCO_3$ represents a net sink of 54.50 kg of carbon per ton of clams and 723.8 tons of carbon in the whole lagoon. Manila clam farming is therefore considered fully sustainable in terms of CO_2 emissions.

In the Sacca di Goro, seed clams of 5–10 mm length are mainly collected from natural nursery areas by the lagoon mouth or outside the lagoon, for re-laying in licensed growing areas where they grow quickly. Since, in the main, the aquaculture operations do not source seed clams from hatcheries, such facilities were excluded from the Turolla et al. (2020) inventory and assessment.

However, following a reduction in wild seed availability, hatchery-reared seed clams now supplement natural seed. Martini et al. (2023) have therefore subjected this Manila clam hatchery production to a similar life cycle assessment covering the three main production phases: microalgal production; broodstock maintenance and conditioning; larval and juvenile rearing to a suitable size for laying. They concluded that the energy demand (required mainly for temperature control and lighting for algal production) contributed 80% of the global warming impact. All other factors together contribute 20%. In relation to global warming specifically, 1 kg of seed production (20,000 individual seed clams) resulted in an impact equivalent to 2.63×10^1 Kg CO_2.

While hatchery production adds significantly to the environmental burden associated with wild seed farming, it also adds to the cost. In this respect, it is noteworthy that Taylor Shellfish, a major clam producer in Washington state, United States, has located its hatchery production facility in Hawaii, where ocean and atmospheric conditions can be judiciously exploited to minimize seed production costs (Chapter 10). Similarly, hatcheries on warmer and sunnier coasts of East Asia, such as in southern China, also reduce costs and significantly enhance environmental performance.

18.3 Benthic impacts and nutrient fluxes

Except for methods involving only the scattering of cultured seed in subtidal areas, as occurs extensively in China, benthic physical disturbance can begin at an early stage in the aquaculture process. Preparation of the intertidal seabed may involve the removal of algal mats and other epiflora, which might smother seed clams. Protecting seed clams from predation by netting also changes the physical and biological character of the seabed. In the Exe estuary, south-west England, Spencer et al. (1996) found that netted plots experienced short-term sediment rates four times higher than control plots, elevating the seabed by around 10 cm. This also led to increases in green algal cover, gastropod herbivores, deposit-feeding polychaete worms, and productivity. Generally similar results were reported in the outer Thames estuary at Whitstable, south-east England, although with species differences (Kaiser et al., 1996).

Work in both the Thames and Exe estuaries after clam harvesting demonstrated that harvesting by suction dredge (Chapter 17) removed the surface sediment sand fraction and greatly reduced the density and diversity of infaunal species. Although recovery time will vary by site, in Whitstable, 7 months later, the sediment structure was largely restored, and no significant difference was found in the infaunal community between experimental and control areas. In both estuaries, it was concluded that clam cultivation did not have long-term effects on the sediment or the benthic community (Kaiser et al., 1996), and that benthic changes as a result of Manila clam cultivation were relatively benign compared to other forms of marine cultivation (Spencer et al., 1997, 1998). In any event, suction dredges are now not used for Manila clam fishing in the Thames due to the unacceptable shell damage they cause (Chapter 17; Humphreys et al., 2026).

Elucidating the chronic long-term effects of persistent clam culture on the seabed is sometimes difficult due to reluctance on the part of fishers to refrain from fishing traditional areas along with the absence of baseline data for already often long-established bivalve fisheries. However, work in the 2000s on the Adriatic lagoons of Italy suggests that such impacts vary on the basis of site-specific factors, including clam population densities and the intensity of exploitation.

In Italy, the Venice lagoon and Sacca di Goro saw a progressive increase in Manila clam production in the 1980s and 1990s. From1990 to 2003, annual landings rose from 1300 tons to over 40,000 tons, most of which was from a largely unregulated wild fishery estimated to involve more than 2000 fishers (Sorokin et al., 1999). Around that time, the Venice Lagoon delivered over 50% of Italian clam production using hydraulic dredges ("rusca" gear: see Chapter 17). In these circumstances, impacts to the seabed included coarsening and erosion of surface sediments and changes to the infaunal community (Pranovi & Giovanardi, 1994). Although more severe, these effects were broadly comparable to those reported in the UK. However, Pranovi et al. (2003, 2004), whose work included trophic energy flow modeling, have provided additional insights.

Pranovi et al. (2003) modeled three basic scenarios: "present" (1998); past (1988), and future (2008). They concluded that large clam population densities alter the functional ecology of the lagoon by greatly increasing the filter-feeding activity in the system. This activity more strongly couples pelagic and benthic subsystems by generating a larger flux of organic matter from the water column into the benthos. They characterized the role of *R. philippinarum* in the lagoon as a "wasp-waist control" system in which the link between lower (primary production) and higher (predator/fishery) trophic levels is dominated, in this case, by a single central component species. Their model estimated that the clam exerts a powerful control on all plankton groups with other consequent effects.

Their 1998 scenario also confirmed earlier work that suspended phytoplankton fell below that necessary to maintain the large Manila clam biomass. Combining this finding with the modeled future scenario in which the cessation of large-scale clam fishing resulted in a 20% reduction in Manila clam biomass, they postulated a "*Tapes* Paradox," involving a positive feedback between clam dredging and clam growth. They attributed this largely to the mobilization of sediment by clam dredging and the resulting resuspension of organic particulates, including benthic diatoms, which the species is known to consume (Chapter 7).

In any event, it is clear from Pranovi et al. (2003, 2004) that while the impacts of mechanized clam dredging had consequences for the lagoonal ecosystem, a gradual cessation of fishing effort would improve the ecological features of the system in comparison to its then degraded state. However, the total clam catch was about 20 times greater in the intensely fished 1998 scenario than in both the past (1988, before the advent of mechanized dredging), or the future scenarios (2008, cessation of mechanized harvesting), a point which illustrates the powerful economic incentives at work. All of which suggests the importance of fishery plans and management for balancing ecological and economic imperatives, as was suggested for the Venice lagoon by, among others (Solidoro et al., 2003; Sorokin et al., 1999) (See Chapter 12).

Along with many other studies at local scales, the analysis of Turolla et al. (2020) suggests that clam culture can improve ecosystem quality by reducing eutrophication through nitrogen and phosphorus sequestration and removal from the system at harvest. Excess nutrient loading of estuarine waters by human activities such as agriculture and sewage discharges is generally damaging to ecosystem function, especially when eutrophication results. Through their suspension-feeding activity, bivalves are known to play a major role in controlling phytoplankton biomass and consequently mitigating the effects of eutrophication in coastal waters. Since bivalve fisheries and aquaculture, in contrast, for example, to salmon farming, involve no feeding regime, anthropogenic sources of nitrogen and phosphorus taken from the water column by phytoplankton are sequestered into bivalve biomass and subsequently removed from the system at harvest.

Among the benefits that accrue are improved light penetration and reduction in water and sediment oxygen deficits that would otherwise be caused by plankton blooms resulting in organic enrichment and microbial degradation (Cranford, 2019). The extent of these eutrophication-related nutrient regulation benefits from bivalve laying has been proposed as an active mitigation measure in itself and a legitimate part of nutrient reduction policies (Petersen et al, 2019), both generally and in estuarine marine protected areas (Gravestock et al., 2020). Thus, in common with other bivalve species, Manila clam farming has been seen to supply both provisioning and regulating ecosystem services (see Chapter 21).

Farmed Manila clams can also play a direct role in nitrogen cycle process rates that eliminate bioavailable nitrogen from estuarine sediments as gaseous nitrogen products. This occurs naturally when clam bioturbation, ventilation, and even oxygen diffusion through siphon walls increase available oxygen in the sediment, facilitating nitrification and denitrification, not least of the phytoplankton-derived ammonia excreted by the clam itself. It is also the case that the hard and soft tissues of clams are colonized by nitrifying and denitrifying bacteria, which can significantly enhance nitrogen fluxes.

In the northern Adriatic Sacco di Goro, at winter water temperatures of 8°C and clam densities of 1440 per m^{-2}, Welsh et al. (2015) calculated that denitrification of sediments in farmed areas could be positively correlated with clam biomass. Consequently, the sediments in clam-farmed areas remained a source of dissolved inorganic nitrogen to the water column. Also in the wider environmental context, any net denitrification benefit should also be balanced by resulting greenhouse (N_2O) gas emissions (Welsh et al., 2015).

Bartoli et al. (2001), reporting earlier work also in the Sacca di Goro, described the heavily exploited Manila clam area as occupying 35%–40% of the seabed with clam densities sometimes exceeding 2000 clams/m^{-2}. Here at summer temperatures above 20°C, fluxes of oxygen, carbon dioxide, ammonium, reactive silica (from diatom consumption) and phosphorus were all increased several-fold by *R. philippinarum* due to reducing conditions caused by the clam's respiration and excretion activities and the accumulation of organic matter produced from clam feces and pseudofeces. Nutrient release increased 6.5 times for NH_4 and 4.6 times for PO_3.

In these more extreme circumstances of extensive cultivation, high clam densities, and high summer water temperatures (27°C in July), taking the lagoon as a whole, they calculated the amount of nitrogen and phosphorus potentially removed from the system as the Manila clam crop was a small fraction compared to that regenerated into the water column in cultivated areas. They concluded that in the Sacco di Goro at that time, the idea that the farming of Manila clams mitigated eutrophication was unrealistic. Nevertheless, where the percentage of cultivated seabed and clam densities is much less, the direct effects of the clams on nutrient dynamics would be negligible or positive in terms of mitigation of eutrophication (Bartoli et al., 2001).

18.4 Disturbance of birds and estuarine amenity

In Europe, conservation directives, including the European Union Birds Directive (Council Directive 2009/147/EC), require an absence of excessive disturbance. This measure *inter alia* applies to over-wintering estuarine birds such as waders

(Charadrii), and wildfowl (Anatidae, including ducks and geese). Many such intertidal estuarine locations are designated marine protected areas. The issue arises in places where the activities of birds and people overlap and relates to whether this causes bird disturbance, significantly reducing overwintering bird fitness and survival, with resulting starvation. Among the human activities that are potential sources of disturbance are dog walking, bird watching, and bivalve fishing (Goss-Custard & Stillman, 2020). The latter either by stock depletion or direct bird disturbance during dredging operations.

While there are examples of detrimental impacts of bivalve fishing on wader mortality due to excessive depletion of the birds' food stocks (e.g., van Roomen et al. (2005) in the Wadden Sea and by Atkinson et al. (2003) in the UK Wash). In the absence of secure scientific information on human disturbance impacts, conservation management practices have been and sometimes remain based on the precautionary principle. This has led to activities being curtailed even when there is little evidence that they have a detrimental effect, while allowing other detrimental activities to continue (Goss-Custard & Stillman, 2020).

Addressing this problem, Goss-Custard and Stillman (2022) have developed modeling approaches elucidating impacts on bird mortality in such a way as to discriminate between damaging and harmless bivalve fishing activity. While acknowledging cases of damage, on the basis of quantitative research, they have also been critical of precautionary management policies which unnecessarily curtail bivalve extraction. In Poole Harbour, their modeling has demonstrated a level of bivalve fishing compatible with high levels of wader survival. Moreover, their work has demonstrated that the controversial introduction and naturalization of the non-native Manila clam (Chapter 24) has increased the carrying capacity of the estuary and resulted in reduced over-wintering mortality of the Eurasian oyster catcher *Haematopus ostralegus* (Caldow et al., 2007).

Subject to Goss-Custard's criticisms are conservation cultures resulting in research too often done to support preconceived concerns and overzealous application of the precautionary principle (Goss-Custard, 2016). Consequent exaggerations of the impact of human activities on birds result, for example, by ignoring their ability to habituate to human activity. It is further suggested that such exaggeration leads to skepticism among the public (and fishers) due to the employment of conservation measures inconsistent with experience, common sense, and scientific evidence (Goss-Custard & Stillman, 2020).

References

Atkinson, P. W., Clark, N. A., Bell, M. C., Dare, P. J., Clark, J. A., & Ireland, P. L. (2003). Changes in commercially fished shellfish stocks and shorebird populations in the Wash, England. *Biological Conservation, 114*(1), 127–141. https://doi.org/10.1016/S0006-3207(03)00017-X.

Bartoli, M., Nizzoli, D., Viaroli, P., Turolla, E., Castaldelli, G., Fano, E. A., & Rossi, R. (2001). Impact of *Tapes philippinarum* farming on nutrient dynamics and benthic respiration in the Sacca di Goro. *Hydrobiologia, 455*, 203–212. https://doi.org/10.1023/A:1011910422400.

Caldow, R. W. G., Stillman, R. A., Dit Durell, S. E. A. L. V., West, A. D., McGrorty, S., Goss-Custard, J. D., Wood, P. J., & Humphreys, J. (2007). Benefits to shorebirds from invasion of a non-native shellfish. *Proceedings of the Royal Society B: Biological Sciences, 274*(1616), 1449–1455. https://doi.org/10.1098/rspb.2007.0072.

Cranford, P. J (2019). Magnitude and extent of water clarification services provided by bivalve suspension feeding. In A. C. Smaal, J. Ferreira, J. Grant, J. Petersen, & Ø Strand (Eds.), *Goods and services of marine bivalves*. Springer https://doi.org/10.1007/978-3-319-96776-9_8.

Goss-Custard, J. D (2016). Mud, birds, and poppycock. *Bulletin of the British Ecological Society, 47*, 26–29.

Goss-Custard, J. D, & Stillman, R. A (2020). How new science should affect the application of protection measures for UK estuarine shorebirds. In J. Humphreys, & R. W. E. Clark (Eds.), *Marine protected areas. Science policy and management* (pp. 525–542). Elsevier.

Goss-Custard, J. D, & Stillman, R. A (2022). Predicting the effect of environmental change on non-breeding shorebirds with individual-based modelling. In J. Humphreys, & S. Little (Eds.), *Challenges in estuarine and coastal science* (pp. 206–222). Pelagic Publishing.

Gravestock, V. J, Nicol, R., Clark, R. W. E., & Humphreys, J. (2020). Assessing the benefits of shellfish aquaculture in improving water quality in Poole Harbour, an estuarine marine protected area. In J. Humphreys, & R. W. E. Clark (Eds.), *Marine protected areas. Science policy and management*. Elsevier.

Humphreys, J., Stuart, K., Wyatt, E., Haupt, P., Clark, R., & Wright, W. (2026). *An emergent Manila clam fishery in the outer Thames estuary, UK, and its role in climate adaptation*. Aquatic Living Resources (Accepted, In Press).

Kaiser, M. J., Edwards, D. B., & Spencer, B. E. (1996). Infaunal community changes as a result of commercial clam cultivation and harvesting. *Aquatic Living Resources, 9*(1), 57–63. https://doi.org/10.1051/alr:1996008.

Krause, G., Vay, L. L, Buck, B. H., Costa-Pierce, B. A, Dewhurst, T., Heasman, K. G., Nevejan, N., Nielsen, P., Nielsen, K. N., Park, K., Schupp, M. F., Thomas, J.-B., Troell, M., Webb, J., Wrange, A. L, Ziegler, F., & Strand, Å (2022). Prospects of low trophic marine aquaculture contributing to food security in a net zero-carbon world. *Frontiers in Sustainable Food Systems, 6*. https://doi.org/10.3389/fsufs.2022.875509.

Martini, A., Aguiari, L., Capoccioni, F., Martinoli, M., Napolitano, R., Pirlo, G., Tonachella, N., & Pulcini, D. (2023). Is Manila clam farming environmentally sustainable? A life cycle assessment (LCA) approach applied to an Italian *Ruditapes philippinarum* hatchery. *Sustainability., 15*(4), 3237. https://doi.org/10.3390/su15043237.

Miguel, E. d. P. (2025). Rethinking material use in low-trophic aquaculture: A global review. *Journal of Environmental Management, 393*(2025), 127091.

Petersen, J. K, Holmer, M., Termansen, M., & Hasler, B. (2019). Nutrient extraction through bivalves. In A. C. Smaal, J. Ferreira, J. Grant, J. Petersen, & Ø Strand (Eds.), *Goods and services of marine bivalves*. Springer https://doi.org/10.1007/978-3-319-96776-9_8.

Pounds, A., & Adamson, E. A. S. (2023). *Aquaculture for a thriving future: Biodiversity, innovation, and economic sustainability in the UK*. The Fish Monger's Company.

Pranovi, F., Da Ponte, F., Raicevich, S., & Giovanardi, O. (2004). A multidisciplinary study of the immediate effects of mechanical clam harvesting in the Venice Lagoon. *ICES Journal of Marine Science, 61*, 43–52.

Pravoni, F., Libralato, S., Raicevich, S., Granzotto, A., Pastres, R., & Giovanardi, O. (2003). Mechanical clam dredging in Venice Lagoon: Ecosystem effects evaluated with a trophic mass-balance model. *Marine Biology, 143*, 393–403.

Pravoni, F., & Geovanardi, O. (1994). The impact of hydraulic dredging for short-necked clams *Tapes spp* on an infaunal community in the Lagoon of Venice. *Scientia Marina, 58*, 345–353.

Solidoro, C., Canu, D. M, & Rossi, R. (2003). Ecological and economic considerations on fishing and rearing of *Tapes philippinarum* in the lagoon of Venice. *Ecological Modelling, 170*, 303–318.

Sorokin, I. I., Giovanardi, O., Pranovi, F., & Sorokin, P. I. (1999). Need for restricting bivalve culture in the southern basin of the Lagoon of Venice. *Hydrobiologia, 400*, 141–148. https://doi.org/10.1023/A:1003707231839.

Spencer, B. E, Kaiser, M. J, & Edwards, D. B (1998). Intertidal clam harvesting: Benthic community change and recovery. *Aquaculture Research*. https://doi.org/10.1046/j.1365-2109.1998.00221.x.

Spencer, B. E, Kaiser, M. J., & Edwards, D. B (1996). The effect of Manila clam cultivation on an intertidal benthic community: The early cultivation phase. *Aquaculture Research, 27*(4), 261–276. https://doi.org/10.1111/j.1365-2109.1996.tb00993.x.

Spencer, B. E., Kaiser, M. J., & Edwards, D. B. (1997). Ecological effects of intertidal Manila clam cultivation: Observations at the end of the cultivation phase. *Journal of Applied Ecology, 34*(2), 444–452. https://doi.org/10.2307/2404889.

Tett, P., Charalambides, G., Franco, S. C., Hughes, A. D., Mikkelsen, E., Nolde Nielsen, K., Arthur Bastos Routledge, E., Nielsen, P., & James, P. (2025). Leaving the niche: Recommendations for mainstreaming low trophic aquaculture in countries around the Atlantic basin. *Marine Policy, 171*, 106475. https://doi.org/10.1016/j.marpol.2024.106475.

Turolla, E., Castaldelli, G., Fano, E. A, & Tamburini, E. (2020). Life cycle assessment (LCA) proves that Manila clam farming (*Ruditapes Philippinarum*) is a fully sustainable aquaculture practice and a carbon sink. *Sustainability, 12*(13), 5252. https://doi.org/10.3390/su12135252.

van Roomen, M., van Turnhout, C., van Winden, E., Koks, B. J., Goedhart, P., Leopold, M., & Smit, C. (2005). Trends in benthivorours waterbirds in the Dutch Wadden Sea 1975-2002. *Limosa, 78*, 21–38.

Welsh, D. T., Nizzoli, D., Fano, E. A., & Viaroli, P. (2015). Direct contribution of clams (*Ruditapes philippinarum*) to benthic fluxes, nitrification, denitrification and nitrous oxide emission in a farmed sediment. *Estuarine, Coastal and Shelf Science, 154*, 84–93. https://doi.org/10.1016/j.ecss.2014.12.021.

CHAPTER

Fishery management and regulation

19

19.1 The legitimacy of inshore fishery management

The extent to which Manila clam fishing is regulated varies. In some countries, the activity is conducted informally without any form of regulation. In East Asia in particular, informal recreational Manila clam gathering is typically done by local women for domestic familial consumption in traditional dishes such as clam soup or noodles (Fig. 16.1). Fishing for Manila clams is sometimes based on a common law system of public rights. Such rights may take the form of informal traditions but were apparently first codified in law in sixth-century Byzantium. In the United Kingdom, the origin of a common law right to fish in territorial tidal waters (including intertidal sediments) is sometimes (incorrectly) attributed to Magna Carta (Appleby, 2005). Unless this public right is legally severed in particular areas (e.g., for aquaculture), then, in theory at least, access remains universal and can even extend to mechanized commercial fishing.

However, in 1995, a UN FAO Code of Conduct for Responsible Fisheries signaled a growing worldwide concern about overfishing. In the United Kingdom, although a public right for intertidal shellfish collection was confirmed by an English court in 1998, it was recognized that such a right, if unrestricted, could threaten stocks and constitute an environmental risk, which therefore could be limited by the UK Parliament (Appleby, 2005). Today, the legislative bodies of central or regional governments create regulatory restrictions. For commercial Manila clam fishing operations, regulations are often comprehensive (see below), but some regulations also apply to recreational Manila clam gathering, such as the minimum size limit in the European Union.

The English common law right to fish removes any need to reserve areas for recreational hand gathering. Consequently, such activities may need to compete with commercial fishing. In contrast, in Canada, specific areas of British Columbia, including Baynes Sound, Nanoose Bay, and Kye Bay, are reserved for the "exclusive use and enjoyment of the public and First Nations for the harvest of shellfish for food, social and ceremonial purposes." Here, commercial harvesting is specifically prohibited. Nevertheless, fishers must have a tidal water recreational harvest license and comply with a daily quota of 60 clams, along with guidelines such as avoiding seagrass areas (IFMP, 2022).

The Canadian limit of 60 clams per day illustrates a potential overlap between the meanings of recreational and subsistence fishing. While commercial fishing is banned

The Manila Clam *Ruditapes philippinarum*: Science and Significance of a Globalised Species
DOI: https://doi.org/10.1016/B978-0-443-36704-5.00024-9
Copyright © 2026 Elsevier Inc. All rights are reserved, including those for text and data mining, AI training, and similar technologies.

in such reserved areas, for some people, hand gathering contributes a seasonal asset in terms of dietary requirements. Since, unlike recreational rod fishing, Manila clam gathering is almost universally done for human consumption, a spectrum exists from forms of subsistence to more organized small-scale informal but commercial activity, to established artisanal fisheries and large-scale commercial enterprises. Each type may have its own regulatory framework.

The modern paradigm of fishery management, which has emerged particularly since the 1990s (Garcia et al., 2014), aims to ensure the sustainable exploitation of finite but renewable resources in the context of biological, socio-economic, and wider environmental priorities. Fishery management therefore involves many processes including stock assessments to inform estimates of potential yield; applying the resulting information to determine sustainable levels of exploitation; establishing effort and gear constraints to ensure both fishery and ecological sustainability; formulating voluntary advice and specific regulations through which fishing can be controlled, and active enforcement of regulations involving fishery patrols and sometimes prosecutions. Not least in coastal fisheries, where water pollution may compromise the quality of the catch, fishery management must also attend to public health (sanitary) risk and standards. These various elements may be encapsulated within a fishery management plan.

In essence, fisheries management involves the formulation and implementation of measures to control fishing activity. Regulations generally include technical measures governing how, when, and where fishers can fish. They may include restricted access to the fishery, catch restrictions, including minimum landing size, and gear restrictions. All of these have been applied in Manila clam fisheries. Regulations may be established in primary legislation by central government or more locally through secondary legislation (sometimes referred to as bylaws) by regional or local government. However, not all active management of fishing involves formal regulation. Voluntary agreements based on advice and best practice may also prove effective. In any event, economic and social objectives are generally thought to be essential considerations in effective inshore fishery management, suggesting that the imperatives of precaution and proportionality should be well balanced.

The legitimacy of fishery management and regulation derives from a combination of ecological, economic, and social considerations. As will be described below, in the absence of effective regulation, the accessibility and value of intertidal Manila clam stocks in particular present both opportunity and incentive for informal, sometimes illegal fishing to an extent that may risk dysfunctional "boom and bust" stock collapses and benthic degradation.

Moreover, the creation and enforcement of regulations in themselves have not always proved effective in rebalancing incentives toward legal fishing, not least as penalties are typically relatively small and prosecutions are resource-intensive for regulators. Therefore, to be effective, fishery management must accommodate the needs of fishers (recreational, artisanal, or industrial) to generate sustenance or income and profit from a high-value resource of limited and naturally variable supply.

As such, fisheries management decisions are not just about fishing stocks or the environment but, above all, the management of people's behavior. Since they have an impact (positive or negative) on the socio-economics of communities for whom fishing is an important part, fisheries policy is also social policy, and social justice is an important consideration (Sinclair, 1990). Fisheries policy and management cannot therefore be fair without due consideration of the people most directly affected and the impacts upon them. Rules considered to be illegitimate are less likely to be complied with. Moreover, a fishery authority's capacity to enforce fishery rules can be made inadequate and ineffective by a disenfranchised community's reluctance to follow them (Clark & Humphreys 2020).

With inshore benthic fisheries such as for the Manila clam, local people and the community as a whole are often generally more engaged with fisheries and fishery rules than with offshore, more remote operations. In this context, fishery rules may be considered illegitimate for various reasons. For intertidal shellfisheries, new rules applied to formerly unregulated informal fisheries or more stringent rules as a consequence of an established fishery falling within a new marine protected area can both be sources of disquiet.

Growing and lucrative fisheries based on a new non-native arrival such as the Manila clam can also draw the interest of large-scale producers, in which case real or perceived unfair access at the cost of long-established coastal communities can delegitimize fishery management rules for small-scale operators. In any event, Agnew et al. (2009) estimated the illegal and unreported catch of clams, cockles, and ark shells worldwide to be less than 10% of the reported catch, which, if true, is toward the lower end of the fishery illegality scale.

Below, examples of Manila clam management and regulation are considered, which illustrate all such difficulties and a number of favored solutions. While most regulated clam fisheries, being in near-shore waters, will most likely always have some level of illegal activity, these examples illustrate approaches by which uncontrolled activity, to the detriment of stocks and habitats, may be significantly reduced. Management approaches that align the interests of stocks, environment, and fishers are legitimized to the extent that they are informed by the involvement of the inshore fishing community. Such approaches often involve "co-management" arrangements as exemplified below.

When new opportunities arise, if left unregulated, even mechanized benthic bivalve fisheries can develop quickly as the capital costs of entry are relatively low, requiring only small boats and initially improvised equipment such as the "rusca" gear initially employed in the Venice Lagoon (Chapter 17). In the optimal conditions of the northern Adriatic lagoons, the arrival and swift naturalization of the species in the 1980s and 1990s created highly lucrative opportunities, which in turn led to an apparently exponential rise in informal and largely unregulated fishing effort. In the Venice Lagoon, an unsustainable and effectively "free access system" for the exploitation of Manila clams (Pranovi et al., 2004) was considered to require a strong clam harvesting management plan and a shift from a wild fishery to spatially controlled aquaculture operations.

In Portuguese estuaries, a similar situation emerged after the Manila clam arrived in the 1980s. In the largest Portuguese estuary of the river Tagus, the naturalization of the species was also followed by an exponential growth in the level of fishing, resulting in an estimated 1700 boats taking an annual catch of up to 17,000 tons with a reported value of Euro 23,000,000. Here, Coelho et al. (2021) reported a general disregard and failure to comply with harvesting legislation, spatial zoning in the estuary, or public health "sanitary" standards. Of the 1700 clam boats working in the Tagus estuary, only 187 had fishing licenses (Mamede et al., 2024).

The case for new forms of regulation that reduce established catches is, for socio-economic reasons, often unpalatable even for regulators. Consequently, the focus in the Venice Lagoon was on reducing benthic damage while maintaining or enhancing the existing Manila clam population. Early initiatives therefore proposed taking small seed clams collected from other parts of the lagoon for growing-on in controlled designated culture areas. Nevertheless, for many opportunistic fishers used to an unregulated and lucrative fishery, such complications were initially unwelcome. By 1999, small areas in the southern basin had already been set aside for Manila clam farming, but the initiative was not yet considered successful due to "difficulties with management and fishermen" (Sorokin et al, 1999).

This illustrates the political reality of establishing regulatory regimes. Even if the objective is to maximize long-term economic benefit through the application of fishery science concepts such as "maximum sustainable yield," fishers tend to be skeptical, and the regulation of entrepreneurial access to income and profit stimulates opposition. Pontecorvo and Shrank (2009) describe commonly held feelings among fishers that fishery managers know less about the reality of the fishery than they do themselves. Also, the history and traditions of fishing communities can give them more political influence than their size or economic significance might suggest. Small-scale inshore fishers common in Manila clam fisheries are often fiercely independent and, like most people, are ingenious at working out methods for getting around restrictions they do not like. Even where fishing restrictions are in force, opportunities often exist for illegal activity.

According to general theories of deterrence, compliance with laws and regulations is most heavily influenced by three key factors: The potential economic gains from the illegal activity, the perceived risk of detection, and the severity of sanctions. Fisheries are no different (Nielsen & Mathiesen, 2000; Sutinen et al., 1990). Manila clams are a high-value seafood, and the economic gains from a day's fishing are substantial. In the United Kingdom, the inshore fishery area divided by the number of fishery patrol boats, implies that each vessel must, on average, patrol 640 km^2 of sea area. Even allowing for new drone technologies and port landing inspections, the probability of detecting illegal fishing is low. Moreover, the value of Manila clam catches can far exceed the penalties for illegal exploitation, which can become little more than a "cost of business" for illegal operators (Clark & Humphreys, 2020). Incentives for illegal fishing are high for those inclined to take the risk.

Below, examples of Manila clam fishery management in England, China, France, and the USA are briefly considered. The first two are in Poole Harbour, Dorset,

England, and the Yalu Estuary, Liaoning Province, China. These sites with widely different contexts, and despite both experiencing significant fishery management issues, both improved in such a way as to be accredited by the Marine Stewardship Council (MSC), an international organization which aims to compensate for failures in fishery management by seeking to mobilize civil society through consumer choice, facilitated by a fishery certification program. To this end, the MSC has established Fisheries Standards designed to prevent overfishing and maintain the structure and function of the host ecosystem (Agnew et al., 2014).

MSC processes are bureaucratic by nature, but their standards have the merit of trying to express global best practice in fishery management through achievable, if challenging, standards. These standards focus largely on the effectiveness of regulation as indicated by regular fishery data gathering and monitoring, rather than on the particular technical measures introduced locally. They require fishery managers to act in response to such data, so as to ensure sustainability. Therefore, MSC standards cover matters including stock status assessment, harvest strategy, bycatch and habitat impact, governance, compliance, enforcement, and monitoring. They include specific standards for "enhanced bivalve fisheries" (involving relaying hatchery produced seed clams on the seabed), which include measures for genetic and translocation management (MSC, 2022).

19.2 Poole Harbour, England

A major part of modern marine conservation strategy is the designation of Marine Protected Areas (MPAs), many of which are in the sorts of estuaries and bays that offer favored habitats for Manila clams (Humphreys & Clark, 2020). Poole Harbour is a multidesignated MPA; a long-established bivalve fishing site; and the location of the country's first licensed Manila clam fishery. The period before 2015 was subject to high levels of illegal activity, and the site illustrates how a laissez-faire fishery in a sensitive marine environment can be brought under control through a range of technical measures along with enhanced enforcement, all with the support of reputable local fishers.

On the relatively warm south coast of England, Poole Harbour became the first site in which the Manila clam naturalized in Britain. By 1994, the first licensed Manila clam fishery had been established; however, it quickly developed a reputation for high levels of illegal activity, with resulting concern both from fisheries and conservation perspectives (Jensen et al., 2004, 2005). Moreover, at that time, the extent of illegal fishing had led to a breakdown between legitimate licensed fishers and fishery managers. The resulting fisheries management project was an attempt to radically improve the reputation and sustainability of the fishery and better conserve the features of the MPA.

The Harbour falls within the responsibilities of the Southern Inshore Fishery and Conservation Authority (SIFCA). SIFCA is one of 10 such regional authorities, each covering a district on the English coast. A range of legal measures is available to such statutory authorities by virtue of their powers to introduce and enforce secondary legislation and prosecute offenders (Clark & Humphreys 2020). The management of the fishery is achieved through measures, as detailed in SIFCA (2023). In summary, they are as follows:

- Restricted entry permit scheme with flexible conditions.

This removed historic regulatory loopholes that provided for illegality. By focusing on the type and use of dredge gear, rather than species it replaced species-specific management measures for two species fished with the same dredge gear (Manila clams and cockles) with a single common regime, which could be adjusted in line with changes in the fishery or environment through annual license conditions (i.e., without changing the byelaw itself). The permit also set out catch restrictions, reporting requirements, gear specifications, spatial and temporal restrictions, including permanently closed areas, and prohibitions on the use or even carriage on board of dredges during the closed season. This restriction on the carriage of dredges closed a loophole through which illegal fishers, when challenged, could claim not to have been using the dredge. Noncompliance with these requirements could lead to loss of permit (Birchenough, 2022).

The idea underlying the new Poole Harbour management regime was to resolve socio-economic and environmental tensions by creating a value to a permit which outweighed the perceived value of illegal fishing (Birchenough et al., 2020). The permit scheme was initially based on a historic track record, therefore limiting permits to all the legitimate fishers who had made catch returns before the new scheme, while thereby also effectively favoring local inshore boats. No such fishers or boats were initially excluded, and a waiting list was subsequently established. Local fishers with permits became incentivized to discourage illegal activity which eroded their own catch potential.

- Minimum landing size

The Minimum Conservation Reference Size Byelaw established 35 mm as the shell length below which Manila clams may not be removed from the fishery, retained on board, trans-shipped, or landed. This size limit originated in European Union (EU) regulations but had to be enshrined in English legislation after "Brexit" (withdrawal of the UK from the EU). It had been reduced from 40mm based on biological evidence on size at sexual maturity, combined with lobbying by European clam fishers (see Arcachon Bay below).

- Other statutory requirements

These included fishing boat registration requirements and maximum fishing boat size restrictions within the Southern IFCA district. Restrictions on hand gathering in Poole Harbour also included a closed season, prohibition from gathering from seagrass beds, and use of a hand rake only. Carrying a hand rake during the closed season also became an offence.

Due to the value of the Manila clam and cockle fisheries in Poole Harbour, it had become a hot spot for illegal activity within the district. Therefore, as part of the overall project and coincident with the various new regulatory measures, the SIFCA invested in increased fishery enforcement patrols within the Harbour, which resulted in a number of early prosecutions.

- Nonstatutory management

The formulation and introduction of the new regulatory regime involved a collaboration between the fisheries authority (SIFCA) and the Poole & District Fisherman's Association (PDFA); environmental NGOs and government officials also acted as advisors. Few fisheries are without some level of illegal activity. However, within a year of the introduction, various metrics indicated that illegal activity within the fishery had declined by 95% (Birchenough, 2022).

- Marine Stewardship Council accreditation

Subsequently, the SIFCA and PDFA set up a project to seek accreditation from the Marine Stewardship Council. This was achieved in 2018 and represented a form of independent framework within which the fishery could continue to develop. It also verified the move away from historic perceptions of an unsustainable fishery driven by illegal activity and thereby opened up new markets for the fishers (Birchenough, 2022).

19.3 The Yalu Estuary, China

The Yalu River marks the boundary between China and North Korea and flows into a biodiverse coastal area of the Yellow Sea, which, like Poole Harbour, has conservation value as a stop-over point for migrating seabirds. On the western side of the estuary lies the Chinese province of Liaoning. The long-established Yalu Estuary Manila clam fishery takes the form of an enhanced "hatch and catch" fishery with clam seed supplied from the more southerly Putian City (Fujian Province), where many hatcheries are based, as climate conditions are conducive to Manila clam seed production.

In Putian on the Taiwan Strait, wild broodstock clams for hatchery operations are fished mainly from shallow subtidal populations by water jet dredge. This broodstock fishery targets larger clams, which are preferred for maximum fecundity. Management involves gear specifications with a mesh size of 40 mm, seasonal closures during the peak spawning period, and a total allowable catch based on standing stock surveys and landings data, among other information. In 2018, a stock survey estimated a standing stock of 33,000 t, and the Putian Clam Association established a maximum sustainable yield of 10,000 t. The Putian wild broodstock fishery is conducted by local fishers with total annual landings of around 4000 t, of which around 2500 t per year is sold to the hatchery as broodstock (MSC, 2021).

From this broodstock, seed clams are produced in large local ponds. Of the 3500ha of ponds used for cultivation in Putian, about 1000ha are used for clams destined for the Yalu. However, most of these ponds are transferred to Putian's intertidal mudflats as "sand seed" (1–4 mm) for on-growing. At around 10 mm shell length, they are harvested using hand raking and simple towed dredges and transported north to Liaoning Province for re-laying subtidally off Donggang on the Yalu estuary. At Market size (around 30 mm), they are harvested again using water jet dredges (or "propellor dredges" similar to those described for the Venice Lagoon in Chapter 17). As such, two fisheries have to be sustainably managed (MSC, 2021).

Initially, MSC certified the fishery in 2021 with conditions requiring improvements in stock, harvesting impact information, and the management capacity to respond effectively to issues raised by such improved monitoring. However, by 2024, the fishery ran into accreditation difficulties due to an insufficiency of fishery monitoring data. Although the wild broodstock fishery was regarded as most likely sustainable, wild stock abundance estimates and landings data were regarded as inadequate by MSC surveyors: For example, infrequent stock surveys and an absence of information on historical annual landings, and fishing effort in terms of the number of boats. All of which prevented secure estimates of stock structure, productivity, and fishery performance. A plan was put in place to rectify matters (MSC, 2024), and the fishery was reinstated, subject to further audit (MSC, 2025).

19.4 Arcachon Bay, France

Arcachon Bay on the Atlantic coast of Aquitaine covers an area of 180 km^2, two-thirds of which are mudflats exposed by low tides. Boats are used to reach much of this area on ebb tides, and hand gathering commences when the boats are grounded, which generally allows 4 h of gathering time. The population is not enhanced by seeding, so the fishery is entirely wild stock. There is no closed season, but fishing is only permitted in daylight hours, and gathering on seagrass beds is prohibited. In 2016, stock declines provoked the creation of three reserve areas, which are left unfished to provide wild broodstock. Stock decline is thought in part to be caused by two non-native arrivals: the Mediterranean gastropod predator *Hexaplex trunculus*, and the Asian date mussel (*Arcuatula senhousia*), which since 2002 has been found attached to live Manila clam shells and is thought to compete for food and space (Montaudouin et al., 2025)

Caill-Milly et al. (2025a) have reviewed the governance of Manila clam fisheries in France along with current knowledge on biology and ecology there. French fishery management is established at three levels: The European Union establishes minimum landing sizes for member states; the National government controls fishing boat permits and sanitary regulations; and regional instruments set spatial and temporal access rules and gear types.

Arcachon Bay has been the focus of research into minimum landing sizes for the Manila clam. In fishery management, minimum landing sizes are designed to maintain sufficient viable breeding stock. This is achieved by ensuring that fishing effort does not take too high a proportion of sexually mature breeding individuals, and that the supply of such individuals from smaller juvenile cohorts is maintained. It is typically set at the size at which 50% of the population reaches sexual maturity, on average, allowing individuals to mature and reproduce at least once before being caught. As such, minimum landing sizes are generally informed by scientific estimates of size at sexual maturity (SSM), which inform a "Minimum Conservation Reference Size" (MCRS).

Under European Union (EU) regulations, as a non-native species, *R. philippinarum* would in normal circumstances be subject to regulations for the prevention, detection, and management of invasive alien species (EU, 2014). However, *R. philippinarum* was exempted from these requirements and those more generally applying to the use of non-native species in aquaculture in 2007 (EU, 2007). Since then, *R. philippinarum* has been managed as a natural rather than a non-native resource (Coelho et al., 2021), and there has been a reduction in minimum landing size (from an original shell length of 40 mm).

In 2020, a group of interested member states submitted a proposal to further reduce the MCRS from 35 to 32 mm in Arcachon Bay. In Arcachon, size at sexual maturity was reported as 26.7 mm, and it was concluded that the impact on juveniles of this reduction was likely to be negligible, and that, while a greater fraction of sexually mature stock would be harvestable, a larger proportion of mature individuals (length 26.7–32 mm) would remain in the population. This MCRS reduction was approved in the context of other management measures, including zones closed to fishing, a system of clam licenses and biennial stock surveys (EU, 2024). Caill-Milly et al. (2025b) provided a review of the biological features considered in a review of the Manila clam's MCRS, with data for Arcachon Bay, including SSM, growth rates, and natural mortality coefficients for both juveniles and adults.

The case of the Manila clam in Europe exemplifies the tensions between ecological and economic priorities in regulatory decisions. While MCRS reductions have been scientifically argued with evidence, the motivation is economic, and the agenda for reducing MCRS has been repeatedly requested by French and other fishers (Caill-Milly et al., 2025b). A call for a 32 mm MCRS was initially rejected, but further scientific work and government-supported industry lobbying brought about its subsequent acceptance.

The extent to which different states prioritize economic arguments varies, among other things, according to the relative strength of industrial and conservation organizations. In France, the fishery and agriculture sectors are traditionally strong and active (note the French lead on MCRS reduction), whereas elsewhere conservation organizations such as environmental NGOs may be more influential on policy. In the Netherlands, for example, two environmental NGOs obtained a ruling from the European Court of Justice relating to the Wadden Sea, which led to changes in the management of UK Manila clam dredging after local challenges by two UK-based NGOs (Clark et al., 2017). At this time, the UK was part of the European Union. At national and regional levels of fisheries governance, management regimes can vary substantially according to prevailing attitudes and the relative strengths of different lobby groups.

19.5 Skokomish tribal territory, Washington, USA

A recognized contributing factor in failed regulatory systems is the extent to which fishers are engaged in the formulation of the regulations with which they have to comply. And in ways which recognize both their often extensive local knowledge and economic needs. The concept of comanagement, while strongly supported on social justice grounds, has the added benefit of reducing the chances of regulatory failure.

Definitions of comanagement vary, but they all require some level of participation of fishers in the processes of rule formulation (Berkes et al., 2001). Many regulatory regimes are now underpinned by some form of comanagement arrangement, including the cases discussed above. For example, Caill-Milly et al. (2021) have described a French comanagement arrangement for Arcachon Bay, which started when fishers became directly involved in stock assessments and in joint working groups. In the United Kingdom, the regional Inshore Fisheries and Conservation Authorities have fisher participation on their governing boards.

A strong social justice argument for comanagement arrangements puts fishing communities at the center of the process. In the tribal lands of Washington State, United States, a strong social justice case, after many years of disputation, resulted in a deep level of comanagement of Manila clam fisheries.

In the mid-19th century, the US Congress negotiated treaties with the coastal First Nation communities (Indian tribes) of Washington, which secured tribal fishing rights on reservations. The Skokomish River passes through such tribal lands, and the Manila clam has been a major part of the catch since its arrival. Until the 1920s, clam fishing was dominated by the tribes, but as tidelands continued to be purchased by "non-Indians," the tribes were slowly excluded from their traditional shellfish harvest areas. However, despite "more than a century of sometimes violent controversy between Indians and non-Indians over fishing rights" (Boldt, 1974), the tribes were tenacious in protecting their rights and sovereignty.

In the 1970s, the US federal government sued Washington State for tribal fishing rights violations. The resulting "Boldt Decision" stipulated that the Skokomish tribe (among others) was, subject to certain conditions, "entitled to exercise its governmental powers by regulating the treaty right fishing of its members without any state regulation thereof," and that the original treaties intended that there should be an equal 50/50 sharing of fish between the tribes and the settlers (Boldt, 1974). A subsequent "Rafeedie Decision" confirmed the same principle for shellfish.

Today, subject only to universal general ordinances, the Skokomish tribe has maintained its extensive treaty rights, within which they have established a tribal Government with a Department of Natural Resources. This governs and supports tribal members' ability to exercise their fishing rights. Tribal responsibilities include stock surveys; the licensing of fishers, buyers, and boats; Manila clam stock enhancements, including clam seed laying. All are directly controlled by the local community (Skokomish Natural Resources Department Habitat Division, 2025a). Specific Manila clam regulations include a minimum landing size (38 mm) and temporal restrictions on subsistence and commercial harvesting (Skokomish Natural Resources Department Habitat Division, 2025b), with management processes including an annual shellfish comanagement meeting and resulting management plans. While the state and other stakeholders remain involved in these processes, the tribal community is now an equal and managing partner, with legally guaranteed shellfish rights (Blair Paul, personal communication).

19.6 Illegality and traceability

The phrase illegal, unreported, and unregulated fishing (IUU) is a common and globally widespread phenomenon. An international action plan on IUU Fishing (IPOA-IUU, 2001), developed by the UN FAO, defines IUU as activities that contravene national and international laws and regulations. Especially for inshore and estuarine fishing, the implications of IUU fishing reach beyond ecological sustainability to the risks that contaminated shellfish pose to the safety of the ultimate consumers. Public health is therefore an essential consideration in fishery regulation, and various measures are taken to ensure the good sanitary status of Manila clams, including the monitoring of contamination; purification when necessary; and full closure of fishing grounds when contamination reaches levels incompatible with human health.

In the European Union, regulations require the classification of designated shellfishing grounds according to the levels of the *Escherichia coli* bacterium found in the flesh and mantle fluids of bivalve samples. In addition, there are labeling regulations to ensure the traceability of each batch of seafood by specifying catch area, production method, and fishing gear (EU, 2013). Illegal fishing and trading can circumvent all such measures, and many aspects of fishery regulation are therefore designed not only to achieve sustainable ecological and economic outcomes, but also to ensure public health.

The public health risks of illegal fishing are experienced, not by illegal fishers, but by consumers, and the provenance of fished shellfish may be lost in supply chains involving fishers, wholesalers, and retailers, unless accurate and honest records are kept. The traceability of Manila clam for consumption is complicated by the fact that the species is one of the most traded bivalves in the world (Mamade et al, 2020). Even within national markets, as a traded commodity, Manila clam supply chains can be complex.

At its simplest, a Manila clam supply chain describes the stages from catch or harvesting to landing port, first sale merchant, final retail market to consumers. Although such simple chains make up a part of the supply chain, in practice, supply chains are often more complex, typically involving an initial merchant who buys from the fisher or aquaculture operator and sells on to a larger distributor before the catch reaches a retail market or restaurant. The chain is further complicated when merchants sell to each other, in response to geographic demand patterns, or when the live product is exported.

Some of the possible additional consequences of international transfers include health risks for both native species and humans, including the introduction of new pathogens, parasites, and inorganic toxins, and with seed clams, new strains of toxic dinoflagellates, etc. Consequently, modern international trade movements may involve stringent sanitary regulations. Nevertheless, fraudulent mislabeling of geographic origin is a widely accepted problem in the seafood trade (Leal et al, 2015).

At a time when the Tagus estuary in Portugal was closed to Manila clam fishing for food safety reasons, Mamade et al. (2020) investigated the harvesting location of four batches of Manila clams suspected of being illegally fished. Based on chemical composition, specimens from three of the four batches tested were found to have a 90% + likelihood of being fished from the closed Tagus estuary. A result that emphasizes the importance of traceability analyses.

Beyond public health considerations, increased consumer interest in the origins and legitimacy of food purchases has also contributed to the need for regulations on the traceability of seafood. The UK Fisheries Act of 2020 defines seafood traceability as the "*ability of any person to discover information about how, where or when the fishery products were (a) caught, harvested or made, or (b) transported, stored or sold.*" For many foodstuffs, some form of "Protected Designation of Origin" (PDO) can add to value in consumer markets by providing an endorsement of authenticity and quality and by reassuring consumers that their purchase does not support IUU activity. Moreover, since such schemes are likely to increase the value of the product, lack of traceability creates conditions in which fraud, mislabeling and IUU fishing can occur (Hopkins et al 2024).

Various techniques have the potential to determine Manila clam origins, including genetic sequencing. However, genetic approaches can be compromised by the mixing of stocks. For example, over 4000 t per year of Chinese seed clams are exported to Korea, with the result that mixing can make the origin of traded clams difficult to determine genetically. Similar export of seed clams and consequent genetic mixing has occurred in parts of Europe. Therefore, although genetics has been used to track the global spread of the species, alternative methods for the determination of the geographical origins of Manila clams have been investigated.

In contrast to the genome, the elemental composition of clams reflects that of the ecosystem in which they live. Therefore, chemical signals from particular estuaries can be compared with the chemical composition of the shell or soft tissue.

In northern Adriatic lagoons where the provenance of local clams has been reported as compromised by "counterfeiting with clams of exotic and dubious provenance" (Bianchini et al., 2021), the regional government has sought to designate the Sacca di Goro as a PDO for Manila clams. These authors have therefore investigated the utility of using stable carbon and nitrogen isotope ratios, determined by Isotope Ratio Mass Spectrometry (IRMS), as a rapid and cost-effective approach to authentication. In the Northern Adriatic lagoons, the environment has low ^{13}C and high ^{15}N values. This isotopic ratio, although somewhat modified biologically and varying over time, is transferred into clam tissues, which Bianchini et al. (2021) have shown can be distinguished from clams originating elsewhere in the world. They have since refined this approach by adding the isotopic ratios of more elements to increase the sensitivity, even to the extent of particular lagoons (Brombin et al., 2022). Won et al. (2021) have also demonstrated the utility of this approach to discriminate between Manila clams from North Korea, China, and South Korea.

For their investigations in Portugal, Mamede et al. (2020) used an alternative biochemical approach based on fatty acid profiles in Manila clam adductor muscle. Here they exploit the variability in fatty acid composition caused by factors including temperature, salinity, and sediment composition. These factors affect the structure and fluidity of cell membranes, characteristics which are themselves determined by combinations of saturated and polyunsaturated fatty acids. For example, high salinity fluctuations and low temperatures result in lower saturated fatty acid content. By comparing the fatty acid content of Portuguese and Spanish Manila clams, Mamede et al. (2020) were able to distinguish Manila clams collected from the Tagus estuary, Portugal, from those taken from other estuaries.

References

Agnew, D. J., Gutiérrez, N. L., Stern-Pirlot, A., & Hoggarth, D. D. (2014). The MSC experience: Developing an operational certification standard and a market incentive to improve fishery sustainability. *ICES Journal of Marine Science, 71*(2), 216–225. https://doi.org/10.1093/icesjms/fst091.

Agnew, D. J., Pearce, J., Pramod, G., Peatman, T., Watson, R., Beddington, J. R., Pitcher, T. J., & Sandin, S. A. (2009). Estimating the worldwide extent of illegal fishing. *PLoS ONE, 4*(2), e4570. https://doi.org/10.1371/journal.pone.0004570.

Appleby, T. (2005). The public right to fish: Is it fit for purpose? *Journal of Water Law, 16*(6), 201–205.

Berkes, F., Mahon, R., McConney, P., Pollnac, R., & Pomeroy, R. (2001). *Managing small-scale fisheries: Alternative directions and methods*. International Development Research Center.

Bianchini, G., Brombin, V., Carlino, P., Mistri, E., Natali, C., & Salani, G. M. (2021). Traceability and authentication of Manila clams from North-Western adriatic lagoons using C and N stable isotope analysis. *Molecules, 26*(7), 1859. https://doi.org/10.3390/molecules26071859.

Birchenough, S. E., Clark, R. W. E., Pengelly, S., & Humphreys, J. (2020). Managing a dredge fishery within a marine protected area: Resolving environmental and socio-economic objectives. In J. Humphreys, & R. W. E. Clark (Eds.), *Marine protected areas. Science, policy, and management* (pp. 459–474). Elsevier.

Birchenough, S. E. (2022). The Manila clam in Poole Harbour. A journey to sustainability. In J. Humphreys, & A. E. Hall (Eds.), *Harbour Ecology, environment and development in Poole Harbour* (pp. 155–170). Pelagic Publishing.

Boldt, G. H. (1974). The Boldt decision. USA v State of Washington. US District Court for the Western District of Washington. Tacoma Division. *Washington*. https://wdfw.wa.gov/sites/default/files/2021-12/boldt_decision.pdf.

Brombin, V., Natali, C., Frijia, G., Schmitt, K., Casalini, M., & Bianchini, G. (2022). Isotope geochemistry for seafood traceability and authentication: The northern Adriatic Manila clams case study. *Foods, 11*(19), 3054. https://doi.org/10.3390/foods11193054.

Caill-Milly, N., Bru, N., de Montaudouin, X., Dang, C., Briaudeau, T., & Sanchez, F. (2025b). Biological features for a minimum conservation reference size revision- case of Ruditapes philippinarum in Arcachon Bay (French Atlantic coast). *Estuarine, Coastal and Shelf Science, 320*, 109288.

Caill-Milly, N., Sanchez, F., Lissardy, M., Bru, N., Kermorvant, C., Montaudouin, X., Lapègue, S., Riquet, F., Bouché, L., D'Hardivillé, C., Lagarde, F., Chambouvet, A., Mayot, N., Dauvin, J.-C., Pezy, J.-P., Basuyaux, O., Guéguen, A., Weiller, Y., Arzul, I., … Carter, C. (2025a). Manila clam (Ruditapes philippinarum) in France: Fishing activity, governance and present knowledge challenges regarding its biology and ecology. *Estuarine, Coastal and Shelf Science, 317*, 109206. https://doi.org/10.1016/j.ecss.2025.109206.

Caill-Milly, N., Sanchez, F., Lissardy, M., de Montaudouin, X., Bru, N., Kermorvant, C., & Ganthy, F. (2021). Drawing lessons from a pluridisciplinary approach associating stakeholders for a better management of a bivalve population (French Atlantic coast)? *Estuarine, Coastal and Shelf Science, 251*, 107194. https://doi.org/10.1016/j.ecss.2021.107194.

Clark, R., Humphreys, J., Solandt, J.-L., & Weller, C. (2017). Dialectics of nature: The emergence of policy on the management of commercial fisheries in English European Marine Sites. *Marine Policy, 78*, 11–17. https://doi.org/10.1016/j.marpol.2016.12.021.

Clark, R. W. E., & Humphreys, J. (2020). Enforcement capabilities and compliance in English marine protected areas: The art of the possible. In J. Humphreys, & R. W. E. Clark (Eds.), *Marine protected areas. Science, policy, and management*. Elsevier.

Coelho, P., Carvalho, F., Goulding, T., Chainho, P., & Guerreiro, J. (2021). Management models of the Manila Clam (*Ruditapes philippinarum*) fisheries in invaded European coastal systems. *Frontiers in Marine Science, 8*. https://doi.org/10.3389/fmars.2021.685307.

EU, 2007 Council Regulation(EC) No 708/2007 of 11 June 2007 concerning use of alien and locally absent species in aquaculture. Official Journal of the European Union. https://eur-lex.europa.eu/eli/reg/2007/708/oj/eng

EU. (2013). Regulation (EU) No 1379/2013 of the European Parliament and of the Council of 11 December 2013 on the common organisation of the markets in fishery and aquaculture products, amending Council Regulations (EC) No 1184/2006 and (EC) No 1224/2009 and repealing Council Regulation (EC) No 104/2000. *Official Journal of the European Union, L354*, 1–21.

EU (2014). Regulation (EU) No 1143/2014 of the European Parliament and of the Council of 22 October 2014 on the prevention and management of the introduction and spread of invasive alien species. Official Journal of the European Union. https://eur-lex.europa.eu/eli/reg/2014/1143/oj/eng.

Garcia, S. M., Rice, J., & Charles, A. (2014). Governance of marine fisheries and biodiversity conservation: A history. In S. M. Garcia, J. Rice, & A. Charles (Eds.), *Governance of marine fisheries and biodiversity conservation: Interaction and coevolution* (pp. 3–17). Wiley.

Hopkins, C. R., Roberts, S. I., Caveen, A. J., Graham, C., & Burns, N. M. (2024). Improved traceability in seafood supply chains is achievable by minimising vulnerable nodes in processing and distribution networks. *Marine Policy, 159*, 105910.

Humphreys, J., & Clark, R. W. E. (2020). A critical history of marine protected areas. In J. Humphreys, & R. W. E. Clark (Eds.), *Marine protected areas. Science, policy, and management*. Elsevier.

IFMP. (2022). *Pacific region. Integrated fisheries management plan. Intertidal clams*. Fisheries and Oceans Canada Catalogue. No. DFO/21-2076 Fs143-3/21-2076E-PDF.

EU (2024). Commission Delegated Regulation (EU) 2024/3204 of 11 October 2024 amending Regulation (EU) 2019/1241 of the European Parliament and of the Council as regards short-necked clam and red seabream. Official Journal of the European Union. https://eur-lex.europa.eu/legal-content/EN/TXT/?uri=CELEX%3A32024R3204&qid=1748507960096.

IPOA-IUU (2001). International plan of action to prevent, deter and eliminate IUU Fishing. *Food and Agriculture Organisation of the United Nations*.

Jensen, A., Humphreys, J., Caldow, R. W. G., Grisley, C., & Dyrynda, P. E. J. (2004). Naturalisation of the Manila clam (*Tapes philippinarum*) an alien species, and establishment of a clam fishery within Poole Harbour, Dorset. *Journal of the Marine Biological Association of the United Kingdom, 84*, 1069–1073.

Jensen, A., Humphreys, J., Caldow, R., & Cesar, C. (2005). The Manila clam in Poole Harbour. In J. Humphreys, & V. May (Eds.), *The ecology of Poole Harbour*. In *Proceedings in Marine Science* (pp. 163–173). Elsevier.

Leal, M. C., Pimentel, T., Ricardo, F., Rosa, R., & Calado, R. (2015). Seafood traceability: Current needs, available tools, and biotechnological challenges for origin certification. *Trends in Biotechnology, 33*(6), 331–336.

Mamede, R., Ricardo, F., Santos, A., Díaz, S., Santos, S. A. O., Bispo, R., Domingues, M. R. M., & Calado, R. (2020). Revealing the illegal harvesting of Manila clams (*Ruditapes philippinarum*) 1 using 2 fatty acid profiles of the adductor muscle. *Food Control, 118*, 107368.

Mamede, R., Santos, A., Ferreira da Silva, E., Patinha, C., Calado, R., & Ricardo, F. (2024). New evidence of fraudulent mislabeling and illegal harvesting of Manila clams (*Ruditapes philippinarum*) through elemental fingerprints of their shells and chemometric analyses. *Food Control, 163*, 110501.

Montaudouin, X. D. e, Baudot, L., Bernard, G., Blanchet, H., Masse, C., & Fouet, M. P. A. (2025). Population dynamics of the potentially invasive Asian date mussel, *Arcuatula senhousia*, in Arcachon Bay, France. *Estuarine, Coastal and Shelf Science, 32*, 109448.

MSC. (2021). MSC fishery assessment report. Public certification report. *MSC fishery assessment report. Public certification report*: 21. Marine Stewardship Council.

MSC 2022. MSC fisheries standard version 3.0 https://www.msc.org/docs/default-source/default-document-library/for-business/program-documents/fisheries-program-documents/msc-fisheries-standard-v3-0.pdf.

MSC. (2024). *Yalu Estuary Manila clam fishery. MSC Notice of Suspension Template v1.2 /SCS Version 1-0 (January 2023*.

MSC. (2025). *Yalu Estuary Manila clam fishery*. Marine Stewardship Council.

Neilsen, J. R., & Mathiesen, C. (2000). Important factors influencing rule compliance in Fisheries: Lessons from Danish Fisheries. In *Proceedings of the IIFET, 2000 Conference* (pp. 10–14).

Pontecorvo, G., & Schrank, W. E. (2009). *Fisheries management: Pandemic failure, workable solutions*. Emerald Publishing Ltd.

Pranovi, F., Da Ponte, F., Raicevich, S., & Giovanardi, O. (2004). A multidisciplinary study of the immediate effects of mechanical clam harvesting in the Venice Lagoon. *ICES Journal of Marine Science, 61*, 43–52.

SIFCA. (2023). *Southern Inshore Fisheries and Conservation Authority* Southern-ifca.gov.uk.

Sinclair, P. R. (1990). Fisheries management and problems of social justice: Reflections on Northwest Newfoundland. *Maritime Anthropological Studies, III*(No. 1), 30–47.

Skokomish Natural Resources Department Habitat Division, 2025a Natural resources. Retrieved May 30, 2025, from https://skokomish.org/natural-resources/#:⟨:text=Management%20services%20include%20licensing%20fishers,include%20clam%20%26%20oyster%20seed%20planting.

Skokomish Natural Resources Department Habitat Division 2025b. Clam regulations. Retrieved May 30, 2025, from https://skokomish.org/tribalregs/.

Sorokin, I., Giovanardi, O., Pranovi, F., & Sorokin, P. I. (1999). Need for restricting bivalve culture in the southern basin of the Lagoon of Venice. *Hydrobiologia, 400*, 141–148.

Sutinen, J. G., Reiser, A., & Gauvin, J. R. (1990). Measuring and explaining non-compliance in federally managed fisheries. *Ocean Development and International Law, 21*, 335–372.

Won, E.-J., Kim, S. H., Go, Y.-S., Kumar, K. S., Kim, M.-S., Yoon, S.-H., Bayon, G., Kim, J.-H., & Shin, K.-H. (2021). A multi-elements isotope approach to assess the geographic provenance of Manila Clams (*Ruditapes philippinarum*) via recombining appropriate elements. *Foods, 10*, 646. https://doi.org/10.3390/foods10030646.

PART

Naturalisation and invasion IV

CHAPTER

Naturalization and dispersal

20

20.1 Diversity and opportunity in Manila clam habitats

Ruditapes philippinarum is a euryhaline species of phylogenetically marine origin which occupies relatively stable sediments. Such habitats are found within the intertidal and shallow subtidal areas of sheltered estuaries, bays, and lagoons. They are also typically "transitional waters" (Basset et al., 2013) subject to varying combinations of tidal and freshwater influence, often including extensive tidal flats. Salinity is a vital environmental factor in such areas, and its instability creates a complex environment in which salinity generally shows an inverse relationship with macrobenthic species diversity. Although these habitats may have high rates of both primary and secondary productivity, they are relatively poor in species in comparison to adjacent fully fresh or fully marine habitats.

This relationship between salinity and species diversity was first formulated by Remane in the Baltic (Remane, 1934; Remane & Schlieper, 1958), and although in some ways limited as a general model, it remains broadly valid for the zoobenthic species on which it was based (Whitfield et al., 2012). The lowest species richness is generally considered to exist in the upper reaches of estuaries with salinities between 5 and 8. Nevertheless, species richness remains suppressed at higher salinities, gradually increasing downstream with prevailing salinity.

Low species diversity in estuarine environments has been accounted for by two categories of explanation: geological and biological. In the first place, the temperate habitats of *R. philippinarum*, while subject to various climatic oscillations operating over different timescales, have been most profoundly influenced by alternating glacial and interglacial periods and resulting changes in the relative elevation of land and sea. As recently as 17,000 years ago, during the last glacial "ice age" with sea levels close to the edge of the continental shelf, many of today's estuaries were fluvial geomorphological features far inland and many in temperate regions, ice-covered. As the icecaps melted during the present interglacial Holocene epoch, sea levels approximating those of the present day created the current coastlines with their estuaries and tidal flats (Fairbridge R W 1980).

The effective destruction of the estuaries and protected coastal flats extant during the last ice age is known to have coincided with the regional extinction of estuarine species such as the bivalve *Mya arenaria,* which disappeared from European waters. Although this species subsequently recolonized from North America, possibly via Viking voyages (Essink & Oost, 2017), it nevertheless illustrates how estuarine

The Manila Clam *Ruditapes philippinarum*: Science and Significance of a Globalised Species
DOI: https://doi.org/10.1016/B978-0-443-36704-5.00028-6

Copyright © 2026 Elsevier Inc. All rights are reserved, including those for text and data mining, AI training, and similar technologies.

fauna would have been lost, and how, as a result, the present-day fauna is largely recently colonized. In this scenario, the geologically recent nature of today's estuarine habitats has allowed little time for adaptive radiation of species into the new Holocene estuaries, not least in the context of their challenging physiological demands.

Moreover, this colonization of emergent estuarine environments will have been additionally physiologically challenging due to the unique demands of temperate estuaries and their intertidal habitats: largely a consequence of their combinations of salinity and aerial exposure regimes. Although the Manila clam can be found in fully marine and sub-tidal habitats, its major success in both native and non-native regions is associated with estuarine (i.e., reduced salinity) sediments (Chapter 11). Estuaries are peculiar in the range and instability of their natural environmental variables: On daily and monthly timescales, tidal cycles with varying amplitude create periods of submergence and aerial exposure, upon which are superimposed fluctuations in weather conditions both locally and in their river catchments. These add stochastic volatility to both salinity and temperature. The resulting unpredictable interactions of environmental variables represent a considerable physiological challenge, which, when combined with the geological recency of these environments, explains species paucity.

Saunders (1968), in his "stability-time hypothesis" predicted that, whereas in a stable benthic environment, many species evolve with narrow niches shaped by competition, less stable and younger environments will have fewer species. Although studies in other habitat types have not always corroborated Saunders' hypothesis (Abele & Walters, 1979; Thistle, 1982), the basic idea remains valid as a general explanation for the paucity of estuarine species (McClusky & Elliott, 2004; Reise, 1985). Moreover, "relative saturation" and "community maturity modeling" by Byers and Noonberg (2003) suggest that lower species diversity in a community reduces species interaction intensity, thereby increasing opportunities for non-native species.

In any event, on tidal flats especially, it cannot be assumed that the arrival of a new species, even when abundant, necessarily causes any corresponding decline in existing species (Reise, 1985). Attrill and Rundle (2002) have described estuaries in terms of two overlapping physicochemical gradients (ecoclines), the first from river to estuary, the second from sea to estuary, each ecocline being inhabited by species of freshwater or marine origin, respectively, both groups at the edge of their tolerance range. This is consistent with arguments that, due to the severity of the physiological stresses they represent, estuaries are, to a significant extent, physically rather than biologically controlled communities and, in this respect, similar to the upper intertidal zone. Based on both empirical evidence and modeling, this category of hypothesis has been suggested by Hurzel et al. (2002) and others to account for the distribution and abundance of *R. philippinarum* and *R. decussatus* in the Bay of Santander (see Chapter 15)

A relatively high frequency of non-native species in estuaries in comparison to open coasts has been observed (Carlton, 1979), and the paucity of estuarine benthic diversity is regarded as a possible factor increasing estuarine susceptibility to non-native arrivals. Wasson et al (2005), in comparing the number of non-native species (including *R. philippinarum*) around Elkhorn Slough, a central Californian estuary, reported an order of magnitude difference between the higher numbers of non-natives in the soft sediments of the estuary compared to the hard structures of the adjacent open coast. Among the possible explanations for this is the lower "biotic resistance" in the form of limited competition with native species in an "under-utilized" habitat.

In any event, that estuarine sediments are relatively susceptible to the establishment of non-natives is well documented (Carlton, 1992; Nichols & Thompson, 1985; Ruiz et al., 1997). Such authors, noting approximately 400 nonindigenous species along US coasts report estuaries to be especially common sites of "invasion."

The idea that habitats with fewer species are more susceptible to nonindigenous species naturalizations can be traced back to Elton's foundational book on the *Ecology of Invasions* (Elton, 1958). Miller et al. (2007) suggested that among the most important species characteristics determining the success of arriving non-native molluscs in estuaries is tolerance to a range of stressors, including low salinities in particular. The paucity of estuarine species has been interpreted in terms of "empty niches" which are available for non-native arrivals to fill (Cohen & Carlton, 1998; Wolff, 1999), not least molluscs. Therefore, for non-native species that have achieved low salinity and other tolerances, entry into estuarine habitats may be facilitated by the weaker competitive and other negative biological interactions that the "empty niche" hypothesis implies.

Other features expediting new estuarine naturalizations include greater retention of larval propagules compared to open coasts, and a concentration of vectors such as shipping or, in the case of *R. philippinarum,* aquaculture operations, (Paavola et al., 2005; Preisler et al., 2009; Wolff, 1999). The observation that most established non-natives occupy the species-poor mesohaline zone of salinity between 5 and 18 suggests that salinity challenge, lower benthic diversity and susceptibility to non-native naturalizations are correlated (Olenin & Mincin, 2024).

For *R. philippinarum,* all this suggests two particular questions. First, the extent to which the Manila clam may present relatively low competitive impact for sympatric indigenous species, second, whether in such circumstances, newly established populations may actually enrich rather than degrade the functional ecology of such systems.

20.2 Processes of naturalization

20.2.1 Propagule pressure and environmental resistance

The naturalization of nonindigenous species requires their introduction into habitats compatible with their niche dimensions. Earlier chapters have considered the extent to which the Manila clam can tolerate a good range of estuarine and inter-tidal variables. However, other species traits expedite naturalization and spread. A readily dispersed life cycle stage and high fecundity are regarded as adaptations associated with the naturalization of new species. These characteristics exert a combined effect referred to as "propagule pressure," defined as the number of individuals released into a region to which they are not native (Lockwood et al., 2005). Higher reproductive effort and effective dispersal mechanisms increase the probability of success by improving the chances that sufficient numbers find suitable habitat and survive predation. Consequently, propagule pressure is regarded as of fundamental importance to range expansion, both generally (Grice, 2009) and in the particular case of marine molluscs in estuarine ecosystems (Miller et al., 2007). As a broadcast spawner with high fecundity and a planktonic larval stage, *R. philippinarum* is equipped for non-native naturalizations (Chapter 10).

However, patterns of distribution are not simply related to areas of high propagule pressure. This is because abiotic and biotic factors, collectively referred to as "environmental resistance" (Williamson, 1996), will tend to limit the success of a new species arrival. For example, larval survival, settlement, and subsequent establishment in bivalves can be significantly affected by predation, food supply, temperature, salinity, pollution, and competition, among other factors. The balance between propagule pressure and environmental resistance will determine the success or otherwise of a new arrival. Therefore, recruitment will not necessarily be directly proportional to larval abundance in a locality, as indicated by Hasegawa et al. (2018), working in northern Japanese estuaries of Hokkaido, where river discharge, salinity, turbidity, and wind direction all influence larval survival and dispersal of Manila clams. Here as elsewhere, an established (naturalized) population will only be maintained when regular successful recruitment episodes are followed by survival and growth leading to sexually mature age cohorts.

In stable ecosystems, environmental resistance will provide a relatively consistent challenge to potential arrivals. However, estuaries, as challenging environments, are prone to wide fluctuations in the abundance of many species (Boesch et al., 1976; McLusky & Elliot, 2004). In a meta-analysis of invasibility, Colautti et al. (2006) found a significant positive association with community disturbance. This suggests that natural volatility in the benthic communities of temperate estuaries must present opportunities to non-native species. The observation by Spencer (2002) that harsh winters can lead to good years for Manila clam settlement by suppressing the abundance of the predator shore crab *Carcinus maenas,* thereby reducing environmental resistance, exemplifies this possibility.

A similar phenomenon has been demonstrated for three bivalve species in the Wadden Sea (Beukema & Dekker, 2014). In Poole Harbour, UK, it has been postulated that the naturalization of the Manila clam in the 1980s may have been expedited by an earlier decline in the abundance of the bivalves *Scrobicularia plana* and *Macoma baltica* attributed to tributyl tin pollution (Caldow et al., 2005; Humphreys et al., 2007). However, since predation is considered a more powerful factor than competition (Chapter 12) and since all three bivalves now coexist, reduction in competition through native species decline appears not to to have been an influential factor in this case.

Since the physical environmental variables responsible for estuarine species paucity may also limit the distribution and abundance of many potential predators, it is possible to conceptualize the challenging estuarine and intertidal habitats favoured by Manila clams as representing refugia from many less tolerant fully marine carnivorous species which would otherwise predate Manila clams (Chapter 15). Among the selective forces that have underpinned this shallow burrowing clam's adaptations for volatile intertidal and estuarine conditions, higher predation pressure in sub-tidal fully marine habitats may have been instrumental. And significant among the explanations for the Manila clams' limited success on the subtidal seabed is the greater abundance and diversity of such predators. Conversely, daylight and low water, especially in intertidal areas, can be times of high vulnerability to the epibenthic predators of Manila clams, as they are themselves then most subject to predation. The Manila clam predators *Carcinus maenas* and the polychaete *Halla okudai* are both known to forage nocturnally.

Therefore, although it may be that the challenges of estuarine and intertidal sediment habitats are balanced by the benefits they represent as forms of refugia, it is clear that becoming established must be dependent on a complex interaction of variables. However, the equilibrium point between propagule pressure and environmental resistance is a hypothetical parameter that, in practice, is virtually impossible to predict in the field.

20.2.2 The relationship between introductions and naturalization

Although difficult to predict, given a basic climate compatibility, the balance between propagule pressure and environmental resistance can be inferred from the relationship between known anthropogenic introductions and subsequent naturalizations. In England, the original naturalization occurred soon after the introduction to Poole Harbour in 1988. No doubt this was expedited by repeated formal aquaculture laying of seed clams, which ensured a significant broodstock, high propagule pressure, and the establishment of a widely distributed feral population.

Moreover, when Manila seed clam laying ceased in 2002, the wild fishery continued to be productive, demonstrating a fully naturalized wild population based on what was then Europe's northern boundary for the species. At the time, it was unclear whether the extensive shallow and relatively warm waters of Poole Harbour, on England's south coast, presented a singular opportunity for Manila clam naturalization, or whether the species might gradually spread to other protected coastal estuaries and bays (Humphreys et al., 2007).

Between 1988 and 2004, government-backed introductions for aquaculture occurred in 19 British estuaries. Of these, by 2010, after all government-sponsored introductions had ceased, the species had naturalized in some but not all of the 19 estuaries to which it had been introduced. Despite introduction, the species had not naturalized in three locations on the southern and south-east coasts of England and four further north in Wales and Scotland. From these findings, it was concluded that the relationship between propagule pressure and environmental resistance was a fine balance which varied by estuary: sometimes favouring naturalization, sometimes preventing it, and sometimes leaving isolated individuals as relics of otherwise unsuccessful settlement events (Humphreys et al., 2015).

Also of interest was the fact that in surveys, the species was found to have naturalized in southern estuaries where no recorded introduction for aquaculture had occurred. This finding suggested an investigation into the mechanisms of introduction and dispersal.

20.3 Natural dispersal to new locations

The significance of anthropogenic factors in the global distribution of the Manila clam is well established (Chapter 3). However, once established in a new region as a non-native, *R. philippinarum* can spread naturally through larval dispersal. This appears to have occurred on the west coast of North America, for example, around Vancouver Island (Gillespie et al., 2012) and in the European Wadden Sea (Reise et al., 2024). Dispersal based on planktonic larval transport is heavily dependent on the hydrodynamics of tidal flows, fluvial discharge, and wind-driven currents, which are in turn influenced by geomorphological features of the coastline and the bathymetry of the seabed. In the tidal estuaries and protected bays occupied by *R. philippinarum,* it is of interest to examine both larval retention within self-recruiting estuarine populations and spread of metapopulations between different coastal estuaries and bays. The first of these is considered in Chapter 10, the latter here.

In estuaries salinity often varies with depth. Consequently, since currents also vary with depth, the depth distribution of larvae is influential on their dispersal (Corell et al., 2012). Larvae of *R. philippinarum* are known to change their vertical swimming activity according to larval stage and in response to salinity gradients (Chapter 10). Various studies have coupled hydrodynamic and behavioural models to predict the transport of Manila clam larvae. Such approaches track the trajectories of larvae in three dimensions based on advection and swimming behavior. Zhong et al. (2022) modeled Manila clam larval dispersal in Southwest Laizhou Bay, China. They reported two potential larval settlement areas and concluded that the dominant controlling factors in that locality were southerly wind- driven currents, with the larvae's response to river plume and estuarine circulation a second factor.

Bidegain and Juanes (2013) modeled dispersal in the Bay of Santander, Spain. Comparisons with *R. decussatus* suggest that *R. philippinarum* is relatively effective in spreading and then maintaining spatially separate spawning locations. While larval behaviour and dispersal are effective in supporting self-recruiting populations, it also supports the persistence of spatially separate meta-populations, albeit largely within large complex and protected coastal features. In comparison to *R. decussatus,* the connectivity between such locations is greater. Self-recruitment grounds, which are less likely to receive larval inputs from other spawning grounds, are more susceptible to decline. In contrast, well-connected grounds can compensate for a deficit with larval transport from elsewhere (Bidegain & Juanes, 2013).

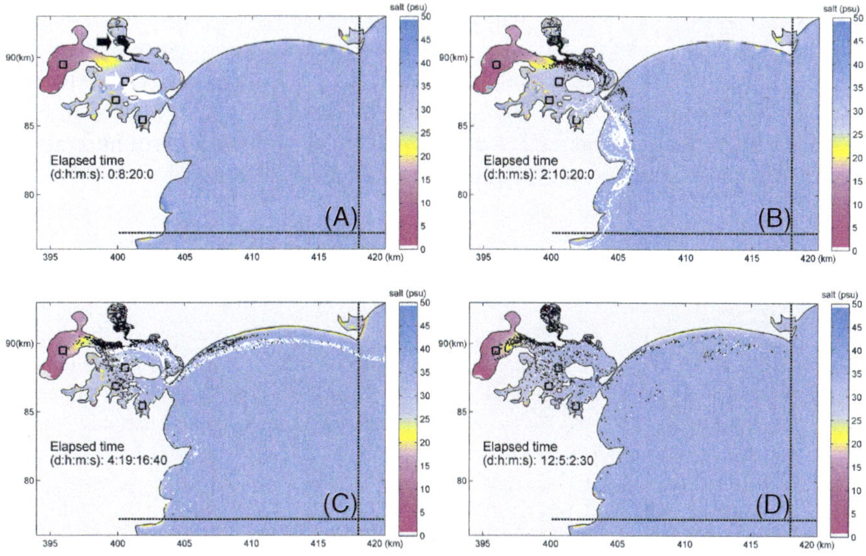

FIGURE 20.1

Modeled Manila clam larval dispersal from two sites (black and white) in Poole Harbour, UK. Both are targeting a salinity of 17 PSU. (A) shows the position after 8 h, (B) after 2.5 days, (C) after 4.75 days, and (D) after 12.25 days. Panels (A) and (B) are mid-ebb tide, whereas (C) and (D) are flood tide. The black particles are strongly retained in the north of the harbor, whereas the majority of the white particles are ejected and move out of the area to the south at first (B) and then the east (C). The background and color bar are modeled on salinity.

From Herbert et al. (2011).

In addition to modeling larval dispersal within Poole Harbour, Herbert et al. (2011) (see Chapter 10) have also modeled the transmission of larvae out into the adjacent coastal areas, with a view to determining the likelihood that the species could "leap-frog" along the coast and into suitable neighbouring estuaries and protected bays. Fig. 20.1 shows the results of this modeling. Although it is estimated that Poole Harbour larvae can reach the entrance of a neighbouring estuary (Christchurch Harbour), which is about 15 km along the coast, it was concluded that the dilution and attenuation of larvae during the journey would result in a propagule pressure unlikely to overcome environmental resistance sufficiently to establish a viable new non-native population. Such conclusions are based on modeled estimations, and it is possible that such transmission could occur if, for example, colder water temperatures prolong larval pelagic life or in particular hydrodynamic circumstances.

Nevertheless, although natural larval dispersal must be significant in spreading the Manila clam within large complex protected coastal features such as the network of estuaries and bays adjacent to the Thames estuary in south-east England, informal anthropogenic introductions provide an alternative explanation for much of the arrival of established populations in English estuaries where no officially sanctioned introductions occurred (Humphreys et al., 2015). Such informal introductions could be purposeful (and in the UK unlawful) by fishers hoping to establish a new fishery, or more likely incidental, for example, from the habit of storing and then landing Manila clams in ports other than where they were caught.

References

Abele, L. G., & Walters, K. (1979). Marine Benthic diversity: A critique and alternative explanation. *Journal of Biogeography, 6*(2), 115. https://doi.org/10.2307/3038047.

Attrill, M. J., & Rundle, S. D. (2002). Ecotone or ecocline: Ecological boundaries in Estuaries. *Estuarine, Coastal and Shelf Science, 55*(6), 929–936. https://doi.org/10.1006/ecss.2002.1036.

Basset, A., Barbone, E., Elliott, M., Li, B.-L., Jorgensen, S. E., Lucena-Moya, P., Pardo, I., & Mouillot, D. (2013). A unifying approach to understanding transitional waters: Fundamental properties emerging from ecotone ecosystems. *Estuarine, Coastal and Shelf Science, 132*, 5–16. https://doi.org/10.1016/j.ecss.2012.04.012.

Beukema, J. J., & Dekker, R. (2014). Variability in predator abundance links winter temperatures and bivalve recruitment: Correlative evidence from long-term data in a tidal flat. *Marine Ecology Progress Series, 513*, 1–15. https://doi.org/10.3354/meps10978.

Bidegain, G., & Juanes, J. A. (2013). Does expansion of Manila clam *Ruditapes philippinarum* cause competitive displacement of the European native clam *Ruditapes decussatus*? *Journal of Experimental Marine Biology and Ecology, 445*, 44–52.

Boesch, D. F., Wass, M. L., & Virnstein, R. W. (1976). *The dynamics of estuarine communities* (pp. 177–196). Estuarine Processes Academic Press.

Byers, J. E., & Noonberg, E. K. (2003). Scale dependent effects of biotic resistance to biological invasion. *Ecology, 84*(6), 1428–1433.

Caldow, W. G., Stillman, R. A., Durell, S. E. l. d., West, A. D., McGrorty, S., Goss-Custard, J. D., Wood, P. J., & Humphreys, J. (2007). Benefits to shorebirds from invasion of a non-native shellfish. *Proceedings of the Royal Society Series B, 274*, 1449–1455.

Carlton, J. (1992). Introduced marine and estuarine mollusks of North America: An end-of-the-20th-century perspective. *J, 11*, 489–505 ournal of Shellfish Research.

Carlton, J. T. (1979). *History, biogeography, and ecology of the introduced marine and estuarine invertebrates of the Pacific Coast of North America.* University of California, Davis.

Cohen, A. N., & Carlton, J. T. (1998). Accelerating invasion rate in a highly invaded estuary. *Science, 279*(5350), 555–558. https://doi.org/10.1126/science.279.5350.555.

Colautti, R. I., Grigorovich, I. A., & MacIasaac, H. J. (2006). Propagule pressure: A null model for biological invasions. *Biological Invasions, 8*, 1023–1037.

Corell, H., Moksnes, P. O., Engqvist, A., Döös, K., & Jonsson, P. R. (2012). Depth distribution of larvae critically affects their dispersal and the efficiency of marine protected areas. *Marine Ecology Progress Series, 467*, 29–46. https://doi.org/10.3354/meps09963.

Elton, C. S. T. (1958). *The ecology of invasions by animals and plants* (Edition 2000). University of Chicago Press.

Essink, K., & Oost, A. P. (2017). How did *Mya arenaria* (Mollusca; Bivalvia) repopulate European waters in mediaeval times? *Marine Biodiversity, 49*, 1–10.

Fairbridge, R. W. (1980). The estuary: Its definition and geodynamic cycle. In E. Olavvson, & I. Cato (Eds.), *Chemistry and geochemistry of estuaries* (pp. 1–35). John Wiley and Sons.

Gillespie, D., Bower, S. M., Marcus, K. L., & Kieser, D. (2012). *Biological synopsises for three exotic molluscs, Manila Clam (Venerupis philippinarum), Pacific Oyster (Crassostrea gigas) and Japanese Scallop (Mizuhopecten yessoensis) licensed for aquaculture in British Columbia*. Ottawa, Ontario: Canadian Science Advisory Secretariat, Fisheries and Oceans Canada.

Grice, T. (2009). Principles of containment and control of invasive species. In M. N. Clout, & P. A. Williams (Eds.), *Invasive species management: A handbook of principles and techniques* (pp. 61–76). Oxford University Press.

Hasegawa, N., Abe, H., Onitsuka, T., & Ito, S. (2018). Association between the planktonic larval and benthic stages of Manila clam *Ruditapes philippinarum* in eastern Hokkaido, Japan. *Fisheries Science., 84*(2), 237–249. https://doi.org/10.1007/s12562-017-1168-z.

Herbert, R. J. H., Willis, J., Jones, E., Ross, K., Hübner, R., Humphreys, J., Jensen, A., & Baugh, J. (2011). Invasion in tidal zones on complex coastlines: Modelling larvae of the non-native Manila clam, *Ruditapes philippinarum*, in the UK. *Journal of Biogeography, 39*(3), 585–597.

Humphreys, J., Caldow, R. W. G., Mcgrorty, S., West, A. D., & Jensen, A. C. (2007). Population dynamics of naturalised manila clams *Ruditapes philippinarum* in British coastal waters. *Marine Biology, 151*(6), 2255–2270. https://doi.org/10.1007/s00227-007-0660-x.

Humphreys, J., Harris, M. R. C., Herbert, R. J. H., Farrell, P., Jensen, A., & Cragg, S. M. (2015). Introduction, dispersal and naturalization of the Manila clam *Ruditapes philippinarum* in British estuaries, 1980-2010. *Journal of the Marine Biological Association of the United Kingdom, 95*(6), 1163–1172. https://doi.org/10.1017/S0025315415000132.

Hurzel, A. H., Hausser, J., Chessel, D., & Perrin, N. (2002). Ecological-niche factor analysis: How to compute habitat-suitability maps without absence data? *Ecology*. https://doi.org/10.1890/0012-9658(2002)083[2027:ENFAHT]2.0.CO;2.

Lockwood, J. L., Cassey, P., & Blackburn, T. (2005). The role of propagule pressure in explaining species invasions. *Trends in Ecology and Evolution, 20*(5), 223–228. https://doi.org/10.1016/j.tree.2005.02.004.

McClusky, D. S., & Elliott, M. (2004). *The estuarine ecosystem. Ecology, threats, and management* (third edition). Oxford University Press.

Miller, A. W., Ruiz, G. M., Minton, M. S., & Ambrose, R. F. (2007). Differentiating successful and failed molluscan invaders in estuarine ecosystems. *Marine Ecology Progress Series, 332*, 41–51. https://doi.org/10.3354/meps332041.

Nichols, F. H., & Thompson, J. K. (1985). Persistence of an introduced mudflat community in South San Francisco Bay, California. *Marine Ecology Progress Series., 24*, 83–97. https://doi.org/10.3354/meps024083.

Olenin, S., & Minchin, D. (2024). Introductions of non-indigenous species to coastal and estuarine systems. In D. Baird, & M. Elliott (Eds.), *Treatise on estuarine and coastal science* (second edition) (6, pp. 259–301). Elsevier. http://dx.doi.org/10.1016/B978-0-323-90798-9.00021-4.

Paavola, M., Olenin, S., & Leppäkoski, E. (2005). Are invasive species most successful in habitats of low native species richness across European brackish water seas? *Estuarine, Coastal and Shelf Science, 64*, 738e750.

Preisler, R. K., Wasson, K., Wolff, W. J., & Tyrrell, M. C. (2009). Invasions of Estuaries vs the adjacent open coast: A global perspective. In G. Rilov, & J. A. Crooks (Eds.), *Biological invasions in marine ecosystems. Ecological studies* (204, pp. 587–617). Springer-Verlag.

Reise, K. (1985). *Tidal flat ecology. An experimental approach to species interactions.* Springer-Verlag.

Reise, K., Wegner, K. M., Borcherding, R., Brand, S., Buschbaum, C., & Waser, A. M. (2024). Manila clams *Ruditapes philippinarum* spreading north and establishing in the European Wadden Sea. *Estuarine, Coastal and Shelf Science, 309*, 108940. https://doi.org/10.1016/j.ecss.2024.108940.

Remane, A. (1934). The fauna of brackish waters. *Verhandlungen der Deutschen Zoologischen Gesellschaft, 36*, 34–74.

Remane, A., & Schlieper, C. (1958). The biology of brackish water. The inland water. *Die Biologie des Brackwassers Die Binnengewasser, 22*, 348.

Ruiz, G. M., Carlton, J. T., Grosholz, E. D., & Hines, A. H. (1997). Global invasions of marine and estuarine habitats by non-indigenous species: Mechanisms, extent, and consequences. *American Zoologist, 37*(6), 621–632. https://doi.org/10.1093/icb/37.6.621.

Sanders, H. L. (1968). Marine benthic diversity: A comparative study. *American Naturalist., 102*(1968), 243–262.

Spencer, B. E. (2002). *Molluscan shellfish farming*. Blackwell Science.

Thistle, D. (1982). The stability-time hypothesis as a predictor of diversity in deep-sea soft-bottom communities: A test. *Deep Sea Research Part A. Oceanographic Research Papers, 30*(3), 267–277.

Wasson, K.United States. Aquatic Nuisance Species Task Force, and National Estuarine Research Reserve System. (2005). *NOAA Technical Report Series 2002: 1*. Silver Spring, MD: National Estuarine Research Reserve System.

Whitfield, A. K., Elliott, M., Basset, A., Blaber, S. J. M., & West, R. J. (2012). *Paradigms in estuarine ecology – A review of the Remane diagram with a suggested revised model for estuaries Estuarine, Coastal and Shelf Science, 97*, 78–90. https://doi.org/10.1016/j.ecss.2011.11.026.

Williamson, M. (1996). *Biological invasions*. Chapman & Hall.

Wolff, W. J. (1999). Exotic invaders of the meso-oligohaline zone of estuaries in the Netherlands: Why are there so many? *Helgo Meeresunters, 52*, 393–400.

Zhong, Y., Zhang, J., Song, D., Zhao, Y., Liu, Y., Wu, W., & Qiao, L. (2022). Numerical study of hydrodynamic effects on Manila clam population distribution and transport in the Southwest Laizhou Bay, China. *Science of The Total Environment, 865*, 161214. https://doi.org/10.1016/j.scitotenv.161214.

CHAPTER

Consequences of naturalization 21

21.1 Introduction: Detriment versus benefit

Most assessments of non-native species risk exclude ecological benefits from consideration. For the avoidance of bias, a balanced assessment of both positive and negative consequences of arrival should arguably be systematically considered. This is particularly necessary in the context of climate change, whose implications for species distribution cannot all be predicted. The beneficial roles of species are commonly expressed as ecological services and sometimes collectively expressed in terms of "natural capital." Among the recognized ecological service categories are "provisioning services," which relate to human food production. In this respect, benefit from the Manila clam is considerable, and in the future it could make an increasing contribution to global food security, with concomitant socio-economic and health benefits (Chapter 26). Since the provisioning services of the Manila clam have been addressed in Chapter 16, these are not considered further in this chapter, which will instead briefly catalogue a range of ecological impacts (positive and negative) that derive from the presence of the species. These relate to bioturbation, nutrient fluxes, competitive impact,and ecological function, all of which apply in the case of naturalized non-native populations.

21.2 Bioturbation and nutrient fluxes

In assessing the consequences of naturalization, it can be difficult to distinguish between the effects of the species *per se* and the effects of fishing and aquaculture practices. The latter have been considered in Chapter 18. A contributing factor to this difficulty is the "Tapes paradox" as elucidated by Pravoni et al. (2003) in which clam dredging itself increases suspended organic particulates such as benthic microalgae, which more strongly connect the pelagic and benthic zones. On the basis of modeling and historic records of biomass under less intense dredging, they estimated that cessation of fishing would result in a 20% reduction of Manila clam biomass.

This "*Tapes* paradox," in which exploitation itself has a positive effect on the exploited resource, is of interest when considering the extent to which the Manila clam, in more "natural," unexploited (or relatively unexploited situations, such as through artisanal fishing), should be considered intrinsically "invasive." Pravoni et al. (2003) were working in an Adriatic Lagoon where an invasive character for the species

The Manila Clam *Ruditapes philippinarum*: Science and Significance of a Globalised Species.
DOI: https://doi.org/10.1016/B978-0-443-36704-5.00027-4

Copyright © 2026 Elsevier Inc. All rights are reserved, including those for text and data mining, AI training, and similar technologies.

has most often been claimed. Therefore, the species might be better considered to be "anthropogenically invasive" i.e., ostensibly invasive in response to anthropogenic alteration of the environment, in this case, intensive dredging. Chapter 22 considers the question of Manila clam invasiveness, including the possibility that invasive tendencies may also be amplified by other anthropogenic factors such as nutrient and organic particulate releases from urban and agricultural effluents.

Lopes et al. (2018) have avoided such confounding dredging-related variables by evaluating bioturbation of sediments and its influence on nutrient cycling in microcosms. Using sediments similar in composition to those found in the Ria de Aveiro lagoon, Portugal, they compared sediment mobilization due to bioturbation by *R. philippinarum* with the deeper burrowing native *Ruditapes decussatus* (*R. decussatus*). Both these clams are known to be active within the sediment. Bioturbation activity was expressed as maximum, median, and mean mix depth of particle redistribution. No significant differences were found between the two species. On this basis, they anticipated no short or long-term change in a scenario where *R. philippinarum* replaced the native species. Similarly, they reported no significant differences between the species in terms of the water concentrations of inorganic nitrogenous and phosphorus compounds.

Nevertheless, for both species, they found significant increases in comparison to no-clam controls with regard to both particle mobilization and nutrient quantities. Thus, exemplifying the fact that benthic infauna are known to exert significant influence over solute fluxes within the sediment and between sediment and water column. Causes include indirect consequences of sediment bioturbation and ventilation due to burrowing and feeding activity, along with direct metabolic activity and chemical transformations of the fauna itself.

Impacts on nutrient fluxes are density-dependent. In Italy, Bartoli et al. (2001) showed the Manila clam to excrete large quantities of ammonium, accounting directly for around 90% of the dissolved inorganic nitrogen released from clam-occupied sediments. Similarly, in farmed deposits of the Sacca di Goro, Welsh et al. (2015), using sediment cores with clams, investigated the relationship between *R. philippinarum* population density and nitrogen cycle processes. At high Manila clam densities (mean 1440 clams m^{-2}), all nitrogen cycle process rates and fluxes between sediment and water column correlated positively with clam biomass. However, this relationship was not primarily due to the clam's creation of favorable conditions for bacterial activity in the sediment, but rather to the direct effect of the clams' own metabolism and of the bacteria that colonize the clams.

This indicates that at high densities of *R. philippinarum* as have occurred in the Adriatic lagoons, bioturbation combined with the clams own excretion of nitrogenous compounds and organic matter through faeces and pseudofeces production, may have a significant effect on system function, but not necessarily all negative: High rates of denitrification and high denitrification efficiency of clam-farmed sediments mitigate the effect of nutrient loading in the polluted waters of the Sacca di Goro lagoon, Italy (Welsh et al., 2015). In Chapter 18 on the environmental performance of Manila clam fisheries and aquaculture, benefits in terms of the removal of excess nutrient loading from eutrophic estuaries, by assimilation followed by harvesting, have been considered. However, this benefit can only accrue in circumstances where the species, when naturalized, is also harvested.

21.3 Trophic adjustment and competitive impacts

Solidoro et al. (2010) have summarized the great changes to the benthic functioning of the Venice Lagoon since the 1930s, including stages of deterioration in which an originally well-diversified community transitioned, such that by the end of the 1980s, 55% of macrobenthic abundance consisted of primary producers, and trophic structure was dominated by herbivores and detritivores. By 1999, after the introduction and spread of the Manila clam, filter feeders became dominant, with the Manila clam generating high levels of secondary production and accounting for about 50% of macrobenthic abundance (Pranovi et al., 2007, 2008).

The implication of the change since the 1990s is that the Manila clam represents a primary cause. However, again, there are good reasons to believe that the mechanisms of Manila clam impact on benthic function may be largely indirect, due to fishing and aquaculture practices rather than natural biological processes consequent on the naturalization of a new species. Manila clam landings peaked in 1999 at a production of 40,000 tons yr^{-1}, although they more than halved by 2003. Pranovi et al. (2003) and Sfriso et al. (2003) regard high-impact Manila clam fishing gear to be a major contributor to sediment erosion and loss of ecological balance.

Solidoro et al. (2010), in their general review, also attribute changes to the Venice Lagoon ecosystem to modern drivers, such as those common to many estuarine and coastal areas. These include nutrient and pollutant load, shipping and tourism, dredging, and the effects of climate change on weather, with fishing and aquaculture among them. In this respect, it is of interest that the benthos of the Venice lagoon was already considered greatly impoverished during the late 1950s to 1960s (Giordani Soika & Perin, 1974). In fact, the Venice Lagoon has undergone frequent profound anthropogenic changes since at least the 13th century (Ravera, 2000).

Comparative studies have been conducted, particularly in non-native areas where competition with indigenous Venerid clams is of conservation and/or economic concern. In indigenous areas, natural competition between sympatric clams is of less concern, and in any event, not easily manageable beyond aquaculture operations. The biogeographical distributions of *Ruditapes aspera* and *Ruditapes bruguieri* overlap with *R. philippinarum* in subtropical regions, but these are essentially tropical species in contrast to the essentially temperate Manila clam.

In Europe, the sympatric venerid species studied with *R. philippinarum* are *Venerupis corrugata* and *Ruditapes decussatus*. Comparative Studies on *V. corrugata* from northeastern Atlantic coasts have confirmed it as an essentially stenohaline species in contrast to the euryhaline *R. decussatus*. Comparisons of the tolerances and distributions of *R. philippinarum* and *R. decussatus* have been considered in Chapter 15.

In North America, *Leukoma staminea* is considered to compete with *R. philippinarum,* and like the European *R. decussatus,* appears to have a similar niche. The native "littleneck clam" *L. staminea* is widely distributed along the west coast of North America from the Aleutian Islands south to Baja, California. Except in the most northerly and southerly parts of its range, it is now frequently sympatric with the non-native Manila clam.

In British Columbia (BC), Canada, Gillespie and Bourne (2005) reported that on a south-to-north progression, the number of Manila clams relative to native littlenecks decreases, reaching parity around Yaculta at 50°N. In BC, the Manila clam is farmed on intertidal seabeds where seeded clams are protected by antipredator netting. Based on field surveys across three regions of, or adjacent to, Vancouver Island, Bendell (2014) reported that average abundances of Manila clams on both farmed and unfarmed (reference) sites were greater than those of the native littleneck. However, the abundance of *L. staminea* increased with the distance between farmed and reference sites, and where the natural dispersal of Manila clams was prevented by colder waters, the native littleneck became the dominant bivalve. Such findings were regarded as an indication that Manila clam introductions caused local declines of native littleneck clams. Based on these site comparisons, Bendell (2014) suggested that, in the absence of Manila clam culture, naturalized populations of Manila clams would be at a density of around 80–120 m^2 (similar to that of wild populations in Poole Harbour, UK).

In the Bay of Santander, Spain, both *R. philippinarum* and *R. decussatus* are present and fished. Investigating six fished zones within the Bay, Juanes et al. (2012) found no significant negative correlation between the abundances of the two species and concluded that competition between them for space or resources is either absent or not intense enough to result in competitive exclusion. Explanations for coexistence or predominance were provided by factors including salinity, hydrodynamics, and proximity to clam culture areas: The native *R. decussatus* dominating in low salinity zones and *R. philippinarum* near culture parks and areas combining significant water flow and eddies. They noted that in areas with mean salinity from 15 to 27.2, lower salinities occurred during heavy rainfall periods (see Chapter 15).

In Arcachon Bay, France, Auby (1993) reported that the Manila clam tended to "supplant" *R. decussatus*, in some areas, the ratio between the two being 6:1 or higher, although, like the Bay of Santander, in other parts of the bay the native species remained dominant. By 2006, the reported numerical dominance of *R. philippinarum* was reported as 90% (Caill-Milly, Duclercq, & Morandeau, 2006), and by 2018, 99%. However, even with a 99% dominance in Arcachon, the density, including small clams, was reported as quite low at 39 clams per m^2 (Sanchez et al., 2018).

The patterns of coexistence and dominance revealed in these various places, from sub-arctic British Columbia to the warmer waters of Santander and Arcachon, suggest that while the Manila clam may be capable of naturalization in a wide range of conditions, its capacity to exclude native equivalents such as *L. staminea* and *R. decussatus* through the appropriation of limited resources of space and food remains largely unevidenced. It appears that coexistence is often the norm, with dominance primarily found in areas where physical conditions are directly less favorable to one or other species rather than indirectly through competition.

In the above cases, densities of the Manila clam, even where dominant, remain relatively low, so the question remains whether, in different conditions and at higher densities, the Manila clam can competitively exclude native equivalents. Two cases of high Manila clam densities elucidate this question: the Californian coast and the Northern Adriatic.

In San Francisco Bay, the Manila clam is one of the most common benthic species and has been reported at densities up to 2000 clams m^2 (Cohen, 2011). Further south in the Colorado Lagoon, Los Angeles County, California (37°N). Burnaford et al. (2011) reported that 6 of the 14 bivalve species collected from intertidal and shallow subtidal areas were non-native, with *R. philippinarum* dominant and making up 87.6% of all collected bivalves. However, in these locations, even when abundant and dominant, the Manila clam coexisted with other bivalves, and the extent to which it competes directly for limited space or food and competitively excludes other species remains unclear.

In the warm Mediterranean waters of the northern Adriatic lagoons, the success of the Manila clam was such as to often be characterised as truly "invasive." The first introduction occurred in 1983 at the southern end of the Venice Lagoon. Following the commercial success of this introduction, seed clams were laid in other northern Adriatic lagoons. Naturalization and natural dispersal were evident by 1986, and in 2002, the most productive areas had densities of >1000 m^{-2}(Breber, 2002).

However, introductions elsewhere in Italy did not result in such high densities, indicating that the northern lagoons offered unusually optimal habitats. The water temperatures in these shallow lagoons respond to air temperatures, which range from around 6°C in winter to 30°C in summer. Breber (2002) attributed the success of the Manila clam in part to their ability to spawn and recruit twice over the summer warm period compared to just a single effective spawning for the native *R. decussatus*. Moreover, the presence of "culture parks" where Manila clams are grown to large size and at high densities ensures sufficient gamete fertilization and larval dispersal to generate strong propagule pressure in surrounding lagoon areas.

In the case of the Sacca di Goro Lagoon, data exist on the densities of the native *R. decussatus* before and after the arrival of the Manila clam. Between 1969 and 1984, records show great fluctuations of mature (>3.5 cm shell length) *R. decussatus* densities between 13 and <1 clam per m^2 independent of any possible Manila clam influence. Breber (2002) found a mean *R. decussatus* density of 0.4 clams per m^2, compared to 384 clams per m^2 for *R. philippinarum*. Clearly, the non-native is dominant, but the native clam was still present at a level not inconsistent with its prior oscillating densities. Breber (2002) attributed the widespread impression that the native had "disappeared" to their similar appearance and considered the Manila clam to be "invasive" but not in the sense of causing the disappearance of *R. decussatus*. This raises questions about the meaning of "invasive," which are considered in Chapter 22.

Therefore, even under high Manila clam densities in California and the northern Adriatic, and despite subjective impressions, secure evidence of competitive exclusion remains lacking. While we cannot rule out the possibility of competition for resources in areas of coexistence, the evidence suggests that competitive exclusion of *R. decussatus* or *L. staminea*, even by high densities of Manila clams, is unlikely. Elsewhere at lower densities, such as in Poole Harbour, the Wadden Sea and Arcachon Bay, the evidence suggests otherwise (Humphreys et al., 2007; Reise, 1985).

21.4 Ecological services from Manila clam naturalization

Although the ecological benefits of wild naturalized populations are clear, as with negative impacts, when wild and farmed populations coexist in an estuary, the positive consequences of each can be difficult to distinguish and are likely to overlap. Such benefits are assumed to be quantitatively reduced as a consequence of the lower population densities of wild populations alone.

Such benefits can be expressed in terms of ecological services, meaning forms of human or environmental advantage. These generally fall into four categories: provisioning, regulating, cultural, and support services. Many of the benefits accruing from Manila clam populations in indigenous areas will also apply to naturalized non-native populations. Native Manila clam populations provide benefits in all four categories of ecological service, and the same can be said of naturalized non-native populations.

Since across most of its worldwide distribution Manila clam populations are fished or farmed, the species' contribution to provisioning services (food) is considerable (see Chapter 16). In respect to the consumption as food, however, recreational gathering may also be seen as a cultural ecological service, especially where local people have a tradition of informal foraging, not least in family groups (Chapter 16).

Another cultural service accrues from the additional non-native food source and consequent reduction in mortality of species of estuarine bird which, by virtue of protected conservation status, are regarded as important in ecological, recreational, and amenity value. In Poole Harbour, Caldow et al. (2007) and Humphreys et al. (2007) have demonstrated that removal of the clams by wild regulated fisheries leaves intact populations sufficient to reduce natural wader overwintering mortality below that which would occur in the absence of the non-native arrival (see Chapter 18).

Regulating services as a category include water purification and climate regulation, both of which Manila clam populations deliver. The former is a consequence of carbon sequestration into shell material. This benefit accrues whether or not the clams are harvested, since even when not harvested, unlike carbon assimilated into the soft tissues, this removal from the short carbon cycle is considered persistent. Farmed clams, by virtue of their population densities, can sequester much more carbon (Chapter 18).

Wild populations that are fished contribute a regulating service by virtue of nutrient removal making a consequent contribution to eutrophication mitigation

(Chapter 18). Since nutrient removal can be seen as contributing to the maintenance of an otherwise degraded system, it can also be categorized as a "supporting" service, since by contributing to a healthier system it supports other ecosystem services. Another such support service has been reported by Li et al. (2017), who observed that the bioturbation of Manila clams can enhance seed burial in the seagrass beds of *Zostera Marina*: habitats known to be significant in terms of estuarine and coastal biodiversity. On the same basis, the reduced wader mortality referred to above can also be seen as a support service.

In terms of a net contribution to ecological services, benefits only accrue when the Manila clam does not competitively replace other species whose ecological service contributions are equal to or greater than those of the Manila clam. However, for reasons explained in Chapter 20, concerning features of clam habitats, including low species saturation and consequently low competitive intensity, in general, Manila clam populations are thought to make a net positive contribution to the range of ecological services. As such, where naturalized as non-natives, they contribute a "biodiversity net gain" in many locations. Nevertheless, in places where it has been described as "invasive," this conclusion can be considered controversial. Therefore, the invasiveness of the species is considered in Chapter 22.

References

Auby, I. (1993). *Technical report*. Laboratory of Biologic Oceanography, Ifremer.

Bartoli, M., Nizzoli, D., Viaroli, P., Turolla, E., Castaldelli, G., Fano, E. A., & Rossi, R. (2001). Impact of *Tapes philippinarum* farming on nutrient dynamics and benthic respiration in the Sacca di Goro. *Hydrobiologia, 455*(1-3), 203–212. https://doi.org/10.1023/a:1011910422400.

Bendell, L. I. (2014). Evidence for declines in the native leukoma staminea as a result of the Intentional Introduction of the non-native *Venerupis philippinarum* in Coastal British Columbia, Canada. *Estuaries and Coasts., 37*(2), 369–380. https://doi.org/10.1007/s12237-013-9677-1.

Breber, P. (2002). Introduction and acclimatisation of the Pacific carpet clam *Tapes philippinarum* to Italian waters. In E. Leppäkoski, S. Gollasch, & S. Olenin (Eds.), *Invasive aquatic species of Europe. Distribution, Impacts and Management.* (pp. 120–126). Springer.

Burnaford, J. L., Henderson, S. Y., & Pernet, B. (2011). Assemblage shift following population collapse of a non-indigenous bivalve in an urban lagoon. *Marine Biology, 158*(9), 1915–1927. https://doi.org/10.1007/s00227-011-1703-x.

Caldow, R W.G., Stillman, R, Durell, S E.A. Le V. dit, West, A.D., McGrorty, S, Goss-Custard, J.D., Wood, P.J., Humphreys, J. (2007). Benefits to shorebirds from invasion of a non-native shellfish. *Proceedings of the Royal Society B - Biological Sciences*, 274 (1616). 1449-1455. doi:10.1098/rspb.2007.0072.

Cohen, A. N. 2011 The exotics guide: Non-native marine species of the North American Pacific Coast. Center for Research on Aquatic Bioinvasions and San Francisco Estuary Institute. Retrieved October 15, 2013, from http://www.exoticsguide.org.

Caill-Milly, N., Duclercq, B., & Morandeau, G. (2006). Campagne d'évaluation du stock de palourdes du bassin d'Arcachon – Année 2006 [Evaluation of clam stocks in the Arcachon Basin, 2006]. Département Ressources Halieutiques, Direction des Ressources Vivantes. Ifremer.

Gillespie, G. E., & Bourne, N. F. (2005). Canadian Manuscript Report of Fisheries and Aquatic Sciences. In *Canadian Manuscript Report of Fisheries and Aquatic Sciences: 2733* (p. 199). Fisheries and Oceans Pêches et Océans Canada.

Giordani Soika, A., & Perin, G. (1974). L'inquinamento della laguna di Venezia: Studio delle modificazioni chimiche e del popolamento sottobasale dei sedimenti lagunari negli ultimi vent'anni [The pollution of the Venice Lagoon: Study of chemical changes and subbasal populations of lagoon sediments in the last twenty years]. *Bollettino del Museo Civico di Storia Naturale di Venezia, 26*, 25–68.

Humphreys, J., Caldow, R. W. G., Mcgrorty, S., West, A. D., & Jensen, A. C. (2007). Population dynamics of naturalised Manila clams *Ruditapes philippinarum* in British coastal waters. *Marine Biology, 151*(6), 2255–2270. https://doi.org/10.1007/s00227-007-0660-x.

Juanes, J. A., Bidegain, G., Echavarri-Erasun, B., Puente, A., García, A., García, A., Bárcena, J. F., Álvarez, C., & García-Castillo, G. (2012). Differential distribution pattern of native *Ruditapes decussatus* and introduced *Ruditapes phillippinarum* clam populations in the Bay of Santander (Gulf of Biscay): Considerations for fisheries management. *Ocean and Coastal Management, 69*, 316–326. https://doi.org/10.1016/j.ocecoaman.2012.08.007.

Li, C. J., Li, W. T., Liu, J., Zhang, X., & Zhang, P. (2017). Zostera marina seed burial can be enhanced by Manila clam *Ruditapes philippinarum*: A microcosm study. *Ocean Science Journal, 52*(2), 221–229. https://doi.org/10.1007/s12601-017-0016-5.

Lopes, L. M., Patrício Rodrigues, J., Crespo, D., Dolbeth, M., Calado, R., & Lillebø, A. I. (2018). Functional traits of a native and an invasive clam of the genus Ruditapes occurring in sympatry in a coastal lagoon. *Nature Scientific Reports, 8*, 16901. https://doi.org/10.1038/s41598-018-34556-8.

Pranovi, F., Da Ponte, F., & Torricelli, P. (2007). Application of biotic indices in the lagoon of Venice: An example by using a time series of macrobenthic community. *Marine Pollution Bulletin, 54*, 1607–1618.

Pranovi, F., Da Ponte, F., & Torricelli, P. (2008). Historical changes in the structure and functioning of the benthic community in the Lagoon of Venice. *Estuarine Coastal Shellfish Science, 76*, 753–764.

Pranovi, F., Libralato, S., Raicevich, S., Granzotto, A., Pastres, R., & Giovanardi, O. (2003). Mechanical clam dredging in Venice Lagoon: Ecosystem effects evaluated with a trophic mass-balance model. *Marine Biology, 143*, 393–403.

Ravera, O. (2000). The lagoon of Venice: The result of both natural factors and human influence. *Journal of Limnology, 59*(1), 19–30. https://doi.org/10.4081/jlimnol.2000.19.

Reise, K. (1985). *Tidal flat ecology*. Springer-Verlag.

Sanchez, F., Caill-Milly, N., & Lissardy, M. M. (2018). Evaluation of clam stocks in the Arcachon basin, 2018. (Campagne d'évaluation du stock de palourdes du bassin d'Arcachon - Année 2018). *Rapport Ifremer R.ODE/LITTORAL/LER AR, 18*(015), 61.

Sfriso, A., Facca, C., Ceoldo, S., Silvestri, S., & Ghetti, P. F. (2003). Role of macroalgal biomass and clam fishing on spatial and temporal changes in N and P sedimentary pools in the central part of the Venice lagoon. *Oceanologica Acta, 26*, 3–13. https://doi.org/10.1016/S0399-1784(02)00008-71313.

Solidoro, C., Bandelj, V., Bernardi, F. A., Camatti, E., Ciavatta, S., Cossarini, G., Facca, C., Franzoi, P., Libralato, S., Canu, D. M., Pastres, R., Pranovi, F., Raicevich, S., Socal, G., Sfriso, A., Sigovini, M., Tagliapietra, D., & Torricelli, P. (2010). Response of the Venice Lagoon ecosystem to natural and anthropogenic pressures over the last 50 years. In *Coastal lagoons: Critical habitats of environmental change* (pp. 483–511). CRC Press. https://doi.org/10.1201/EBK1420088304.

Welsh, D. T., Nizzoli, D., Fano, E. A., & Viaroli, P. (2015). Direct contribution of clams (*Ruditapes philippinarum*) to benthic fluxes, nitrification, denitrification and nitrous oxide emission in a farmed sediment. *Estuarine, Coastal and Shelf Science, 154*, 84–93. https://doi.org/10.1016/j.ecss.2014.12.021.

CHAPTER 22

Invasion and the Manila clam

22.1 General "invasive" traits in the Manila clam

The characteristics of successful invasive species are widely stated to include fast growth, high fecundity, relatively short generation time, high dispersal potential, and some form of human association. Additional features often attributed to invasive species include "pioneer" characters which enable colonization of disrupted habitats, such as phenotypic plasticity enabling adaptive responses to new conditions and food types. Permutations of such attributes are considered to provide a competitive advantage over indigenous species. Moreover, native communities and habitats themselves vary in their susceptibility to invasion as a function of such features as low native biodiversity and disturbance as a consequence of human activity (CBD, 2002; Kolar, 2001).

Previous chapters of this volume demonstrate that the Manila clam possesses several traits frequently attributed to invasive species, and that the habitats within which it becomes naturalized share features associated with susceptibility to invasion (Chapter 20). In this context, it is not surprising that the Manila clam has been described as invasive in such classic sites as the lagoons of the northern Adriatic and in some estuaries of the Iberian Atlantic coast.

However, equally apparent is the absence of any attribution of invasiveness to Manila clam arrivals in many other places. This raises three questions: First, why, despite its various ostensibly invasive characteristics, is the species clearly not universally considered invasive? Second, what is it about certain locations that triggers high population densities and assertions of invasive species status? Third, to what extent are such assertions legitimate even in these locations?

In order to address these questions, it is first necessary to clarify the meaning of ecological invasion and its relation to the naturalization of non-native species.

22.2 Concepts of invasion and invasive species

The phrase invasive species has achieved wide usage in the literature, not least as some species undoubtedly pose a major threat to biodiversity and indigenous ecosystems. However, in seeking to understand the meaning of the term, a difficulty arises due to differences in definitions as used by different authorities, along with usage changes over time. Larson (2011) identified two universal defining attributes in all

The Manila Clam *Ruditapes philippinarum*: Science and Significance of a Globalised Species.
DOI: https://doi.org/10.1016/B978-0-443-36704-5.00021-3

Copyright © 2026 Elsevier Inc. All rights are reserved, including those for text and data mining, AI training, and similar technologies.

definitions of "invasive species": First, they are non-native, and second, anthropogenically introduced, intentionally or otherwise. Beyond these, however, definitions then dichotomize, with some usages requiring only arrival with a tendency to abundance, while others also explicitly require harm, often expressed in terms of the biodiversity of the recipient community.

An early ambiguity in the use of the term invasion can be traced to a seminal text by Elton, 1958, in which, while at the outset he describes his book as about "ecological explosions" (p. 15), he later refers to "invasions that never happen…" (p. 109), i.e., are unsuccessful. In this usage, all non-native species arrivals are invasions. In a later influential book, Williamson (1996) continued this usage while emphasizing that "most invasions fail" or, even after arrival, spread and naturalization "have minor consequences."

Today, the phrase invasive species is still often used as a synonym for non-native species, regardless of impact. However, the term is increasingly, and more legitimately, used in the more restrictive sense of arrivals with significant, normally deleterious, ecological consequences, i.e., not all non-native species that become established are invasive. For example, the term "invasive alien species" has been used to distinguish between noninvasive and invasive non-native species (Clout & Williams, 2009). Maynard and Nowell (2009) therefore define invasive species as those that "can spread aggressively, to intrude or overwhelm other organisms." Similarly, the UN Convention on Biological Diversity (CBD) refers to measures to deal with "Alien species which threaten ecosystems, habitats or species" (CBD, 2002).

Therefore, although it is recognized that all newly naturalized non-indigenous species must find an ecological role in their recipient indigenous community, it is now widely accepted that this may not always be harmful. Two major types of harm associated with "successful" invasions were identified by Williamson (1996): the depression or even extinction of native species and the restructuring of native ecosystems. The latter takes the form of significant changes to the structure or functional ecology of the system. Moreover, invasive risk goes beyond the direct ecological effects of a species to include the introduction of associated non-native pathogens and parasites along with any concomitant impacts of fishing and aquaculture activities. Therefore, while true invasion has negative consequences ranging from biodiversity loss and compromised ecosystem services to human health effects, secondary parasite and pathogen introductions and fishing impacts, while detrimental, do not derive directly from the non-native species to which they may be attributed. The case of the Manila clam introduction to Britain, for example, was conducted only after scrupulous preparation of pathogen-free stock.

Conversely, examples exist of newly established arrivals that coexist within local communities rather than exerting detrimental effects such as species displacement. In many locations, the Manila clam falls into this category (Reise et al., 2017). Such species may strengthen ecological function and add to biodiversity. Various positive ecological impacts have been reported for *Ruditapes philippinarum,* including new food sources for predatory fish and birds and increased biofiltration in eutrophic waters (Chapter 21). Katsanevakis et al. (2014) calculated that 35% of the non-native marine species they assessed in Europe were reported to have a positive impact on other species. A complicating factor is climate change, which may, over time, shift the risk/benefit balance of particular new arrivals (Chapter 23). While warming waters may conceivably change a benign non-native into an "invasive" species, conversely, the northerly retreat of colder water species may reduce the biodiversity, productivity, and resilience of native temperate communities, if new warmer water species are not introduced by natural spread or anthropogenic factors.

Therefore, in practice, a spectrum of non-native species impacts exists from truly invasive to negligible to ecologically advantageous. Where a non-native species such as the Manila clam falls on this spectrum varies with locality and professional judgement. In the implementation of policy on non-native arrivals, assessing the potential and actual effects of an arrival or introduction becomes an important exercise, not least for the deployment of limited management resources. In this context, the scientific subtleties of a spectrum of effects from malign to benign to advantageous often give way to a dichotomy, and professional judgement must decide between one of two polarized categories, using such alternative phrases as "Invasive Alien Species" (IAS) in contrast to simply nonindigenous species (NIS).

However, while general definitions of invasiveness are useful for some purposes, the extent of competitive or functional effects necessary to categorize a species as invasive is generally not well defined. For example, while "local extinction" may be clearly apparent, is competitive "decline" measured in terms of abundance or distribution? And if the latter, to what extent does "invasiveness" include some reduction of a realized niche if healthy populations of a native species remain extant and some form of coexistence is the result, as is often the pattern with Manila clam naturalizations (Chapter 21)? Moreover, how might such effects be balanced against socio-economic considerations? Despite much study of invasion and invasive species since the 1950s, when, where and whether a species becomes "invasive" remains a subjective question, resolved in practice by "expert judgement" and consensus among subsets of scientists. One difficulty that this presents is that, even when expert judgement is based on peer-reviewed primary field reports (which is often not the case), different subsets of scientists may reach very different conclusions (Chapter 24).

Sandvik (2023) has questioned the "misconception" that the use of qualitative data in invasion biology is justified by uncertainty, suggesting instead that this widespread practice actually increases uncertainty. Valid quantitative criteria are defined as numerically defined values of empirically measurable quantities. Conversely, qualitative criteria involve a scale of subjective terms even if they refer to categories defined by numbers (e.g., 0 = low, 1 = high, etc). Non-native risk assessments always involve uncertainty, but causes may be epistemic (e.g., observation error) or semantic (e.g., linguistic vagueness). As Sanvik points out, unlike semantic uncertainty, which is unscientific, epistemic uncertainty can be corrected with evidence, thereby achieving a benchmark of science. Moreover, attempts to improve the situation with frameworks of standards for non-native species risk assessments, (Roy et al., 2018), do not solve this problem themselves. Unfortunately, semantic vagueness to which the concept of invasion is itself vulnerable is also commonly apparent in risk assessments, and no strictly quantitative assessments have been conducted for the Manila clam.

22.3 Assessing the invasiveness of the Manila clam

The achievement of non-native naturalization involves various stages from first arrival to establishment and spread. In general, the ratio between each stage is quite small. Williamson (1996), for example, proposed a "rule of ten" in which only 10% of non-native arrivals become established and only 10% of those become invasive (Williamson & Brown, 1986). Despite the occurrence of numerous exceptions to this "rule" and whatever the validity of its generalization for coastal habitats, it has the merit of drawing attention to the distinction between non-native and invasive and encourages an examination of the reasons for the difference in terms of the characteristics of particular species and the features of particular recipient communities.

As a basis for conclusions on the invasive potential of species, methods have been developed in which the focus of professional judgement is steered through the application of established criteria of invasiveness, relating to both species and habitats. Jensen et al. (2023) have applied one such approach as a screening method on 46 nonindigenous species in Danish waters, using expert judgements to assess six invasiveness criteria, using a scale of 0–3, with 3 being the highest risk. *R. philippinarum* was assessed as 3 for dispersal potential, 2 for conservation value of habitat, 1 each for impact on native species, impact on ecosystem function, and economic impact, and 0 for health impacts. For a species to be evaluated as invasive, the following four criteria and scores must apply: "*...the species shall reproduce and spread in the new habitat, i.e., the score for dispersal must not be 0. Additionally, the sum of the scores for both dispersal and the nature quality of invaded habitats has to be ≥3; the species must have an impact on native species, i.e., the score for impact on native species must not be 0; and, in addition, the sum of scores for impacts on indigenous species and functioning of invaded ecosystems must also be ≥3*" Jensen et al. (2023). The resulting judgements put *R. philippinarum* near the bottom of a risk analysis for "not yet established" marine species (although the species is now found in Danish waters, Chapter 3).

Such approaches are applied for the purposes of implementing the policy on non-native species. From a purely scientific standpoint, the limitations of such analyses are apparent. In the first place, the general characteristics of invasive species employed can also be found in noninvasive species. For example, "an ability to reproduce and spread" does not necessarily reliably discriminate between noninvasive and invasive species. Therefore, while the Manila clam scores 3 for dispersal, there is no attention given to the fact that propagule pressure is an adaptive advantage mainly because predation of propagules can be as prodigious as their production. Nevertheless, it should be recognized that this method was intended only as a screening process, i.e., to identify species meriting further investigation.

In the context of the European Union Alien Species Regulations, and using a more detailed assessment, (the Aquatic Species Invasiveness Screening Kit or AS-ISK), Copp et al. (2016a) included *R. philippinarum* as a trial species with the coast of France as the risk assessment area. This method involved 49 questions on the species' biogeographical and historical traits and biological and ecological interactions. One of these questions was "Is it likely to disrupt food web structure/function...?," the answer being "Yes," with a comment citing (Bonham et al., 2015; Copp et al., 2016b; Goulletquer, 1997) to the effect that *"Changes in nutrient cycles, sedimentation rates, stronger benthic-pelagic coupling and benthic fauna have been reported as well as reduced resilience. Increasing population densities are likely to lead to further disruptions.* While such effects may simply be an inevitable feature of many infaunal species, the degree of impact necessary for them to be "disruptive" is unclear. Moreover, whether the changes outlined are the result of the Manila clam *per se* or a consequence of fishing and aquaculture practices is neglected. An illustration of the subjective nature of such assessments is shown in the comment elsewhere that the Manila clam is *"already well known to be invasive,"* even though the opposite is arguably true (Humphreys et al., 2007). It is also of interest that the analysis requires that the "benefits of non-native species should not normally be included in the risk assessment" (Copp et al., 2016a).

A more discerning assessment lies in various other answers in AS-ISK where the Manila clam's capacity to replace native species ranges from "partly" in the Venice lagoon (Q. CCA 10) to "not" in France (Q. CCA 15). These two comments illustrate the importance of climate and local conditions in the clam's capacity to represent risk. Nevertheless, the fact that in this analysis "undesirable traits" included the Manila clam's *"adaptive capacity as an euryhaline, eurythermal species and also an active and competitive filter-feeder,"* without indicating why these traits are undesirable in a largely coastal intertidal species, or any specification of the degree of competition that filter feeding generates in such habitats.

In summary, approaches to assessing a species' potential to become invasive through the application of generalized criteria as proxies for "invasiveness" are unreliable. This is partly because they cannot securely represent the ecological complexities of different habitats, and secondly because they provide an ostensibly objective gloss to an essentially subjective process. While they may satisfy certain policy needs, they are not a rigorous basis for scientific conclusions.

In general, whereas naturalization is a function of distribution, invasiveness is a function of abundance. While there are locations where the biomass and densities of non-native Manila clam populations have reached levels claimed to equate with one or more defining attributes of invasiveness (Chapter 21), the unscientific nature of the concept impedes definitive judgements. For example, where the Manila clam has been reported as responsible for "supplanting" the European native, this has never been shown to be by extirpation, but rather by confining it to more restricted areas than might otherwise be the case.

In British Columbia, where the Manila clam is seeded under antipredator netting (Bendell, 2014), several lines of evidence have been reported, which Bendell argues suggest that the Manila clam is replacing the native venerid clam *Leukoma staminea*. These include finding greater numbers of the non-native in areas where farming is most intense, and the Manila clam becomes dominant. However, she also reports that in sites where the Manila clam is excluded by temperature, *L. staminea* is dominant. One difficulty with such assessments occurs when the densities of native species before the arrival of the Manila clam are unclear, and Bendell's lines of evidence alone do not demonstrate replacement of the native. Certainly, in some locations such as the Tagus estuary, Portugal, the abundance of the Manila clam has been described as invasive (Correia et al., 2025; Pranovi et al., 2006), and its arrival has reportedly coincided with a significant decrease in *R. decussatus* (Moura et al., 2007). Other reportedly invasive cases include the Venice lagoon (Pranovi et al, 2003) and sites on the French Atlantic coast such as Arcachon Bay (Flye-Sainte-Marie et al., 2007).

Elsewhere, however, such as in Poole Harbour, UK, and the Bay of Santander in Spain, the two clam species coexist, and there is no evidence that the Manila clam supplants or restricts the distribution of the native clam, which appears to be determined primarily by physico-chemical tolerances (Humphreys et al., 2015; Juanes et al., 2012). This raises the question of the factors that determine the level of success and therefore the "invasiveness" of the Manila clam in different estuaries. In the first place, repeated introductions of Manila clam seed for fishery enhancement or formal aquaculture operations will directly increase population densities and, through enhanced propagule pressure, support the proliferation and resilience of populations in the wider environment. However, of greater utility in understanding the invasive tendencies of the Manila clam *per se* are indirect effects of human activity, which create anthropogenically altered environments well suited to the particular traits of the species.

The first such anthropogenic alteration derives from pollution. The Manila clam is known to thrive in nutrient-rich eutrophic environments, which supply enhanced quantities of food in the form of phytoplankton, benthic microalgae, and often also as nonliving organic particulates. In this context, the wide food spectrum, high filtration and food assimilation rates, growth rate, and reproductive output of the species (Chapters 7 and 10) enable the efficient exploitation of such anthropogenically stimulated resource abundances. In its native range, the Manila clam attains its highest abundances in eutrophic environments (Lee, 1996), and the significance of such conditions for Manila clam abundance is corroborated by the collapse of indigenous East Asian populations after improved water treatment changed eutrophic to oligotrophic conditions (Chapter 14).

A second cause of habitat alteration has been described in the Venice lagoon. Here, Breber (2002) noted that the spatial distribution of clams coincided with intertidal flats where light is strong, wave action is weak, and benthic diatoms flourish. However, trophic mass-balance modeling by Pranovi et al. (2003) has also shown that the energy required to sustain "invasive" levels of Manila clam abundance is greater than that available in phytoplankton biomass. In investigating this, they found that the resuspension of benthic food material by clam dredging creates a positive feedback loop between clam fishing effort and clam standing stock. Their model predicted a 20% decline in Manila clam biomass when clam dredging is reduced to zero. This work provided a quantitative elucidation of the "*Tapes* paradox" originally reported in Libralato et al. (2002).

Therefore, human activity in the form of both nutrient pollution and mechanical dredging fundamentally alters the energy dynamics of Manila clam habitats and the structure and function of the ecosystem. In this context, the effectiveness and efficiency with which the Manila clam can exploit these alterations provide the basis for a level of success that has been regarded as invasive. Sites such as the Venice Lagoon and Tagus estuary, where the character of invasiveness has often been attributed to Manila clam populations, have both experienced particularly intense levels of clam dredging, and both are eutrophic. Synergistically interacting impacts of multiple anthropogenic habitat alterations are not unique to the case of the Manila clam. For example, Byrnes et al. (2007), in a wide-ranging analysis, identified a "trophic skew" in which combinations of human-caused extinctions typically at higher trophic levels (e.g., predators) and non-native introductions, typically at lower trophic levels (such as filter feeders and detritivores), reshaped the marine food webs of the Wadden Sea.

In summary, it can be argued that the Manila clam is not an inherently invasive species. It only achieves densities sufficient to provoke claims of invasion, in specific combinations of circumstances where it is anthropogenically introduced to areas where optimal climatically and locally driven patterns of temperature and salinity combine with anthropogenic alterations of the habitat, which in turn enable it to maximally exploit its advantageous traits. Beyond such circumstances in less altered habitats, one or other limiting factor acts to constrain densities to a level allowing spatial coexistence with native bivalves. Consequently, any invasiveness claimed for the Manila clam might at most be described as anthropically induced: that is to say, any "invasiveness," rather than being an inherent feature of the species, is restricted to and triggered by habitats significantly altered by human activity.

It follows that anthropically induced "invasiveness" implies that solutions to "invasive" densities of Manila clams regarded as undesirable lie in addressing the causes rather than the effects, namely addressing nutrient pollution levels and/or effectively managing fishing effort and aquaculture operations.

22.4 Conservation policy and the Manila clam

The arrival and naturalization of any non-native species will have some level of influence on the recipient system. From a strict conservation viewpoint, all non-native arrivals are in principle negative since by their presence they represent a change to the indigenous system, i.e., the system is changed rather than conserved, even if the change is of only low significance in terms of ecological function or biodiversity. In many cases of incipient or recent arrival of new species, introduction has been inadvertent, in which case conservation responses include assessments of "invasiveness" and management measures focusing on prevention or mitigation remedies such as containment or eradication.

However, the situation is more compounded for species that have been, or are proposed to be, introduced purposefully, with motivations involving benefit to human populations. In these circumstances, conservation and economic imperatives interact. *R. philippinarum* has been a case in point, and as such provides an opportunity to consider both sides of the benefit-detriment equation in such a way as to reflect on the decision to introduce.

Several conservation policies are generally influential in determining responses to non-native Manila clams. Many are underwritten by international agreements and treaties governed by organizations such as the United Nations, whose general overarching commitments are enshrined in the UN Sustainable Development Goals (SDGs). These include SDG 14: To conserve and sustainably use the oceans, seas and marine resources for sustainable development. Other goals relevant to Manila clam introduction, aquaculture, and fishing relate to economic growth (SDG 8), good health (SDG 3), and climate action (SDG 13). Specific policies related to these general goals (although sometimes preceding them) include the widespread adoption of the "precautionary principle," policies concerning non-native species, and policies on marine protected areas (MPAs). Tensions between such parallel conservation and socio-economic priorities are sometimes apparent in a dichotomy of attitudes to the Manila clam, which have been manifest since Phase 2 of its globalization commencing in the 1970s (Chapter 3). The controversy around the UK introduction of the Manila clam is a case in point (see Chapter 24).

Ostensibly, resolution of the tensions that exist between conservation and socio-economic goals is achieved through the concept of sustainable development, which has been widely adopted since its introduction to the international policy arena in 1987. Enshrined in this idea is the politically attractive premise that there need not be a trade-off between economic growth and the environment (Humphreys & Herbert, 2018). An overarching statement of this principle is, for example, found within the European Union "Ocean Pact," whose six key priorities include "Boosting the sustainable competitiveness of the blue economy" (European Commission, 2025)

Its preeminent global position among exploited mollusks indicates the importance of the Manila clam in terms of socio-economic goals. This value, combined with its inshore rather than deep-sea habitats, also offers particular beneficial contributions to relatively disadvantaged coastal communities, even in otherwise affluent countries. As a high-value seafood, the species also makes contributions in terms of "decent work" (SDG 8), good health (SDG 3), and gender equality (SDG 5, see Chapter 16). Moreover, from the conservation and sustainability standpoints, the Manila clam has merit derived from its status as a low-trophic aquaculture product and relatively low carbon impact from farming (Chapter 18). As such, when properly managed and regulated (Chapter 19), Manila clam farming and fishing can reasonably be argued as a prime example of sustainable development.

However, despite the fact that many of the habitats that offer opportunities for Manila clam naturalization and exploitation are already greatly altered by human impacts, the arrival of the species does generate conservation concerns. Aside from general attitudes to non-native species, these relate most directly to the impacts of fishing and aquaculture practices on the seabed and consequences for the overlying water column. Particular issues arise in the context of the growing proportion of estuaries and tidal flats designated as marine protected areas (MPAs). Since the 1990s, area-based conservation has become one of the main strategies for marine conservation (Humphreys & Clark, 2020). In 2022 and after ratification in 2025, an international global biodiversity framework agreed under the auspices of the CBD committed signatory nations to the effect that at least 30% of all sea areas should fall within some form of MPA. The effect of this and earlier CBD targets has been the proliferation of MPAs, not least in inshore areas.

In England, for example, by 2025, 51% of inshore waters fell within the MPA network (Collin et al., 2025). A significant proportion of estuaries worldwide now fall under such designations. These operate under several different statutes and normally seek to conserve specific conservation features such as species or habitats. Often, among such features are the range of intertidal substrates including "coarse sediments; sandy and muddy sediments; mud sediments; and mixed sediments" (JNCC, 2019). Many intertidal sediments are protected for marine wading birds under conservation directives such as the international Ramsar Convention, but can also accumulate various other national or regional designations.

Therefore, many forms of MPA adopt feature-based approaches in which specific named features within the MPA boundary are protected, in contrast to a "whole-site" approach. Within an MPA, goals typically involve the achievement through restoration or maintenance of "favourable condition" of the features. Features-based approaches are consistent with the principle of sustainable development as they are more conducive to the coexistence of ecological and economic goals. However, the adoption of whole-site approaches, which seek to conserve everything within an MPA boundary, has long been and continues to be called for by conservation organizations (Clifford 2025). The arguments deployed to this end are often legitimate assertions of the crucial links between specific features and their environments. To some extent, this is recognized when a specific named feature is a habitat rather than a species.

However, the general goals of restoration are commonly expressed in terms of the achievement of "favourable condition" meaning "healthy stable habitats" where the "natural range" is stable (Department for Environment, Food & Rural Affairs (DEFRA) et al., 2021). The question then becomes more obscure: restoration to what condition? The transitional habitats of Manila clams are geomorphologically protected from high-energy wave action. Such coastal features have been used by human communities since the early Holocene. Archaeological research in Japan, North America, and Europe, for example, demonstrates prehistoric uses of the coastal zone for bivalve fishing, not only for subsistence but also for trading

(Easton et al., 2021; Shimoyame et al., 2015; Winder & Campbell, 2022; and Chapter 1). Such sites have long been in continuous use by coastal communities to the present day (Humphreys, 2022).

In this context, what do phrases like "favourable condition" and "natural range" mean in practice? Arguably, it is reasonable to include human activity as a legitimate part of a baseline "natural range" coastal system, so long as it remains compatible with the conservation of highly valued or rare specific named features. If so, then certain "whole site" conservation approaches, in so far as they imply a cessation of long-established clam fishing and harvesting, could be disproportionate. Conversely, "feature specific" approaches to marine protected areas are often deemed compatible with properly managed sustainable exploitation, which balances ecological and economic goals.

In any event, many estuaries and other transitional locations have long-since ceased being "wilderness" environments where human impact is historically minimal. Where this is the case, by comparison with terrestrial systems, some such locations can be more accurately perceived as analogous to semiurban farming land, which, although needing good environmental practices and careful management, is thought compatible with many types of terrestrial protected areas. The problem of establishing legitimate baselines from which to assess marine non-native impacts has been addressed by Carlton (2009) and Ojaveer et al. (2018), among others. In their review, Ojaveer et al. (2018) recommend that effective management should recognize the interactions between non-natives and other environmental stressors and also consider shifting global patterns, including climate change (Chapter 23).

All this suggests that attitudes to the introduction or natural arrival of the Manila clam, or proposals for new aquaculture operations, merit a balanced and dispassionate consideration. However, consensus on such matters can be difficult to achieve. Not least as they are essentially political in nature, involving questions similarly posed by re-wilding introductions, or genetically engineered crop varieties. For non-native species, decisions can be contentious and difficult to resolve.

The UK plans for the sustainable development of the "blue economy," including growth in sustainable aquaculture. However, the development of new Pacific oyster farms has been a matter arguably unresolved for many years (Humphreys et al., 2014, 2021). It is correctly assumed that new farms would cause an increase in naturalized populations. However, there are good reasons to believe that restrictions on Pacific oyster aquaculture will not prevent further inevitable naturalization, especially in warming waters (Herbert et al., 2012; Shakspeare et al., 2024). It has been argued that UK policy processes, ostensibly motivated by conservation concerns, have been driven by a political need to be seen to be doing "conservation," rather than based on any scientifically understood relationship between cause and effect, thereby contributing little or no material conservation advantage (Humphreys, 2022).

Below, the introduction of the Manila clam to English waters is considered in such a way as to elucidate how deeply held convictions, combined with the political nature of such decisions, can lead to flawed evidence selection and an unproductive polarization of scientific opinion at the interface between science and policy.

References

Bendell, L. I. (2014). Evidence for declines in the native *Leukoma staminea* as a result of the intentional introduction of the non-native *Venerupis philippinarum* in coastal British Columbia, Canada. *Estuaries and Coasts., 37*(2), 369–380. https://doi.org/10.1007/s12237-013-9677-1.

Bonham, V., Humphreys, J., Yang, N., Haiying, O., & Caiping, Y. (2015) Datasheet report for *Ruditapes philippinarum* (Japanese carpet shell). CABI Retrieved October 22, 2015, from www.cabi.org/isc/datasheetreport?dsid=61697.

Breber, P. (2002). Introduction and acclimatisation of the Pacific Carpet Clam Tapes philippinarum to Italian waters. In E. Leppäkosld, et al. (Eds.), *Invasive Aquatic Species of Europe* (pp. 120–126). Kluwer Academic Publishers.

Byrnes, J. E., Reynolds, P. L., Stachowicz, J. J., & Lusseau, D. (2007). Invasions and extinctions reshape coastal marine food webs. *PLoS One, 2*(3), e295. https://doi.org/10.1371/journal.pone.0000295.

Carlton, J. T. (2009). Deep invasion ecology and the assembly of communities in historical time. In G. Rilov, & J. A. Crooks (Eds.), *Biological invasions in marine ecosystems* (pp. 13–56). Springer-Verlag.

CBD 2002. COP 6 Retrieved October 1, 2025, from https://www.cbd.int/decision/cop?id=7197.

Clifford, D. (2025). Reality check for marine protection, *The marine biologist* (36, pp. 10–13). Marine Biological Association.

Clout, M. N., & Williams, P. A. (Eds.). (2009). *Invasive species management: A handbook of principles and techniques*. Oxford University Press.

Copp, G. H., Godard, M. J., Russell, I. C., Peeler, E. J., Gherardi, F., Tricarico, E., Miossec, L., Goulletquer, P., Almeida, D., Britton, J. R., Vilizzi, L., Mumford, J., Williams, C., Reading, A., Rees, E. M. A., & Merino-Aguirre, R. (2016a). A preliminary evaluation of the European on-native species in aquaculture risk assessment scheme applied to species listed on Annex IV of the EU Alien species regulation. *Fisheries Management and Ecology, 23*(1), 12–20. https://doi.org/10.1111/fme.12076.

Copp, G. H., Vilizzi, L., Tidbury, H., Stebbing, P. D., Tarkan, A. S., Miossec, L., & Goulletquer, P. (2016b). Development of a generic decision support tool for identifying potentially invasive aquatic taxa: AS-ISK. *Management of Biological Invasions, 7*, 343–350.

Correia, M. J., Chainho, P., Goulding, T., Carvalho, F., Cabral, S., Gomes Ferreira, F., & Vasconcelos, L. (2025). Participatory action research supporting adaptive governance of Manila clam fisheries. *Marine Policy, 174*, 106605. https://doi.org/10.1016/j.marpol.2025.106605.

Collin, S., Smith N., Pearson Z. (2025). Review and evaluation of the marine protected area networks in England and Northern Ireland. A report commissioned by the Office for Environmental Protection. https://www.theoep.org.uk/sites/default/files/investigations-files/MPA%20Review%20and%20Evaluation_FINAL.pdf.

Department for Environment, Food & Rural Affairs, Welsh Government, Natural England and Natural Resources Wales, 2021. Guidance: Duty to protect, conserve and restore European sites. Competent authorities must take action to help protect, conserve and restore the protected habitats and species of European sites. Retrieved November 07, 2025, from https://www.gov.uk/guidance/duty-to-protect-conserve-and-restore-european-sites#:(:text=By%20meeting%20the%20objectives%2C

Easton, N. A., Moore, C., & Mason, A. R. (2021). The archaeology of submerged prehistoric sites on the North Pacific Coast of North America. *The Journal of Island and Coastal Archaeology, 16*(1), 118–149. https://doi.org/10.1080/15564894.2020.1785061.

Elton, C. S. T. (1958). *The ecology of invasions by animals and plants*. University of Chicago Press.

European Commission, 2025 The European ocean pact. Retrieved November 07, 2025, from https://oceans-and-fisheries.ec.europa.eu/european-ocean-pact_en

Flye-Sainte-Marie, J., Jean, F., Paillard, C., Ford, S., Powell, E., Hofmann, E., & Klinck, J. (2007). Ecophysiological dynamic model of individual growth of *Ruditapes philippinarum*. *Aquaculture, 266*(1-4), 130–143. https://doi.org/10.1016/j.aquaculture.2007.02.017.

Goulletquer, P. 1997 A bibliography of the Manila clam *Tapes philippinarum*. IFREMER, RIDRV-97.021RN: 122 pp. http://archimer.ifremer.fr/doc/00000/3221.

Herbert, R. J. H., Humphreys, J., Davies, C. J., Roberts, C., Fletcher, S., & Crowe, T. P. (2012). Ecological impacts of non-native Pacific oysters (Crassostrea gigas) and management measures for protected areas in Europe. *Biodiversity and Conservation, 25*(2016), 2835–2865. https://doi.org/10.1007/s10531-016-1209-4.

Humphreys, D. A. (2022). An industrial history of Poole Harbour. In J. Humphreys, & A. E. Hall (Eds.), *Harbour ecology: Environment and development in Poole Harbour*. Pelagic Publishing.

Humphreys, J., Syvret, M., Horsfall, S., Williams, C., Woolmer, A., & Adamson, E. A. S. (2021). Why we should learn to love Pacific Oysters. *The marine biologist* (pp. 10–11). Marine Biological Association.

Humphreys, J., Caldow, R. W. G., Mcgrorty, S., West, A. D., & Jensen, A. C. (2007). Population dynamics of naturalised Manila clams *Ruditapes philippinarum* in British coastal waters. *Marine Biology, 151*(6), 2255–2270. https://doi.org/10.1007/s00227-007-0660-x.

Humphreys, J., & Clark, R. W. E. (2020). A critical history of marine protected areas. In J. Humphreys, & R. W. E. Clark (Eds.), *Marine protected area. Science policy and management* (pp. 1–12). Elsevier.

Humphreys, J., Harris, M. R. C., Herbert, R. J. H., Farrell, P., Jensen, A., & Cragg, S. M. (2015). Introduction, dispersal and naturalization of the Manila clam *Ruditapes philippinarum* in British estuaries, 1980-2010. *Journal of the Marine Biological Association of the United Kingdom, 95*(6), 1163–1172. https://doi.org/10.1017/S0025315415000132.

Humphreys, J., & Herbert, R. J. H. (2018). Marine protected areas: Science, policy & management. *Estuarine, Coastal and Shelf Science., 215*, 215–218. https://doi.org/10.1016/j.ecss.2018.10.014.

Humphreys, J., Herbert, R. J. H., Roberts, C., & Fletcher, S. (2014). A reappraisal of the history and economics of the Pacific oyster in Britain. *Aquaculture, 428-429*, 117–124. https://doi.org/10.1016/j.aquaculture.2014.02.034.

Humphreys, J. (2022). Marine conservation: Smoke and mirrors in the coastal zone. In J. Humphreys, & S. Little (Eds.), *Challenges in estuarine and coastal science*. Pelagic Publishing.

Jensen, K. R., Andersen, P., Andersen, N. R., Bruhn, A., Buur, H., Carl, H., Jakobsen, H., Jaspers, C., Lundgreen, K., Nielsen, R., Strandberg, B., & Stæhr, P. A. U. (2023). Reviewing introduction histories, pathways, invasiveness, and impact of non-indigenous species in Danish marine waters. *Diversity, 15*(3). https://doi.org/10.3390/d15030434.

JNCC, 2019 Description of biotope or habitat type: Littoral sediment. Joint Nature Conservation Committee, Retrieved November 05, 2025, from https://mhc.jncc.gov.uk/biotopes/jnccmncr00000274

Juanes, J., Bidegain, G., Echavarri-Erasun, B., Puente, A., García, A., García, A., Bárcena, J. F., Álvarez, C., & García-Castillo, G. (2012). Differential distribution pattern of native *Ruditapes decussatus* and introduced *Ruditapes phillippinarum* clam populations in the Bay of Santander (Gulf of Biscay): Considerations for fisheries management. *Ocean & Coastal Management, 69*, 316–326. https://doi.org/10.1016/j.ocecoaman.2012.08.007.

Katsanevakis, S., Wallentinus, I., Zenetos, A., Leppäkoski, E., Ertan Çinar, M., Oztürk, B., Grabowski, M., Golani, D., & Cardoso, A. C. (2014). Impacts of invasive alien marine species on ecosystem services and biodiversity: A pan-European review. *Aquatic Invasions, 9*(4), 391–423. https://doi.org/10.3391/ai.2014.9.4.01.

Kolar, C. S., & Lodge, D. M. (2001). Progress in invasion biology: Predicting invaders. *Trends in Ecology & Evolution, 16*, 199–204. https://doi.org/10.1016/S0169-5347(01)02101-2.

Larson, B. M. H. (2011). Embodied realism and invasive species, *Philosophy of ecology: Volume 11 in handbook of the philosophy of science* (11, pp. 129–146). https://doi.org/10.1016/B978-0-444-51673-2.50005-4.

Lee, S. Y. (1996). Distribution pattern and interaction of two infaunal bivalves, *Tapes philippinarum* (Adams and Reeve) and *Anomalocardia squamosa* (Linnaeus), (Bivalvia: Veneridae). *Journal of Experimental Marine Biology and Ecology, 201*(1-2), 253–273. https://doi.org/10.1016/0022-0981(96)00015-9.

Libralato, S., Pastres, R., Pranovi, F., Raicevich, S., Granzotto, A., Giovanardi, O., & Torricelli, P. (2002). Comparison between the energy flow networks of two habitats in the Venice Lagoon. *Marine Ecology, 23*(s1), 228–236. https://doi.org/10.1111/j.1439-0485.2002.tb00022.x.

Maynard, G., & Nowell, D. (2009). Biosecurity and quarantine for preventing invasive species. In M. N. Clout, & P. A. Williams (Eds.), *Invasive species management: A handbook of principles and techniques* (pp. 1–18). Oxford University Press.

Moura, P., Garaulet, L. L., Vasconcelos, P., Chainho, P., Costa, J. L., & Gaspar, M. B. (2007). Age and growth of a highly successful invasive species: The Manila clam Ruditapes philippinarum (Adams & Reeve, 1850) in the Tagus Estuary (Portugal). *Aquatic Invasions, 2*(2), 133–146.

Ojaveer, H., Galil, B. S., Carlton, J. T., Alleway, H., Goulletquer, P., Lehtiniemi, M., Marchini, A., Miller, W., Occhipinti-Ambrogi, A., Peharda, M., Ruiz, G. M., Williams, S. L., Zaiko, A., & Bianchi, C. N. (2018). Historical baselines in marine bioinvasions: Implications for policy and management. *PLOS ONE., 13*(8), e0202383. https://doi.org/10.1371/journal.pone.0202383.

Pranovi, F., Franceschini, G., Casale, M., Zucchetta, M., Torricelli, P., & Giovanardi, O. (2006). An ecological imbalance induced by a non-native species: The Manila clam in the Venice Lagoon. *Biological Invasions, 8*, 595–609. https://doi.org/10.1007/s10530-005-1602-5.

Pranovi, F., Libralato, S., Raicevich, S., Granzotto, A., Pastres, R., & Giovanardi, O. (2003). Mechanical clam dredging in Venice Lagoon: Ecosystem effects evaluated with a trophic mass-balance model. *Marine Biology, 143*, 393–403.

Reise, K., Buschbaum, C., Büttger, H., & Wegner, K. M. (2017). Invading oysters and native mussels: From hostile takeover to compatible bedfellows. *Ecosphere, 8*(9). https://doi.org/10.1002/ecs2.1949.

Roy, H. E., Rabitsch, W., Scalera, R., Stewart, A., Gallardo, B., Genovesi, P., Essl, F., Adriaens, T., Bacher, S., Booy, O., Branquart, E., Brunel, S., Copp, G. H., Dean, H., D'hondt, B., Josefsson, M., Kenis, M., Kettunen, M., Linnamagi, M., … Zenetos, A. (2018). Developing a framework of minimum standards for the risk assessment of alien species. *Journal of Applied Ecology, 55*(2), 526–538. https://doi.org/10.1111/1365-2664.13025.

Sandvik, H. (2023). For quantitative criteria in alien species assessment. *Frontiers in Environmental Science, 11*. https://doi.org/10.3389/fenvs.2023.1119094.

Shakspeare, A., Cameron, T. C. and Steine, M. 2024 Restrictions on UK aquaculture of Pacific oyster (*Magallana gigas*) will not prevent naturalised spread but suppress ecological and economic benefits to coastal communities. (Preprint) https://doi.org/10.32942/X2PP60.

Shimoyama, S., Ichihara, T., Tsukano, K., Kabashima, M., Momoshima, N., Komorita, T., & Tsutsumi, H. (2015). Historical occurrence of the short-neck clam, *Ruditapes philippinarum* (Adams & Reeve, 1850), on the sandy flats of Ariake Bay, Kyushu, western Japan. *Plankton Benthos Research, 10*(4), 202–214.

Williamson, M. H., & Brown, K. C. (1986). The analysis and modelling of British invasions. *Philosophical Transactions of the Royal Society of London. B, Biological Sciences, 314*(1167), 505–522. https://doi.org/10.1098/rstb.1986.0070.

Williamson, M. (1996). *Biological invasions*. Chapman & Hall.

Winder, J., & Campbell, G. (2022). The archaeology of the marine shells of Poole Harbour. In J. Humphreys, & A. E. Hall (Eds.), *Harbour ecology: Environment and development in Poole Harbour*. Pelagic Publishing.

CHAPTER

Climate change: Effects and implications

23

23.1 The anthropogenic context

The coastal zone has been described as an epicenter of climate change impacts. These include rising sea levels, an increase in water temperature, and a decrease in average salinity through precipitation. In addition, changes in carbon chemistry result in the acidification of seawater. Moreover, transitional areas such as estuaries and lagoons can amplify the salinity and temperature changes expected for the open sea (Ferrarin et al., 2014; Humphreys & Little, 2022; IPCC, 2014). Ocean warming and acidification are already adversely affecting food production from shellfisheries; however, such effects are not occurring uniformly (IPCC, 2023). With salinity, for instance, the balance between precipitation and evaporation varies with latitude, and in higher latitudes, ice melt reduces salinity.

The impacts of climate change on the Manila clam operate directly through physiological and metabolic processes such as growth and reproduction, and indirectly through chronic or acute changes to the biological and physical environment. While some effects, such as ocean acidification, represent a general risk to all bivalves, the impact of others, including sea temperature change, will vary from negative to positive, for example, according to latitude, thus also enabling northerly range expansion. In addition, shallow coastal and intertidal zones in particular are vulnerable to the increasing frequency and severity of acute weather events such as storms and heat waves.

As part of their framework for assessing climate change, the Intergovernmental Panel on Climate Change (IPCC) produces different concentration pathways (RCPs) representing future greenhouse gas concentrations, along with shared socioeconomic pathways (SSPs) representing possible future emissions scenarios, ranging from sustainability to further fossil fuel development (Riahi et al., 2017). They also support a Coupled Model Intercomparison Project (CMIP) to compare models in such a way as to advance climate modeling and promote collaboration (Eyring et al., 2016). These various initiatives and facilities have been widely applied in the context of experimentally determined or geographically inferred tolerances to predict the potential effects of climate change on species such as the Manila clam.

The Manila Clam *Ruditapes philippinarum:* Science and Significance of a Globalised Species.
DOI: https://doi.org/10.1016/B978-0-443-36704-5.00019-5
Copyright © 2026 Elsevier Inc. All rights are reserved, including those for text and data mining, AI training, and similar technologies.

23.2 Ecological and biological effects

Most Manila clam climate-related work has focused on temperature. Above 25°C, Manila clam growth rate declines as the species experiences physiological and metabolic disturbances (Isono et al., 1998, see Chapter 8). As an intertidal species, a level of temperature tolerance will serve as a "preadaptation" to the thermal effects of climate change in more northerly latitudes, but further south, impacts such as heat waves are already considered to be implicated in population declines. Ponti et al. (2017) have suggested that increased surface sea temperature has already contributed to stock decline in the relatively warm lagoons of the Adriatic. Bertolini and Pastres (2021) have developed a thermal tolerance landscape model for *R. philippinarum*, which accommodates both heatwave intensity and duration in Adriatic lagoons. They found that between 2008 and 2018, "mortality thresholds" were exceeded in 3 years over a 10 year period

Climate effects can cause changes in both food availability and predator activity for bivalves, including the Manila clam. The multifactor nature of climate change can sometimes make it difficult to establish unequivocal causal linkages between Manila clam abundance and indirect climate-related environmental changes, such as in phytoplankton or predator distribution. Dethier et al. (2019) have shown that predator impact on Manila clams is context-dependent. In Washington State, United States, mortality is highest in high salinity low intertidal conditions, most likely due to greater marine predator access during longer immersion times. Whereas in lower salinities, the relative effect of predation on mortality was lower. Since both predator distribution and salinity are affected by climate change, complex synergies are likely to be at work with differing results locally.

Tezuka (2025), in reviewing the possible causes of Manila clam declines in Japan, has suggested that the impact of warm-water predators such as the Black Seabream has increased with coastal warming. Conversely, it might be supposed that some Manila clam predators extant in current climate conditions may, in the future, decline in abundance.

Among biotic influences on survival, competition for food is often regarded as being a lesser forcing process compared to predation, as food supply, whether planktonic, benthic, or detrital, is not typically a limiting factor. Yoon et al. (2013), working in Akkeshi Lake, a sub-Arctic lagoon in Japan's most northerly main island, modeled the impact of a 2°C water temperature increase on food availability (planktonic and benthic) on Manila clam growth. Their model predicted an increase in clam growth due to this indirect effect of warming. A corresponding decline in food availability was attributed to ingestion by filter feeding, but in the warming scenario, food was never a limiting factor on clam growth.

23.3 Synergistic pollution impacts

Oxidative stress is a response to chemical toxicity in marine animals. Under warming conditions (17°C–21°C) when exposed to caffeine pollution,

R. philippinarum avoids significant cellular damage by activating antioxidant and detoxification mechanisms (Cunha et al., 2025, Chapter 11). However, there is a limit to such tolerance mechanisms. Lompre et al. (2025) investigated the interaction between metal toxicity and temperature using chromium and the "rare earth" terbium at two temperatures (18°C and 21°C). At the higher temperature, they reported an inhibition of antioxidant and biotransformation enzymes, suggesting that oxidative stress impaired enzyme activity, especially when both metals were present, as is likely in electronic waste. They postulated a general increase in risk from metal toxicity under climate change.

Pharmaceutical pollutants include the synthetic hormone 17-gamma-ethinylestradiol (EE2), used variously from oral contraceptives to prostate therapy. This compound is chemically stable and is known to accumulate in aquatic animals. Consequently, it has been included in the European Union Watch list for contaminants of emerging concern. EE2 is known to induce negative cellular effects, including oxidative stress in *R. philippinarum* (Silva et al., 2022). Rodrigues et al. (2023) have elucidated the complex metabolic responses of *R. philippinarum* when exposed to different concentrations at 17°C and 21°C. When exposed to higher concentrations of EE2 (625 ng/L) at the higher temperature, these responses, while they may enhance antioxidant properties, also have degradation effects on cell membrane stability.

Despite an innate resilience to oxidative stress from pollutants, it seems that the higher temperatures predicted in climate change scenarios can significantly reduce the Manila clam's ability to deal with pollutant stressors, to an extent that has led to calls for better management of pollutant releases in the context of climate change.

23.4 Acidification

Anthropogenic ocean acidification refers to a reduction of the pH of seawater due to increased CO_2 emissions. When dissolved, CO_2 becomes a weak acid, which generates changes in the carbon chemistry of seawater, including the equilibrium between $CaCO_3$ precipitation and dissolution. The phrase acidification is misleading insofar as it implies that sea surface waters will necessarily become acidic. Since pH will likely remain above 7.0 until at least 2100, ocean acidification might alternatively be termed "carbonation" as it represents an increase in the concentration of dissolved inorganic carbon (Gattuso & Hansson, 2011).

Under RCP8.5, a high-end scenario, IPCC (2019) projects ocean water pH averages to decrease from the current 8.1 to 7.7. However, even in the open ocean, pH varies naturally with other variables, including temperature, and in coastal bays and estuaries can fall as low as 7.0 due to dilution with fresh water and sometimes H_2S release (Sverdrup et al., 1942). On Mediterranean coasts, for example, pH has been reported to vary seasonally from 7.9 to 8.3 (Prado et al., 2016). It is therefore likely that estuarine and coastal species will have evolved metabolic and physiological responses conferring a level of natural tolerance, which in turn will offer some level of "preadaptation" for anthropogenic acidification.

Nevertheless, such changes in carbonate chemistry are sufficiently serious to represent a risk to various biological processes, from photosynthesis to molluscan shell formation. Like sea warming, acidification will impact directly on metabolism, growth, and reproduction, and indirectly by trophic changes relating to photosynthetic plankton and predator species.

At a "control condition" of pH 7.8 and salinity 28, representing the Tagus estuary in Portugal, Velez et al. (2016) reported that *R. philippinarum* maintained both physiological condition and biochemical performance for at least the 28-day experimental duration. Moreover, combining lower pH (7.3) with lower (14 and 25) salinities indicated that pH has a lesser impact than salinity at these levels.

When exposed to seawater with pH reduced from 8.1 to 7.7, Manila clams exhibit significantly increased carbonic anhydrase activity, an enzyme involved in calcification, along with depletion in tissue lipid and glycogen levels. Moreover, there is a transgenerational benefit to reduced pH exposure, in the sense that clams whose parents experienced lower pH showed a significantly higher carbonic anhydrase response, faster shell growth, and higher Condition Index at the lower pH, compared to clams spawned from parents exposed to the higher pH of currently ambient seawater. This transgenerational acclimation effect applies to both juvenile and adult Manila clams (Zhao et al., 2018, 2019) Observing that the clam's energy budget was lower under the lower pH, and that no similar transgenerational benefit accrued in relation to energy-metabolizing enzymes, these authors suggested that the most critical impact of acidification may be on energy budget.

However, it also seems that after transgenerational acclimation, the Manila clam can adopt a less energetically costly strategy to mitigate water acidification: Using stable carbon isotopic signatures, Zhou et al. (2018) have also demonstrated that in acclimated clams, up to 61% of shell carbonate comes from carbon produced during metabolism, rather than from seawater. This also applies to nonacclimated clams but to a lesser extent (45%). Extraction from seawater is known to be more energetically demanding. Based on their findings, Zhao et al. (2018) stated that the Manila clam may be more resilient to acidification than is sometimes suggested.

Yuan et al. (2020) have reached a similar general conclusion when investigating the effects of a range of salinities (14, 28, and 35) and pH (7.8 and 7.3), alone and in combination, on the physiological condition and oxidative stress markers of *R. Philippinarum* over 28-day periods. The 28-day responses at the lowered pH were the same as at the control (7.8) pH, and changes in physiological condition and biochemical performance could be explained by salinity alone. They also concluded that predicted acidification would be of less consequence for Manila clams than salinity shifts.

Nevertheless, pH change can destabilize acid-base homeostasis. In experiments subjecting Manila clams to CO_2-induced pH shifts from 8.10 (the control pH) to 7.48, Jiang et al. (2023) found increases in foot contraction and movement but also impeded burrowing time and depth. These behavioral responses coincided with increased intracellular bicarbonate ions and decreased pH. Almeida et al. (2022) investigated the combined responses of Manila clams to lowered pH (8–7.6) and exposure to antiepileptic and antihistamine pharmaceutical pollutants. After 28 days' exposure, they reported that the clam's response to these drugs changed with pH. Lowered pH alone resulted in cellular damage, but this was not enhanced when combined with the pharmaceutical agents. Indeed, lower pH seemed to increase activation of the drug elimination processes. De Marchi et al. (2017) found that under lowered pH, the toxicity of *R. philippinarum* to carbon nanoparticles was the same as at the control pH.

Modeling the effect of pH and temperature on growth, Maynou et al. (2020) predicted that while a temperature increase of 2°C–3°C would increase length and weight by 2%–3%, a reduced pH of 7.3 would reduce shell length by 2%–5%. Decreased pH would also negatively affect "reproductive potential," but this was more than compensated for by the temperature increase aggregated over 10 years.

In summary, the evidence suggests that currently predicted anthropogenic acidification will, in itself, be within the tolerances of *R. philippinarum*, although it may contribute to increased oxidative stress and energetic demand, especially when in combination with commonly occurring pollutants.

23.5 Predicted distribution change

Jung et al. (2025) have used Surface Sea Temperature (SST) data from an ensemble of eight climate system models provided by Phase 6 of CMIP, under four IPCC SSPs from SSP1-2.6 (sustainable development) to SSP5-8.5 (fossil fuel-dependent rapid economic growth). To predict future Manila clam distribution across seven sites in North and South Korea, they defined vulnerable conditions for Manila clam populations as a sequence of eight or more consecutive days with water temperatures exceeding 26°C. Except for the two most southerly Korean peninsula sites, all other sites showed an increase in high temperature days (over 25°C) with corresponding greater Manila clam vulnerability over time.

Their results show that patterns of ocean circulation, rather than just latitude, are an important factor in determining Manila clam vulnerability. By 2075, Wonson, the most northerly and only east coast site, had the fewest high-temperature days (3.75), while Cheongcheon, with a similar latitude on the west coast, had 47.25 high-temperature days. All other sites exceeded 65.75 days, with Geoje on the southern coast predicted to have the highest (88.25 high-temperature days). These results were consistent with the oceanographic influence of the warm Kuroshio Current, which forms the western side of the North Pacific Ocean gyre, carrying water from the tropics. At all sites, the period from 2050 to 2075 is predicted to suffer the greatest increase in high-water temperature days.

Based on their results, Jung et al. (2025) suggest that the appropriate habitat for the Manila clam will move north and that the north-eastern coast represents a greater future cultivation potential. At central sites, production could be maintained under mid-range scenarios if adaptation includes changing the farming calendar and using deep water extraction for cooling systems in hatcheries and nurseries, as has been developed in Hawaii and Japan (Nakasone & Akeda, 1999). However, in the southernmost site at Geoje, it is predicted that the existing Manila clam industry will become unsustainable.

In the Venice Lagoon, modeling also based on IPCC scenarios, predicted changes in larval abundance and the timing of spawning and settlement from Spring/Summer to Winter/Autumn (Ghezzo et al., 2018). These authors also predict changes in both spatial distribution and the absolute quantity of settled larvae in the lagoon, first positively and over time negatively.

In a preliminary assessment of Manila clam distribution changes at a global scale, Johnson et al. (2025) used maximum entropy modeling to predict worldwide distribution in 2100 under a worst-case IPCC climate change scenario, in which global greenhouse gas emissions continue to rise and global average temperature rises exceed 4°C by 2100. They modeled published tolerances of the species against two climate scenario variables, temperature and salinity. The model predicted a northerly poleward increase in suitable habitat reaching as high as 70°N in the North Atlantic, giving new opportunities for aquaculture production in Scandinavia and also in Alaska. Conversely, their modeling suggested a decrease in habitat suitability in Asia between latitudes 20° and 40°N and on Italy's Mediterranean coast.

This study by Johnson et al. (2025) exemplifies a difficulty of model construction when cross-validation uses present-world distribution data. Although this was taken to indicate good model performance, the accuracy of current global Manila clam distribution data is problematic. In this case, some presence data was taken from the Global Biodiversity Information Facility (GBIF), which (along with other online datasets) is questionable. In particular, with regard to the inclusion of an ostensibly tropical distribution in South-East Asia, which has not stood up to scrutiny. Tropical distribution reports, even including the supposed Philippine type locality, are now thought to be erroneous due to a combination of persistent cataloguing errors and misidentifications with the similar but tropical species *Ruditapes aspera* (Humphreys, 2025, and Chapter 2) .Johnson et al. (2025) have suggested further work to verify their findings.

In the context of increases in water temperature and a reduction of sea ice, a commissioned assessment of the risk of non-native species in the Bering Sea put *R. philippinarum* in the top ten of a list of "species of concern" (Reimer et al., 2017).

In summary, changes in local, regional and global distribution will undoubtedly occur in response to climate change, and predictions of global range expansion are not only plausible, but also consistent with reports of northerly spread of Pacific populations and, in Europe, the species' arrival in the Swedish Skagerrak (Greeve et al., 2025 and see Chpater 3).

References

Almeida, Â, Calisto, V., Esteves, V. I., Schneider, R. J., Soares, A. M. V. M., & Freitas, R. (2022). Responses of *Ruditapes philippinarum* to contamination by pharmaceutical drugs under ocean acidification scenario. *Science of The Total Environment, 824*, 153591. https://doi.org/10.1016/j.scitotenv.2022.153591.

Bertolini, C., & Pastres, R. (2021). Tolerance landscapes can be used to predict species-specific responses to climate change beyond the marine heatwave concept: Using tolerance landscape models for an ecologically meaningful classification of extreme climate events. *Estuarine, Coastal and Shelf Science, 252*, 107284. https://doi.org/10.1016/j.ecss.2021.107284.

Cunha, M., Raimundo, S., De Marchi, L., Leite, C., Pretti, C., Meucci, V., Soares, A. M. V. M., Antunes, S. C., & Freitas, R. (2025). Biochemical dynamics of the invasive clams *Ruditapes philippinarum* and *Corbicula fluminea* in response to caffeine concentrations under present and future climate scenarios. *Estuarine, Coastal and Shelf Science, 318*, 109217. https://doi.org/10.1016/j.ecss.2025.109217.

De Marchi, L., Neto, V., Pretti, C., Figueira, E., Chiellini, F., Morelli, A., Soares, A. M. V. M., & Freitas, R. (2017). The impacts of seawater acidification on *Ruditapes philippinarum* sensitivity to carbon nanoparticles. *Environmental Science: Nano, 4*(8), 1692–1704. https://doi.org/10.1039/C7EN00335H.

Dethier, M. N., Kobelt, J., Yiu, D., Wentzel, L., & Ruesink, J. L. (2019). Context-dependence of abiotic and biotic factors influencing performance of juvenile clams. *Estuarine, Coastal and Shelf Science, 219*, 201–209. https://doi.org/10.1016/j.ecss.2019.02.013.

Eyring, V., Bony, S., Meehl, G. A., Senior, C. A., Stevens, B., Stouffer, R. J., & Taylor, K. E. (2016). Overview of the Coupled Model Intercomparison Project Phase 6 (CMIP6) experimental design and organization. *Geoscientific Model Development, 9*(5), 1937–1958. https://doi.org/10.5194/gmd-9-1937-2016.

Ferrarin, C., Bajo, M., Bellafiore, D., Cucco, A., De Pascalis, F., Ghezzo, M., & Umgiesser, G. (2014). Toward homogenization of Mediterranean lagoons and their loss of hydrodiversity. *Geophysical Research Letters, 41*(16), 5935–5941. https://doi.org/10.1002/2014gl060843.

Gattuso, J.-P., & Hansson, L. (2011). *Ocean acidification* (pp. 1–20). Oxford University Press.

Ghezzo, M., Pellizzato, M., De Pascalis, F., Silvestri, S., & Umgiesser, G. (2018). Natural resources and climate change: A study of the potential impact on Manila clam in the Venice lagoon. *Science of The Total Environment, 645*, 419–430. https://doi.org/10.1016/j.scitotenv.2018.07.060.

Greeve, Y., Lindegarth, M., Panova, M., & Strand, Å. (2025). First report of the Manila clam *Ruditapes philippinarum* on the Swedish Skagerrak coast. *Estuarine Coast Shelf Science, 325*, 109487.

Humphreys, J. (2025). Revisiting the indigenous distribution of the "Manila clam" *Ruditapes philippinarum*: Mechanisms of persistent error. *Estuarine, Coastal and Shelf Science, 314*, 109152. https://doi.org/10.1016/j.ecss.2025.109152.

Humphreys, J., & Little, S. (2022). Trajectories and Challenges in estuarine and coastal sciences. In *Challenges in estuarine and coastal science* (p. 264). Pelagic Publishing.

IPCC, 2014 Climate change 2014: Impacts, adaptation and vulnerability. Part A global and sectoral aspects. Contribution to the fifth assessment report of the Intergovernmental Panel on Climate Change. Cambridge University Press

IPCC. (2019). Changing ocean, marine ecosystems and dependent communities. In N. L. Bindoff, W. W. L. Cheung, & J. G. Kairo (Eds.), *IPCC special report on the ocean and cryosphere in a changing climate*. Cambridge University Press https://www.ipcc.ch/srocc/download-report.

IPCC., 2023: Climate change 2023: Synthesis report. Contribution of working groups I, II and III to the sixth assessment report of the intergovernmental panel on climate change H., Lee and J., Romero (Eds.) IPCC, 184 pp., https://doi.org/10.59327/IPCC/AR6-9789291691647.

Isono, R. S., Kita, J., & Kishida, C. (1998). Upper temperature effect on rates of growth and oxygen consumption of the Japanese littleneck clam, *Ruditapes philippinarum. Nippon Suisan Gakkaishi (Japanese Edition), 64*(3), 373–376. https://doi.org/10.2331/suisan.64.373.

Jiang, W., Fang, J., Rastrick, S. P. S., Samuelsen, O. B., Liang, B., Mao, Y., Strand, Ø., Fang, J., & Jiang, Z. (2023). CO2-Induced ocean acidification alters the burrowing behavior of Manila clam *Ruditapes philippinarum* by reversing GABAA receptor function. *Environmental Science and Technology, 57*(24), 8921–8932. https://doi.org/10.1021/acs.est.3c00707.

Johnson, A. S., Bridges, A. E. H., & Knights, A. M. (2025). Predicting the future distribution of a commercially important clam (*Ruditapes philippinarum*) in a changing climate. *Estuarine, Coastal and Shelf Science, 320.* https://doi.org/10.1016/j.ecss.2025.109307.

Jung, Y., Tae, S., Nam, J., & Kim, Y. H. (2025). Projected thermal stress and regional vulnerability of Manila clam production in Korean waters under climate change scenarios. *Estuarine, Coastal and Shelf Science, 318*, 109209.

Lompre, J. S., Cunha, M., Pinto, J., Soares, A. M. V. M., Pereira, E., & Freitas, R. (2025). Cellular responses of *Ruditapes philippinarum* clams to global changes: Assessing the risks of rising temperatures and E-waste. *Estuarine, Coastal and Shelf Science, 319*, 109262.

Maynou, F., Galimany, E., Ramón, M., & Solé, M. (2020). Impact of temperature increase and acidification on growth and the reproductive potential of the clam *Ruditapes philippinarum* using DEB. *Estuarine, Coastal and Shelf Science, 247*, 107099. https://doi.org/10.1016/j.ecss.2020.107099.

Nakasone, T., & Akeda, S. (1999). The application of deep sea water in Japan. *Proceedings of 28th UJNR Aquaculture Panel Symposium, 28*, 69–75 UJNR Technical Report.

Ponti, M., Abbiati, M., Ceccherelli, V. U., Gamba, E., Ragazzoni, A., & Castellini, A. (2017). Decline of the Manila clam stocks in the northern Adriatic lagoons. *Acta Adriatica, 58*, 89–104. https://doi.org/10.32582/aa.58.1.7.

Prado, P., Roque, A., Pérez, J., Ibáñez, C., Alcaraz, C., Casals, F., & Caiola, N. (2016). Warming and acidification-mediated resilience to bacterial infection determine mortality of early Ostrea edulis life stages. *Marine Ecology Progress Series, 545*, 189–202. https://doi.org/10.3354/meps11618.

Reimer, J. P., Droghini, A., Fischbach, A., Watson, J. T., Bernard, B., & Poe, A. (2017). Assessing the risk of non-native marine species in the Bering Sea. NPRB Project 1523. Alaska Center for Conservation Science, University of Alaska. 39 pp.

Riahi, K., van Vuuren, D. P., Kriegler, E., Edmonds, J., O'Neill, B. C., Fujimori, S., Bauer, N., Calvin, K., Dellink, R., Fricko, O., Lutz, W., Popp, A., Cuaresma, J. C., KC, S., Leimbach, M., Jiang, L., Kram, T., Rao, S., Emmerling, J., … Tavoni, M. (2017). The Shared Socioeconomic Pathways and their energy, land use, and greenhouse gas emissions implications: An overview. *Global Environmental Change, 42*, 153–168. https://doi.org/10.1016/j.gloenvcha.2016.05.009.

Rodrigues, J., Bispo, D. S. C., Silva, M., Araújo, R., Soares, A. M. V. M., Freitas, R., & Gil, A. M. (2023). Impact of sea warming and 17-α-ethinylestradiol exposure on the lipid metabolism of *Ruditapes philippinarum* clams. *International Journal of Molecular Sciences, 24*(11), 9485. https://doi.org/10.3390/ijms24119485.

Silva, M., Esteves, V. I., Meucci, V., Battaglia, F., Soares, A. M. V. M., Pretti, C., & Freitas, R. (2022). Metabolic and oxidative status alterations induced in *Ruditapes philippinarum* exposed chronically to estrogen 17α-ethinylestradiol under a warming scenario. *Aquatic Toxicology, 244*, 106078. https://doi.org/10.1016/j.aquatox.2022.106078.

Sverdrup, H. U., Johnson, M. W., & Fleming, R. H. (1942). *The oceans, their physics, chemistry and general biology*. Prentice Hall.

Tezuka, N. (2025). Increased impact of predators and coastal warming on *Ruditapes philippinarum* populations in Japan. *Estuarine, Coastal and Shelf Science, 317*, 109201. https://doi.org/10.1016/j.ecss.2025.109201.

Velez, C., Figueira, E., Soares, A. M. V. M., & Freitas, R. (2016). Combined effects of seawater acidification and salinity changes in *Ruditapes philippinarum*. *Aquatic Toxicology, 176*, 141–150. https://doi.org/10.1016/j.aquatox.2016.04.016.

Yoon, S., Abe, H., & Kishi, M. J. (2013). Responses of Manila clam growth and its food sources to global warming in a subarctic lagoon in Japan. *Progress in Oceanography, 119*, 48–58. https://doi.org/10.1016/j.pocean.2013.06.005.

Yuan, H., Xu, X., Yang, F., Zhao, L., & Yan, X. (2020). Impact of seawater acidification on shell property of the Manila clam *Ruditapes philippinarum* grown within and without sediment. *Journal of Oceanology and Limnology, 38*(1), 236–248. https://doi.org/10.1007/s00343-019-8281-z.

Zhao, L., Yang, F., Milano, S., Han, T., Walliser, E. O., & Schöne, B. R. (2018). Transgenerational acclimation to seawater acidification in the Manila clam *Ruditapes philippinarum*: Preferential uptake of metabolic carbon. *Science of The Total Environment, 627*, 95–103.

Zhao, L. (2019). Transgenerational biochemical effects of seawater acidification on the Manila clam (*Ruditapes philippinarum*). *Science of The Total Environment*. https://doi.org/10.1016/j.scitotenv.2019.136420.

CHAPTER

The politics of invasion: From "alien monster" to climate adaptation

24

24.1 Socio-economics and conservation

In this chapter, the case of the Manila clam in Britain is examined with the purpose of elucidating aspects of the politics and psychology of intentional non-native introductions, not least among both politicians and scientists. The UK has a strong conservation lobby, including both government agencies and environmental non-governmental organizations. Consequently, the 1980s introduction of the Manila clam was highly contentious. Moreover, this government-sponsored introduction for aquaculture occurred during a key period in the emergence of conservation policy and invasion science.

The UK's long and complex coastline offers great potential for the farming of bivalve mollusks, but these opportunities were then (as now) significantly unrealized. In the late 1980's, the dominant cultivated bivalve species was the mussel *Mytilus edulis,* which, along with the native oyster *Ostrea edulis*, was on-grown from juvenile stock taken from natural habitats and transferred to leased areas of sheltered seabed. However, at that time, over 90% of UK bivalve production came from fisheries rather than cultivation (MAFF, 1989), and native oyster fisheries had long been in decline, probably though a combination of overfishing and disease (Humphreys, 2022). Aware of the potential economic benefits for coastal communities, the UK government was encouraging aquaculture enterprises and assessing the potential of non-native species such as the Pacific oyster, *Magallana gigas,* and the Manila clam (Humphreys et al., 2014, 2010).

The earlier French introduction and a growing European market provided the rationale for the introduction. With the species not commercially exploited, annual farmed production from native clam species was estimated at only 25 tons (MAFF, 1989). Without the Manila clam, the UK industry was regarded as disadvantaged in the European free market (Utting & Spencer, 1992).

The organizational structures involved in the policy process at that time included the Ministry of Agriculture, Fisheries, and Food (MAFF), which made all policy decisions. This ministry was supported by the Nature Conservancy Council, established as an independent agency to provide conservation advice, and the government's network of Fisheries Research Stations, including one in Conwy, Wales, with a focus on shellfish of commercial importance. Several eNGOs also became involved in the contention, including the Royal Society for Nature Conservation (then the UK

The Manila Clam *Ruditapes philippinarum:* Science and Significance of a Globalised Species.
DOI: https://doi.org/10.1016/B978-0-443-36704-5.00022-5
Copyright © 2026 Elsevier Inc. All rights are reserved, including those for text and data mining, AI training, and similar technologies.

division of the World Wildlife Fund) and the Marine Conservation Society. The British introduction occurred before the precautionary principle had become enshrined in legislation.

In 1980, the first UK Manila clams were imported from the USA to the Fisheries Station at Conwy. However, in the following year, new British legislation for the control of non-native species was introduced as the Wildlife and Countryside Act 1981. Section 14 established strict legal controls on "*release or escape into the wild*" of species "*not ordinarily resident*" in UK waters. Since the Manila clam was clearly "not ordinarily resident" in UK waters, the new Act appeared to preclude the Manila clam as a solution to the government's economic ambitions. Consequently, early introductions of the Manila clam sparked a controversy centered on the legal question of whether the introduction contravened the legislation. In the resulting dispute, whether the species could successfully reproduce and establish wild populations in British waters became a key point of contention. Government fishery scientists from the Conwy Station did not believe it could, but conservation scientists, both governmental (NCC) and from the eNGOs, believed it would. In covering the controversy, a national newspaper described the species as an "alien monster" which "could destroy native marine life" (Telegraph, 1989).

The 1981 Act had created a similar difficulty with the non-native Pacific oyster (*Magallana gigas*). Although initially introduced in 1965 for aquaculture, the continued annual laying of hatchery-reared seed oysters was also now ostensibly a violation of the statute. The government's solution was to declare the Pacific oyster to be "ordinarily resident" and therefore exempt from the prohibition of release to the wild. In 1982, a general license to "Release or allow to escape into the wild" was issued (The Gazette, 1982).

Such legal maneuvers are not unusual. France also declared the Pacific oyster a resident species and subsequently applied industry-sympathetic aspects of the European Union Aquaculture Alien Species regulation (Regulation (EC) No 708/2007) to it as a long-used non-native species (Humphreys et al., 2014). The Manila clam, however, provided no similar route to exploitation, since there could be no plausible argument that the species might be considered resident. Therefore, the other major component of the 1981 Act was brought to bear, namely the prohibition of "release or escape to the wild." To this end, the government decided that seed clams introduced for aquaculture should be confined to nets to prevent escape. This requirement had the added advantage of protecting from predators.

24.2 Evidence and consequences

As a broadcast spawner, larvae enter the plankton during which time they are widely dispersed before settling (Herbert et al., 2012, 2016). To circumvent this difficulty, a second matter was therefore raised in an effort to claim that the introduction would be legal. Although juveniles of the species were known to be able to grow well in

British conditions, whether the more temperature-sensitive life cycle stages of gametogenesis and spawning could be completed in Britain became the key issue. If these reproductive stages could be achieved in UK climatic conditions, then laying under netting would not prevent "release or escape," and introductions would be illegal in any sensible interpretation of the 1981 Act. Moreover, if they could reproduce, they could most likely establish non-native "wild" populations. The scientific dispute that followed, therefore, focused on the clam's ability to successfully reproduce in UK climatic conditions.

Early work at the government's Conwy fishery station found no evidence that clams could spawn in British conditions (Millican & Williams, 1985). In 1995, after 5 years of Manila clam trials in the Exe Estuary, Devon, the UK fisheries ministry reported that natural spatfalls had not been recorded (MAFF, 1995). However, in other countries where the Manila clam had been introduced, detailed studies on temperature and reproduction were also available at the time. In Washington State, United States, Mann (1979) reported successful spawning from 15°C to 21°C. With ripe gonadal material evident at all temperatures from 12°C. On the French Atlantic coast around La Rochelle, the species was known to be able to spawn Goulletquer et al. (1987), and Beninger and Lucas (1984) reported seasonal spawning patterns in Brittany. In Arcachon Bay, after its introduction in 1980, the Manila clam was known to have rapidly naturalized (Dang et al., 2010).

Papers from Japan (Ohba, 1959), the USA, (Holland & Chew, 1974), and Conwy (Liang et al.,1987), among others, showed that gonad activity could commence at around 8°C, gamete ripening at 12°C, spawning and larval growth around 14°C or 15°C. However, once settled, juvenile growth could be achieved at 10°C. Therefore, assuming sufficient food, it appeared that a sustained progressively warming season of sea temperatures between 8°C and 14°C of sufficient duration would be needed for successful larval settlement in the wild (see also Chapter 10).

Other relevant information available at the time concerned the distribution of wild populations elsewhere in the world. Northeast Asian indigenous populations were known to extend to Vostok Bay in Russia, the northern Japanese island of Hokkaido, and as far north as the Amur estuary and the Strait of Tartary (Chapter 2). Moreover, in the large, shallow, lagoon-like conditions of Poole Harbour on the coast of southern England, an early site of introduction, annual variation in mean water temperatures was similar to the Gulf of Morbihan in Brittany, France, where the species had naturalized (Caldow et al., 2007).

Nevertheless, in 1984, commercial trials commenced, and bivalve hatchery techniques developed in the United States and France were refined in Conwy for Manila clam production and introduced to commercial hatcheries at Reculver on the outer Thames estuary and elsewhere (Humphreys et al., 2015).

In Poole Harbour, by 1986, a catastrophic outbreak of *Bonamia* infection in oysters resulted in the closure of all commercial bivalve activity (Humphreys, 2022). In response, MAFF scientists suggested the Manila clam as an alternative to oysters. In spring 1988, the first seed clams were introduced. These Manila clam deposits were described as conducted under shellfish legislation designed to control the introduction of diseases. MAFF also considered the requirements of a 1979 European "Revised Code of Practice to Reduce the Risks of Adverse Effects Arising from Introduction of Marine Species." This document included a general recommendation that "The probable effects of introduction into the new area should be assessed carefully, including examination of the effects of any previous introductions of this or similar species in other areas" (ICES, 1979). It is noteworthy that this code emphasized probable effects and risk assessment rather than the hard prohibition of the new UK legislation, which was already in force. The first Manila clam crop was successfully harvested in 1990.

The 1980s Manila clam trial sites included a designated marine reserve. This triggered the "full-scale row" between the fisheries ministry and conservation organizations, including the NCC, as reported in the national press (Telegraph, 1989). The government's environment agency lead officer on marine was quoted as saying that "It is our view that such introductions are unwelcome… and we would like to see it stopped," adding that the NCC had made this clear to the Ministry. Similarly, several environmental NGOs were reported as describing the government's position on "containment" under netting in terms such as "non-sensical" and "making a mockery of the law" (Telegraph, 1989).

Specific concerns about the introduction both at the NCC and the NGOs related to the protection of native infaunal communities from competition for food and space, creating risks for indigenous species and the functional ecology of the benthos. This attitude to the Manila clam was a specific case of a more general concern about the negative impacts of "alien" invasive species, which at the time was of increasing scientific interest. In 1985, the matter was raised in Parliament, and the government confirmed its position that, since successful recruiting of wild populations could not be established, the introductions were not a contravention of legislation (Humphreys, 2025).

After its introduction to Poole Harbour for aquaculture in 1988, wild Manila clams were soon appearing in the catches of cockle fishers, and it was clear that the species had successfully naturalized (Jensen et al., 2004). At the time it was uncertain whether Poole Harbour conditions might be unusually conducive to Manila clam naturalization, but by 2010 the species, having been introduced with government support in 19 British estuaries was known to have naturalized in at least 11 locations in southern and south-east England (Humphreys et al., 2015) Since then, it has spread to more estuaries, not least the extensive outer Thames complex of estuaries in the counties of Kent and Essex (Stuart, 2025). Clearly, government scientists were wrong in their conclusions.

On the other hand, the worst fears of conservationists have not materialized. UK population densities remain low relative to parts of East Asia or warmer regions of Europe (Humphreys et al., 2007). In Poole Harbour, the native sympatric *Ruditapes decussatus* is mixed in with Manila clam catches, indicating the two can coexist. Additionally, the Manila clam has proved a trophic asset for protected migrating waders whose estimated over-winter mortality is reduced because of the new food source (Caldow et al., 2007). While the clam itself has no measurable detrimental impact, its harvesting has been shown to affect sediment composition and increase species richness, although to an extent insufficient to require a reclassification of the biotope (Clarke et al., 2018). In particular, there is no evidence that the Manila clam has become "invasive" (Chapter 22). Moreover, the value of the Manila clam is such as to have provided a sustainably managed fishery, providing benefits to coastal communities (Birchenough et al., 2020). In impartial assessments, it is reasonable to argue that the establishment of wild populations has been both ecologically and economically beneficial. An outcome that neither conservation nor fisheries scientists anticipated.

24.3 The psychology of the "alien monster"

Analysis of the policy process associated with the UK Manila clam introduction has shown how reputable scientists may unintentially but problematically select, interpret, and communicate evidence when engaged at the interface of science and policy, how this can facilitate a polarization of scientific judgements, and how such polarization impedes rather than benefits accurate prediction. For example, it appears that fishery scientists favored their own experimental evidence to a degree that effectively excluded other evidential sources (Humphreys, 2025).

A similar phenomenon has been reported at the science/policy interface in other contexts. The epidemiologist Woolhouse (2022) analyzed failed scientific predictions about the course of the COVID-19 pandemic in 2020. He attributed these errors to the conflation of evidence generation and evidence interpretation. The scientists generating evidence were also advising the government. It appears that scientists interpreting their own evidence tended to overstate its efficacy as an exclusive basis for action. Cognitive psychologists seeking to understand attitudes at the intersection of science and policy (not least on climate change) have investigated how even well-informed protagonists make mistakes involving judgements informed by preconceived ideas or knowledge, resulting in the polarizations of scientific opinion.

While forms of bias are thought to be a universal human phenomenon, "confirmation bias" is associated with professional activities involving the application of expertise. It typically involves scientists acting in good faith, making incorrect inferences or assessments influenced by prior dispositions (Guiney, Goodfellow, Timothy, & Canfield, 2020; Kahan, 2012; Kahan et al., 2012). In the polarized case of the UK Manila clam controversy, the possibility of confirmation bias may have been at work on both sides of the argument. Whilst the fishery scientists were wrong in their conclusions, it is also fair to acknowledge that in the event, various conservation scientists were prone to overstate the consequences of naturalization, not least when suggesting that the clam could "destroy…the area's unique wildlife," "destroy native marine life by competing for food and breeding sites" and put at risk the diversity of invertebrates on which "spectacular flocks of wildfowl and wading birds depend" (Telegraph, 1989). In the latter case, the Manila clam introduction has proved advantageous to wading birds (Caldow et al., 2007).

The UK Manila clam introduction occurred at a time when scientific attitudes to "alien" species were still coalescing under the ecological sub-division of "invasion biology." Davis (2006) has described how this subject developed into a distinct field in the 1980s, as it shifted from an earlier preoccupation with population dynamics and community ecology to a discipline increasingly informed by conservation values. Warren et al. (2017) examined the language used to frame non-native species arrivals to investigate the possibility of bias. They analyzed 651 peer-reviewed papers on non-natives and concluded that a lack of objectivity in the form of "context bias" was a widespread phenomenon. Davis et al. (2011) suggested in *Nature* that "biological bias" is a contributing factor in attitudes to non-native species. This famously provoked a vigorous response from no less than 141 scientists in the next issue (Simberloff, 2011).

Others have criticized how terminology such as " invasion," "alien" and even "non-native" used by scientists is itself pejorative by nature: To an extent that Larsson (2011) has felt the need to distinguish between the reality of invasive species (in a culturally-independent physical world), and the conceptual scheme applied to them that may involve tendencies and commitments not always borne out in reality and are, in general, culturally dependent (Larson, 2011).

Reise et al. (2023), in noting a tendency to consider non-native species as a kind of biological pollution, Elliott, 2003 has suggested that greater differentiation about the effects of introductions can meet with strong opposition, including assertions of "invasive species denialism" (Elliott, 2003; Ricciardi & Ryan, 2018). This allegation and its refutations (Crowley et al., 2017) have been the subject of too much debate to be reviewed here (Davis, 2020). Reise et al. (2023) have responded with a dispassionate habitat-specific assessment of the effects of over 100 introductions to the Wadden Sea coastal sediment ecosystem, mostly from warmer waters over a period of a century.

24.4 Postscript: The Manila clam as climate adaptation

While noting negative effects from some non-native arrivals, Reise et al. (2023) observed that no resident Wadden Sea populations had become extinct due to introduced species and that, rather than degrading the ecosystem, introduced species had increased the capacity of the system to follow environmental change, not least by facilitating the inevitable transformations of global warming. Some, for example, promote stabilization of sediments and mud accretion, which could help tidal flats keep up with rising sea levels. Others have diversified lower food webs and benefited foraging birds, as reported also by Caldow et al. (2007). For tidal ecosystems of low biodiversity, such as those typically occupied by the Manila clam Chapter 20), it is suggested that the common general condemnation of non-native species should be reconsidered (Reise et al., 2023).

Beyond such ecological considerations, the arrival of the Manila clam is increasingly seen in similarly positive socio-economic terms, not least as a climate adaptation measure. Since the 1980s introduction, an initially small population adjacent to a hatchery on the outer Thames estuary, UK, has slowly become widespread to the point where cockle fishers finding clams in their dredges began to report an increasing clam presence to the regional Kent & Essex Inshore Fisheries and Conservation Authority. The Authority, after monitoring the distribution and abundance of the species, established a managed trial fishery in 2024 (Humphreys et al., 2026; Stuart, 2025).

This initiative has been motivated in part as a strategy to maintain a declining inshore fishery fleet and associated communities at a time when global warming has started to reduce the commercial viability of traditional fin fisheries in the southern North Sea, and in anticipation of adverse impacts of severe heatwaves on intertidal indigenous bivalve fisheries (Coulthard et al., 2025; Verdelhos et al., 2015). In particular, it is thought that a cockle fishery of great historical importance in the region may decline with warming waters, and it is anticipated that the Manila clam will contribute to the continuity of the inshore fleet. Regional socio-economic initiatives encouraged by the inshore fisheries authority include related local facilities such as purification and wholesale enterprises (Will Wright, personal communication, and Stuart, 2025).

In the UK, therefore, the Manila clam never came to resemble an "alien monster" and instead has offered both ecological and economic benefits. In both respects, the case supports the assertion by Reise et al. (2023) that the general condemnation of non-native species should be reconsidered for tidal sediment ecosystems.

References

Beninger, P. G., & Lucas, A. (1984). Seasonal variations in condition, reproductive activity, and gross biochemical composition of two species of adult clam reared in a common habitat: *Tapes decussatus* L. (Jeffreys) and *Tapes philippinarum* (Adams & Reeve). *Journal of Experimental Marine Biology and Ecology, 79*(1), 19–37. https://doi.org/10.1016/0022-0981(84)90028-5.

Birchenough, S. E., Clark, R. W. E., Pengelly, S., & Humphreys, J. (2020). Managing a dredge fishery within a marine-protected area: Resolving environmental and socio-economic objectives. In *Marine protected area, science policy and management* (pp. 459–474). Elsevier.

Caldow, R. W. G., Stillman, R. A., Dit Durell, S. E. A. L. V., West, A. D., McGrorty, S., Goss-Custard, J. D., Wood, P. J., & Humphreys, J. (2007). Benefits to shorebirds from invasion of a non-native shellfish. *Proceedings of the Royal Society B: Biological Sciences, 274*(1616), 1449–1455. https://doi.org/10.1098/rspb.2007.0072.

Clarke, L. J., Esteves, L. S., Stillman, R. A., Herbert, R. J. H., & Kenchington, E. (2018). Impacts of a novel shellfishing gear on macrobenthos in a marine protected area: Pump-scoop dredging in Poole Harbour, UK. *Aquatic Living Resources, 31*. https://doi.org/10.1051/alr/2017044.

Coulthard, S., Hatt, A., Lewis, P., Stewart, B. D., Roach, M., Clark, R., Fanshawe, S., White, C. S., Urquhart, J., Percy, J., Gray, T., Bulled, E., Richards, J., Turner, R., Baker, E., Evans, L., Chaigneau, T., Hooper, T., Longsden, J., & Anbleyth-Evans, J. (2025). Recognising and protecting the national benefit of sustainable fisheries in the UK. *Fish and Fisheries, 25*(5), 1–15. https://doi.org/10.1111/faf.12898.

Crowley, S. L., Hinchliffe, S., Redpath, S. M., & McDonald, R. A. (2017). Disagreement about invasive species does not equate to denialism: A response to Russell and Blackburn. *Trends in Ecology and Evolution, 32*(4), 228–229. https://doi.org/10.1016/j.tree.2017.02.004.

Dang, C., de Montaudouin, X., Gam, M., Paroissin, C., Bru, N. oë, & Caill-Milly, N. (2010). The Manila clam population in Arcachon Bay (SW France): Can it be kept sustainable? *Journal of Sea Research, 63*(2), 108–118. https://doi.org/10.1016/j.seares.2009.11.003.

Davis, M. A. (2006). Invasion biology 1958–2005: The pursuit of science and conservation. In M. W. Cadotte, S. M. McMahon, & T. Fukami (Eds.), *Conceptual ecology and invasion biology*. Springer.

Davis, M. A. (2020). Let's welcome a variety of voices to invasion biology. *Conservation Biology, 34*(6), 1329–1330. https://doi.org/10.1111/cobi.13608.

Davis, M. A., Chew, M. K., Hobbs, R. J., Lugo, A. E., Ewel, J. J., Vermeij, G. J., Brown, J. H., Rosenzweig, M. L., Gardener, M. R., Carroll, S. P., Thompson, K., Pickett, S. T. A., Stromberg, J. C., Tredici, P. D., Suding, K. N., Ehrenfeld, J. G., Philip Grime, J., Mascaro, J., & Briggs, J. C. (2011). Don't judge species on their origins. *Nature, 474*(7350), 153–154. https://doi.org/10.1038/474153a.

Elliott, M. (2003). Biological pollutants and biological pollution—An increasing cause for concern. *Marine Pollution Bulletin, 46*(3), 275–280. https://doi.org/10.1016/s0025-326x(02)00423-x.

Goulletquer, P., Lombas, I., & Prou, J. (1987). Influence of the immersion time on the reproductive activity and growth of the Japanese clam Ruditapes philippinarum and the Japanese oyster Crassostrea gigas. *Haliotis, 16*, 453–462.

Guiney, P., Goodfellow, W., Timothy, J., & Canfield, T.J. (2020). An overview of confirmation bias in science: Examples and opportunities for improvement. Washington D.C. Society of Environmental Toxicology and Chemistry (SETAC):. SETAC North America 41st Annual Meeting.

Herbert, R. J. H., Humphreys, J., Davies, C. J., Roberts, C., Fletcher, S., & Crowe, T. P. (2016). Ecological impacts of non-native Pacific oysters (*Crassostrea gigas*) and management measures for protected areas in Europe. *Biodiversity Conservation, 25*, 2835–2865. https://doi.org/10.1007/s10531-016-1209-4.

Herbert, R. J. H., Willis, J., Jones, E., Ross, K., Hübner, R., Humphreys, J., Jensen, A., & Baugh, J. (2012). Invasion in tidal zones on complex coastlines: Modelling larvae of the non-native Manila clam, *Ruditapes philippinarum*, in the UK. *Journal of Biogeography, 39*(3), 585–599. https://doi.org/10.1111/j.1365-2699.2011.02626.x.

Holland, D. A., & Chew, K. K. (1974). Reproductive cycle of the Manila clam (*Venerupis japonica*) from Hood Canal, Washington. *Proceedings of the National Shellfish Association, 64*, 53–58.

Humphreys, J. (2022). Ecology and exploitation of Poole Harbour oysters. In J. Humphreys, & A. E. Hall (Eds.), *Harbour ecology*. Pelagic Publishing.

Humphreys, J. (2010). The introduction of the Manila clam to British coastal waters. *Biologist, 57*, 134–138.

Humphreys, J. (2025). The politics of an Alien Monster: Retrospective analysis of the use of evidence at the science-policy interface. *Ocean and Coastal Management, 266*. https://doi.org/10.1016/j.ocecoaman.2025.107693.

Humphreys, J., Caldow, R. W. G., Mcgrorty, S., West, A. D., & Jensen, A. C. (2007). Population dynamics of naturalised Manila clams *Ruditapes philippinarum* in British coastal waters. *Marine Biology, 151*(6), 2255–2270. https://doi.org/10.1007/s00227-007-0660-x.

Humphreys, J., Harris, M. R. C., Herbert, R. J. H., Farrell, P., Jensen, A., & Cragg, S. M. (2015). Introduction, dispersal and naturalization of the Manila clam *Ruditapes philippinarum* in British estuaries, 1980-2010. *Journal of the Marine Biological Association of the United Kingdom, 95*(6), 1163–1172. https://doi.org/10.1017/S0025315415000132.

Humphreys, J., Herbert, R. J. H., Roberts, C., & Fletcher, S. (2014). A reappraisal of the history and economics of the Pacific oyster in Britain. *Aquaculture, 428–429*, 117–124. https://doi.org/10.1016/j.aquaculture.2014.02.034.

Humphreys, J., Stuart, K., Wyatt, E., Haupt, P., Clark, R., & Wright, W. (2026). An emergent Manila clam fishery in the outer Thames estuary, UK, and its role in climate adaptation. Aquatic Living Resources. (In press)

ICES. (1979). *Report of the ICES working group on the introduction of non-indigenous marine organisms* https://imr.brage.unit.no/imr-xmlui/bitstream/handle/11250/103373/CM_1979_E_22.pdf?sequence=1&isAllowed=y.

Jenson, A. C., Humphreys, J., Caldow, R. W. G., & Dyrynda, P. E. J. (2004). Naturalisation of the Manila Clam *Tapes philippinarum,* an alien species, and establishment of a clam fishery within Poole Harbour, Dorset. *Journal of the Marine Biological Association of UK, 84*, 1069–1073.

Kahan, D. M. (2012). Cognitive Bias and the Constitution of the Liberal Republic of Science. Yale Law School Public Law Research Paper No. 270. SSRN. 1–36. https://doi.org/10.2139/ssrn.2174032.

Kahan, D. M., Peters, E., Wittlin, M., Slovic, P., Ouellette, L. L., Brahaman, D., & Mandel, G. (2012). The polarizing impact of science literacy and numeracy on perceived climate change risks. *Nature Climate Change, 2*, 732–735.

Laing, I., Utting, S. D., & Kilada, R. W. S. (1987). Interactive effect of diet and temperature on the growth of juvenile clams. *Journal of Experimental Marine Biology and Ecology, 113*, 23–28.

Larson, B. M. H. (2011). Embodied realism and invasive species. In *Philosophy of ecology: Volume 11 in handbook of the philosophy of science: 11* (pp. 129–146). https://doi.org/10.1016/B978-0-444-51673-2.50005-4.

Mann, R. (1979). The effect of temperature on growth, physiology, and gametogenesis in the Manila clam *Tapes philippinarum* (Adams & Reeve, 1850). *Journal of Experimental Marine Biology and Ecology, 38*(2), 121–133. https://doi.org/10.1016/0022-0981(79)90016-9.

MAFF, 1995. Clam Cultivation: Results of an Experiment in the River Exe, Devon (1991-1995).

Millican, P. F., & Williams, D. R. (1985). The seasonal variation in the level of meat content, lipid and carbohydrate in Mercenaria mercenaria and Tapes semidecussatus grown in fertilised and unfertilised. *water*: CM2985/K.51. Copenhagen: International Council for the Exploration of the Sea.

MAFF. (1989). Fish farming in the UK: Memorandum submitted to the Agriculture Select Committee. Minutes of evidence, 1 November 1989. House of Commons Papers, Session 1988–89 (HC 594-i). London, UK: UK Parliament Archives.

Ohba, S. (1959). Ecological studies in the natural population of a clam, *Tapes japonica*, with special reference to seasonal variations in the size and structure of the population and to individual growth. *Biological Journal of Okayama University, 5*, 13–47.

Reise, K., Buschbaum, C., Lackschewitz, D., Thieltges, D. W., Waser, A. M., & Wegner, K. M. (2023). Introduced species in a tidal ecosystem of mud and sand: Curse or blessing? *Marine Biodiversity, 53*(1). https://doi.org/10.1007/s12526-022-01302-3.

Ricciardi, A., & Ryan, R. (2018). The exponential growth of invasive species denialism. *Biological Invasions, 20*(3), 549–553. https://doi.org/10.1007/s10530-017-1561-7.

Simberloff, D. (2011). Non-natives: 141 scientists' object. *Nature, 475*, 36–37.

Stuart, K. (2025). *Manila Clam Handbook*. Ramsgate, UK: Kent and Essex Fisheries and Conservation Authority www.kentandessex-ifca.gov.uk.

Telegraph. (1989). Alien invasion warning. *Weekend Telegraph*, 9–10.

The Gazette (1982). Official Public Record: General. Licence permitting the release or escape of the Pacific oyster into the wild. *London*.

Utting, S. D., & Spencer, B. E. (1992). Introductions of marine bivalve molluscs into the United Kingdom for commercial culture - Case histories. *ICES Marine Science Symposium, 194*, 84–91.

Verdelhos, T., Marques, J. C., & Anastacio, P. (2015). Behavioural and mortality responses of the bivalves *Scrobicularia plana* and *Cerastoderma edule* to temperature, as indicator of climate change's potential impacts. *Ecological Indicators, 58*, 95–103. https://doi.org/10.1016/j.ecolind.2014.11.022.

Warren, R. J., King, J. R., Tarsa, C., Haas, B., & Henderson, J. (2017). A systematic review of context bias in invasion biology. *PLoS ONE, 12*(8). https://doi.org/10.1371/journal.pone.0182502.

Woolhouse, M. (2022). *The year the world went mad*. Sandstone Press Ltd.

PART

Preadaptation for pre-eminence

V

CHAPTER

Explaining the ascent of the Manila clam

25

25.1 Introduction

Previous chapters of this book have detailed the globalization of *Ruditapes philippinarum* from Indo-Pacific origins to its current cosmopolitan distribution, and how as a consequence, the species has achieved prominence in both economic and ecological spheres. Examination of the science of the species has revealed some general and persistant issues from the amplification of erroneous distribution reports to how scientific polarization can operate at the science-policy interface. Above all, however, the most significant aspect of the species is its current global preeminence as a consequence of the economic activities of human communities, from local to global scales. In this final Part E, two concluding chapters reflect on the nature and implications of this interspecies relationship. First, in this chapter, summarizing the permutation of characters and circumstances which have provided for the extraordinary ascent of this quintessential globalized species on the basis of a mutualistic Manila clam-human synergy. Second, in the final chapter, the meaning of humanity's relationship with *R. philippinarum* for conventional paradigms of conservation in a changing environment is considered, as are the opportunities presented by the species in years to come.

25.2 The basis of global pre-eminence

An interplay of terrestrial, fluvial, marine, atmospheric, and geomorphological processes ensures that the estuaries and coasts inhabited by the Manila clam are naturally volatile and challenging environments. Here also changes in water temperatures and pH combine with sea level rise to signal profound anthropogenic impacts. Moreover, many of the habitats of the Manila clam are located adjacent to cities and towns, which expose them to high levels of effluent and multiple other industrial pollutants. Such "urban sea systems" (Greene & Delany, 2025) exhibit highly altered environmental variables and therefore present new evolutionary landscapes and selective regimes for coastal marine organisms (Alter et al., 2020). Human activity has led to anthropogenic niche formation and niche revision, phenomena which, although commonly used in relation to terrestrial domestication and mutualism, can also be applied in coastal transitional habitats (Boivin et al., 2016).

Such "anthropogenic environments" are characterized by high levels of inorganic toxins, pharmaceutical compounds, organic particulates and inorganic nutrients, with

The Manila Clam *Ruditapes philippinarum*: Science and Significance of a Globalised Species.
DOI: https://doi.org/10.1016/B978-0-443-36704-5.00031-6
Copyright © 2026 Elsevier Inc. All rights are reserved, including those for text and data mining, AI training, and similar technologies.

multiple consequences for biodiversity and ecological function (Hulme-Beaman et al., 2016). Extractive activities such as fishing and aquaculture are often the most ancient of these anthropogenic stressors and, in some cases, reshaped coastal marine ecosystems early in their postglacial histories (Jackson, 2001). A range of port activities also contribute to ecosystem change, from non-native species arrivals to benthic sediment alteration. It is in such radically altered transitional habitats that *R. philippinarum* reaches its highest abundances.

In seeking to explain the ascent of the species, it must be recognized at the outset that it is not some sort of super-species (in the literal rather than taxonomic sense). Like many bivalves, it is vulnerable particularly to stochastic change and can suffer both chronic and acute population declines. Nevertheless, the species is highly effective within its particular niche. So, what attributes provide the explanation for *R. philippinarum* as a global phenomenon? In fact no single attribute answers this question. While *R. philippinarum* shares many of its characteristics and abilities with other species, it is a unique combination of traits from the physically mundane to the physiologically sophisticated that provides the explanation for its globalized pre-eminence. Summarizing from the chapters of this book, this permutation includes the following traits:

- *Nutritive versatility*

The clam's anatomical and physiological adaptations provide for a wide range of food types from diverse long-chain phytoplankton species and benthic pico-microalgae to organic detritus. Directly advantageous both biologically and economically, this versatility also enables the species to benefit from a variety of anthropogenically derived organic particulates such as are routinely discharged from agricultural catchments and sewage outflows.

- *Energetic prioritization*

Its energetic prioritization of growth and reproduction makes the species highly effective in its natural habitats, but also attractive as a wild or cultured fishery species. A capacity for high productivity is particularly apparent in fertile waters such as those exhibiting eutrophication. Uncontrolled metabolic prioritization may lead to "anthropically-induced invasiveness," which can be of great socio-economic benefit to coastal communities. That improvements in wastewater treatment creating oligotrophic conditions can contribute to population decline further elucidates the clam's affinity with anthropogenic environments.

- *Connectivity, persistence, and natural dispersal*

Ruditapes philippinarum is effective in maintaining and expanding the extent and number of its spawning locations. Larval behavior is effective both in perpetuating self-recruiting populations and ensuring the development and persistence of spatially dispersed meta-populations. In comparison to some sympatric Venerid populations' connectivity is greater. Self-recruitment grounds, which are less likely to receive larval inputs from other spawning grounds, are more susceptible to decline. In contrast, well-connected grounds can compensate for a local deficit with larval transport from elsewhere.

- *Pollution, acidification, and infection tolerance*

The species possesses effective metabolic and physiological adaptations to natural transitional and intertidal habitats prone to high volatility in physical and chemical variables. These same adaptations offer tolerances to a range of pollutants such as metal toxins, pharmaceutically active pollutants, and declining pH as a result of carbon emissions. Also, among the challenges of infaunal estuarine life is the pathogen and parasite-rich nature of sediments. Expansion of immunity-related genes has been interpreted as an adaptation to such environments. Although *R. philippinarum*, like many bivalves, can sometimes succumb to disease, the high natural microbial content of sediments has provoked tolerances which improve resistance to anthropogenically expedited pathogens.

- *Temperate distribution and climate adaptation*

R. philippinarum is an eurythermal species found in northern hemisphere temperate waters with wide seasonal fluctuations. It also has a natural biogeographical range from the subarctic to the subtropical. Temperate estuarine and intertidal habitats demand good tolerance of short-term temperature fluctuations and aerial exposure. These adaptations, along with its capacity to fully close and seal its valves are advantageous in terms of commercial storage and bulk transportation. Estuarine locations in temperate coastal regions are largely a feature of the northern hemisphere. This locates the species alongside human populations inclined to exploit it and developed trade routes facilitating its use as a lucrative commodity. Effective adaptations for its natural habitats also allow it to survive acute episodes of extreme heat and cold, increasingly associated with climate change.

- *Suitability for small and large-scale, cost-effective production*

An infaunal sediment habit makes *R. philippinarum* susceptible to a wide range of exploitation approaches, from simple recreational and subsistence fishing to mass culture and mechanized harvest analogous to intensive arable agricultural production on land. The species has the additional advantage of being a high-value appetizing meat product. Although epibenthic bivalves such as mussel and oyster species are quite widely distributed and susceptible to aquacultural production, their culture and harvest technologies tend to be labor and cost intensive relative to industrialized clam farming. Nevertheless, like those species, clam farming involves no costly nutrients to be applied. While hatchery production is more costly, the value of on-grown clams for the market more than compensates. Successful Manila clam farming is a lucrative enterprise.

- *Positive feedback loops from infaunal harvesting*

Clam fishing and aquaculture generate various positive feedback loops that increase production. These include enhancement of food availability and therefore productivity by the resuspension of benthic organic matter by the action of clam dredge gear (the Tapes paradox). Secondly, clam dredging can change the nature of benthic substrates through a process of "gravelization." While the species has a wide sediment range, it benefits from the presence of coarse sand and gravel particles to which it can more securely attach during settlement. Gravelization can also suppress the effectiveness of some predators, and accumulating shell debris from enhanced fisheries or aquaculture operations is similarly beneficial.

Although when exploited for human benefit, *R. philippinarum* can be associated with indirect changes to habitats of conservation concern, to some extent, the seabed impacts of bottom-towed fishing gear are mitigated because the species occupies the top sediment layer. Benthic biotope damage from bottom-towed fishing gear used in sediment habitats is not comparable to that caused on rock seabeds. Shallow penetrating dredges are compatible with the maintenance of species assemblages and, therefore, ecological function in historically fished areas.

Finally, despite the traits that make it an attractive seafood crop, the Manila clam is not intrinsically invasive. Anthropically-induced invasiveness, as described in Chapter 22, is a consequence of habitat alteration rather than an inherent character of the species. Normally, *R. philippinarum* occupies a place in the intertidal zones of coastal transitional waters with little or no material risk to indigenous species or ecological function. In such circumstances, both ecological and economic benefits can accrue.

In comparison to most other forms of food production, the Manila clam is highly efficient and relatively benign in terms of conservation and carbon emissions. As a low trophic aquaculture species, it has the potential to deliver significant socio-economic benefits from local to global scales, while simultaneously mitigating eutrophication in coastal waters and avoiding the levels of carbon emissions and pollution associated with other cultured species such as the Atlantic salmon. Moreover, it does not exert the same degree of environmental deficit as is associated with terrestrial protein production. All these various features of the Manila clam add up to a unique combination of assets which collectively provide the basis for its close association with human communities.

25.3 The nature of the *Homo sapiens/Ruditapes philippinarum* synergy

Employing the perspective of evolutionary theory, to what extent can we say that the adaptations for a volatile and challenging natural environment "preadapted" *R. philippinarum* for the additional demands of human exploitation and anthropogenic change? Preadaptation requires that adaptive traits evolved for purposes other than those for which they are currently useful. In such circumstances, the fitness conveyed by a characteristic is a by-product of an original adaptation. After Gould and Lewontin's (1979) critique of the term "pre-adaptation," Gould and Vrba's (1982) introduction of the concept of "exaptation" has been widely used for such "co-opted adaptations." A classic example is feathers, which are thought to have been an adaptation for thermal regulation but with the potential to become a preadaptation (exaptation) for flight. In metabolic systems, Barve and Wagner (2013) have shown that one adaptive pathway can entail multiple potential exaptations conveying a "latent" capacity for evolutionary innovations.

Many of the above characteristics of *R. philippinarum*, which equip it to be an anthropogenically expedited species, act in ways similar to their original function, although in a different context. For example, some biochemical and physiological stress mitigation responses now prove effective against new anthropogenic stressors such as pollutants. Whether or not such characteristics can be considered preadaptations *sensu stricto* is debatable; however, the term preadaptation more broadly applied has explanatory utility in this case. Essentially, adaptations for volatile and challenging natural habitats have preadapted *sensu lato R. philippinarum* for many of the additional challenges of anthropogenic change.

Therefore, many of those features that emerged during the evolutionary history of the species, providing an advantage in the wild ecological niche, also preadapt the Manila clam for a relationship with *Homo sapiens*. Consequently, and above all, we must recognize the role of human agency in the success of a species which is now cosmopolitan and has contributed so much to human food security and socio-economic benefit.

If we consider *R. philippinarum* and *Homo sapiens* as no more than just two biological species, we can also define their relationship in terms of ecological theory. To what extent do conventional ecological concepts such as "predator-prey," "commensalism," or mutualism, apply? In exclusively wild fisheries, only a predator-prey relationship exists with humans. However, in enhanced fisheries and aquaculture, *R. philippinarum* moves its ecological relationship with humans to something more complex.

Commensalism is defined as a relationship between two species in which one benefits, and the other receives neither positive nor negative impact. In their framework for assessing types of commensalism in anthropogenic environments, Hulme-Beaman et al. (2016) distinguish between "obligate commensals," which are entirely reliant on, and restricted to, the anthropogenic environment, and "anthropophilic taxa," which will sometimes exploit, but are not dependent on anthropogenic resources, and can successfully occupy both anthropogenic and nonanthropogenic niches. Wholly unexploited populations of *R. philippinarum*, which benefit from enhanced food supplies as a consequence of pollutants such as organic particulates or algal nutrients and eutrophication (see above), might be considered "anthropophilic" (or facultative) commensals. However, when fished or cultured, commensalism does not fully describe the relationship since both species may benefit from the relationship. In such circumstances and more especially when the more sophisticated forms of aquaculture are applied, mutualism is arguably a better descriptor of the human–clam relationship.

A classic form of mutualism is exemplified by the widespread distribution of aphid-farming ant species. In this case, both species benefit from a relationship in which the aphids get protection from predators, while the ants get nutrition in the form of the aphids' honeydew secretion (Way, 1963). A parallel with the human–clam relationship becomes more apparent when it is also recognized that the ants will also consume whole aphids, which thereby also become a protein source (Ivens & Kronauer, 2022). For some ant and aphid species, such interactions can be described as facultative rather than obligate mutualism, when neither is exclusively dependent on the other, being also capable of living independent lives.

A key point of the ant-aphid relationship is that it is considered to be mutualism even though individual aphids are also consumed by the ants. On this basis, it is clear that mutualism is seen as a population-level phenomenon (as is commensalism). In the aquaculture relationship, while the Manila clam is beneficial to humans through the consumption (and related economics) of a high-value protein food source, *R. philippinarum* also receives benefits such as one or more of enhanced reproduction, dispersal, and often protection from predators. The clam–human relationship can therefore be posited as a form of facultative mutualism.

The emphasis on population rather than individual organism level benefit in types of symbiotic relationship is ecologically important as it raises the question of selection pressures and the evolution of new adaptive traits. In ant-aphid mutualism, for example, behavioral adaptations have evolved that facilitate the relationship (Depa et al., 2020). This selective process could be considered analogous to the artificial selection of color forms in cultured Manila clams, but in any event, it is evident that populations of *R. philippinarum* are now genetically distinguishable to an extent that genomes provide the basis of tracking the pattern of worldwide introduction.

Whether or not in strict scientific terminology, *Ruditapes philippinarum* can be legitimately considered preadapted to thrive in anthropogenically altered environments and in a form of facultative mutualistic relationship with *Homo sapiens*.

References

Alter, S. E., Tariq, L., Creed, J. K., & Megafu, E. (2020). Evolutionary responses of marine organisms to urbanized seascapes. *Evolutionary Applications, 2021*(14), 210–232.

Barve, A., & Wagner, A. (2013). A latent capacity for evolutionary innovation through exaptation in metabolic systems. *Nature, 500*(7461), 203–206. https://doi.org/10.1038/nature12301.

Boivin, N. L., Zeder, M. A., Fuller, D. Q., Crowther, A., Larson, G., Erlandson, J. M., Denhami, T., & Petraglia, M. D. (2016). Ecological consequences of human niche construction: Examining long-term anthropogenic shaping of global species distributions. *Proceedings of the National Academy of Sciences of the United States of America, 113*(23), 6388–6396. https://doi.org/10.1073/pnas.1525200113.

Depa, Ł, Kaszyca-Taszakowska, N., Taszakowski, A., & Kanturski, M. (2020). Ant-induced evolutionary patterns in aphids. *Biological Reviews, 95*(6), 1574–1589. https://doi.org/10.1111/brv.12629.

Gould, S. J., & Lewontin, R. C. (1979). The spandrels of San Marco and the Panglossian paradigm: A critique of the adaptationist programme. *Proceedings of the Royal Society of London. Series B: Biological Sciences, 205*, 581–598.

Gould, S. J., & Vrba, E. S. (1982). Exaptation – A missing term in the science of form. *Palaeobiology, 8*(1), 4–15.

Green, H. G., & Delaney, J. R. (2025). Urban sea systems and why they are important: A preliminary characterization. *Estuarine, Coastal and Shelf Science, 322*, 109274.

Hulme-Beaman, A., Dobney, K., Cucchi, T., & Searle, J. B. (2016). An ecological and evolutionary framework for commensalism in anthropogenic environments. *Trends in Ecology and Evolution, 31*(8), 633–645. https://doi.org/10.1016/j.tree.2016.05.001.

Ivens, A. B. F., & Kronauer, D. J. C. (2022). Aphid-farming ants. *Current Biology, 32*(15), R813. https://doi.org/10.1016/j.cub.2022.06.072.

Jackson, J. B. C. (2001). What was natural in the coastal oceans? *Proceedings of the National Academy of Sciences, 98*(10), 5411–5418. https://doi.org/10.1073/pnas.091092898.

Way, M. J. (1963). Mutualism between ants and honeydew-producing homoptera. *Annual Review of Entomology, 8*(1), 307–344. https://doi.org/10.1146/annurev.en.08.010163.001515.

CHAPTER

26 Alternative futures

26.1 Introduction

The characteristics and processes that lie behind the ascent of the Manila clam will ensure its continued significance, both in its native East Asia and across the northern hemisphere. Recently discovered populations in Scandinavia and on western Atlantic coasts (Chapter 3) are highly significant. The former likely involves natural dispersal, but the latter is far removed and inconsistent with any plausible natural explanation. A powerful combination of natural dispersal and human agency will ensure that the global spread of the species continues. Predictive modeling confirms this likelihood, and its facilitation by climate change. However, aspects of this future for the Manila clam remain to be determined.

In East Asia, the species will most likely continue to grow in importance. However, for places where the clam is a non-native, the future is likely to vary according to social and cultural attitudes and, consequently, prevailing policy priorities. Its potential value to these societies may be optimized in line with their own interpretations of sustainable development or attenuated by norms inconsistent with the fulfilment of its socio-economic and nutritive potential. In such places, a dialectic between percieved environmental concerns and socio-economic priorities will influence its future.

26.2 The Anthropocene environment

Over the approximately 12,000 years since the commencement of the current post-glacial Holocene epoch, humans have transformed the planet. Since the Neolithic (agricultural) and then industrial "revolutions," this process has progressed to the point where quantifiable human impacts can be seen in almost every facet of the Earth system (Chure et al., 2022). In the last 50 years, the impacts of greenhouse gas emissions have been recognized as manifest in climate change and ocean acidification, but human impacts go beyond these, and the extent of widespread alteration of the world's environment and ecosystems is sometimes represented in broad general concepts such as "anthropogenic planet" (Chure et al., 2022) and "anthromes" (anthropogenic biomes). Such ideas seek to encapsulate the profound geographical range, scope, and significance of human influence on the Holocene world. Transitional coastal waters, so often integrated with and intimately connected to human settlements and urban landscapes, are arguably aquatic "anthromes" *par excellence*.

The Manila Clam *Ruditapes philippinarum*: Science and Significance of a Globalised Species.
DOI: https://doi.org/10.1016/B978-0-443-36704-5.00030-4
Copyright © 2026 Elsevier Inc. All rights are reserved, including those for text and data mining, AI training, and similar technologies.

In the early 21st century, recognition of the extent of these changes led to the idea that the Earth has transitioned into a new Anthropocene age (Crutzen, 2002). As a formal geological epoch, the Anthropocene was rejected in 2024 but is thought to retain utility as an informal concept representing the widespread and irreversible impact of humanity on the planet (Adam, 2024). Consequently, it remains in use in environmental scientific discourse with a number of scientific journals dedicated to it (Malhi, 2017). The Anthropocene concept refers to a planet that has not only transitioned from its Holocene stable state, but also that this transition is now irreversible. The global success of *Ruditapes philippinarum* is a quintessential phenomenon of the Anthropocene.

With its many iterations, the concept of conservation can be difficult to pin down (Ray & McCormick-Ray, 2014). The United States Conservation Service describes conservation as the protection, management or restoration of natural environments. But what are "natural environments," and to what historic condition are we seeking to restore them? Moreover, in what circumstances can we "protect" against the "irreversible changes" of the Anthropocene? For example, in contrast to some other impacts, climate change is a pernicious problem, since, as a planetary-scale phenomenon, it is difficult or impossible to protect species and habitats from all its effects (Cahill et al., 2013).

Arguably, some of the implicit tenets and dogmas underlying conventional conservation theory and practice involve conserving the Holocene world. Is it realistic, or even desirable, to seek in principle to conserve a Holocene world which in many ecological respects has already vanished, or should the strategic goals of conservation shift to accommodate and optimize the consequences of irreversible change?

The problem of "Shifting Baseline Syndrome" is recognized as an obstacle to addressing global environmental issues through the normalization of incrementally degraded ecosystems (Soga & Gaston, 2018). But, establishing recent or present condition as the baseline to which conservation goals should aspire implies a stasis in environments which have been changing for millennia and will, continue to do so. Moreover, in the context of irreversible changes such as planetary-level system impacts, for many conservtion projects, there is a risk that they become performative, with no real chance of long-term success or a reasonable return on the investment they involve (Humphreys, 2022). Yet largely performative responses may be a contributing factor in the widespread failure of conservation when measured as global outcomes, rather than process metrics.

It is possible, for example, that some efforts on the restoration of European native oysters will never achieve their goals, as complex and interacting causes of decline may ultimately prove intractable. In such circumstances, the earlier introduction to Europe of the non-native Pacific oyster may unintentionally have provided a form of solution, maintaining both local biodiversity and ecological function. But in Britain, if not always elsewhere, a strong conservation lobby has created a persistent reluctance to accept the Pacific oyster as part of the British marine fauna, despite a long-established presence (Humphreys et al., 2021; Yonge, 1960).

The idea of "The Sixth Extinction" has entered popular culture (Kolbert, 2009), and few would argue the veracity of species and genetic losses and the risks they present. But what part does the Manila clam play in this crisis? Once the general premise is accepted that non-native species are a universal problem, then scientific reasoning gives way to false logic, the deduction being: "Non-natives are bad; this is a non-native; therefore, this is bad." In the English introduction of *R. philippinarum*, this invalid syllogism appears to have been demonstrable not just in deductive logic, but in what subsequently transpired (Chapter 24).

Despite references to non-native species as "biological pollution," *R. philippinarum* demonstrates that new arrivals vary greatly in their effects, some of which include benefits. Arguably, it is misleading to place all such fauna in a single category except with less pejorative terms such as "neobiota." This term originated in Germany in part to avoid the negative connotations and value-laden associations of labels such as alien (Kowarik & Starfinger, 2009). Aside from avoiding implied universally negative associations, the term also has the merit of emphasizing what these species are, rather than what they are not. However, despite the existence of an interdisciplinary scientific community and a journal of that name, "neobiota" has not yet replaced "non-native" and similar implicitly perjoritive terms, except in some parts of Europe.

At least such species are sometimes distinguished by alternative terms such as noninvasive and invasive non-natives. However, this distinction is generally motivated by a desire to focus conservation resources rather than representing the necessary paradigm shift, which accommodates a range across the full spectrum of consequences, including beneficial. Neither does it imply that impartial risk assessments should include benefits in their deliberations. Conversely, it leaves such species in the same category as pollution, over-fishing, and habitat destruction, as one of the four major widely promulgated global threats.

In Europe on marine coasts, an acceleration of arrivals in the 1960s and 1970s had already resulted in over 800 recorded non-native marine species by 2004, the majority of which were benthic animals (Streftaris et al., 2005). Change rather than stasis is the norm. In Poole Harbour, UK, of around twelve extant species of bivalve, six are considered non-native. One of which was probably reintroduced by Vikings from America in the 13th century (Eno et al., 1997). These species have become essential ecological assets, and without them, impoverished ecosystems would be manifest. The generalization that neobiota represent a risk to diversity or ecological function is a relatively recent development, and, since the impacts of non-natives are known to be context-dependent, in coastal environments at least, is misleading.

Yet such generalizations are still pervasive in the policy and practice of conservation, obscuring important benefits. Consequently, beyond its native regions, an arrival will typically be assessed in terms only of its propensity to cause ecological damage. Arguably, in practice, an implicit desire to conserve the Holocene remains a priority in an irreversibly altered Anthropocene world, and in some quarters, it remains contentious to suggest that, despite its "non-native" status, there is little evidence that *R. philippinarum* is a direct conservation risk, in contrast to the degraded systems within which it can thrive. Fortunately, in many countries where it is both indigenous and not-so, research programs continue to enhance our understanding of the species. Moreover, it appears that international collaborations on this globalized bivalve are strengthening (Humphreys et al., 2025; Riquet et al., 2026), and there is now clear evidence that it can be both ecologically and socio-economically beneficial.

26.3 Blue transition

The global food system falls far short of achieving goals either for human health or the environment. The agricultural sector is reported to account for up to 27% of global greenhouse-gas emissions and is a major cause of water pollution, soil degradation, and biodiversity loss (Hirvonen et al., 2020; Hughes et al., 2022). Seafood provides an important supply of protein for billions of people worldwide and is generally rich in protein and micronutrients. Bivalves in particular are relatively rich in vitamins such as A and B12 and in minerals including calcium, iodine, iron, and zinc, compared to various terrestrially sourced animal foods. Nevertheless, in comparison to terrestrial food, seafood production is often undervalued, consequently receiving suboptimal investment and policy attention (Golden et al., 2021).

Guidelines from the Committee on World Food Security indicate that policy is needed to transform food systems by prioritizing aquatic-sourced foods (Hirvonen et al., 2020). However, finfish capture fisheries production has stagnated for decades (FAO, 2024) and is, in any event, prone to unsustainable overfishing, not least due to large-scale illegal, unreported and unregulated fishing activity. Alternatives include managed coastal invertebrate fisheries and aquaculture. Golden et al. (2021) have calculated that investment in aquaculture production sufficient to significantly increase the world's production of aquatic animal-sourced food would decrease prices, increase consumption, and reduce the consumption of less healthy and less sustainable red and processed meats, while also benefiting human health through improved micronutrient intake. The UN Food and Agriculture Organization has identified a "blue transition" to sustainable aquaculture growth as a priority action (FAO, 2024). But what is the most sustainable form of aquaculture?

A major factor in the sustainability of aquaculture production is the trophic level of the cultivated species. In Chapter 18, the sustainability of Manila clam production has been examined: High-trophic aquaculture involves the cultivation of carnivorous animals, which are two or more food-chain steps away from primary production. Examples of high-trophic aquaculture include salmon, which require a food supply that is itself derived from smaller fish or equivalent feed. In contrast, "low trophic aquaculture" (LTA) includes the production of primary consumers such as filter-feeding bivalves, which consume microalgae. LTA has a higher efficiency of natural resource use and lower environmental impact, with lesser greenhouse gas emissions and water contamination per unit weight of food produced (Tett et al., 2025).

Animal seafood aquaculture production is largely focused on finfish, crustaceans, and mollusks. Approaches vary according to the extent to which they remain linked to a wild capture system. This linkage can be due to the need for direct feeding inputs when feed is derived from wild-captured smaller animals. Feeding also results in pollution in the water column, on the seabed below culture pens, or in wastewater. Of 203 marine aquaculture species reviewed by Froehlich et al. (2023), 66% required direct feed inputs. The remaining 33% were all molluscan filter feeders.

A second type of wild capture-aquaculture link exists when the farmed species must first be harvested from the wild at early life-cycle stages for growing-on to marketable size. This approach has been termed capture-based aquaculture. While it can reduce impacts on wild stocks compared to purely wild capture fisheries, it can nevertheless exacerbate pressure on such populations. While in the United Kingdom and elsewhere, some generally smaller-scale translocations of this sort are practiced with *R. philippinarum*, the approach is not a feature of the great majority of Manila clam farming practices, which, even in indigenous regions, largely rely on seed clams from hatcheries.

Sustainable seafood production is most effectively achieved by approaches that break the link between wild fisheries and aquaculture. Hatchery-based "domesticated aquaculture" of *R. philippinarum* involves closed production cycles relying much less, if at all, on wild inputs (Chapter 17). In its purest form, this does not deplete natural stocks of any species either for food or broodstock (Froehlich et al., 2023). The prodigious fecundity of the clam ensures that the number of adult clams taken for broodstock is small, and in some Chinese Manila clam hatcheries, broodstock is taken from and then returned to the farmed populations after a few spawning cycles. Hatchery-based production can also compensate for reduced wild catch due to overfishing and stock collapse, and there is evidence that such approaches are expediting the "blue transition" in which aquaculture overtakes wild capture production (Stoll et al., 2019).

The practice of enhancing natural Manila clam stocks with hatchery-reared seed also achieves this end. Although hatchery seed production can be energy-intensive, this challenge may be reduced by situating hatcheries in warmer climates than the cooler seabed areas used for on-growing, as is the case in southern China. To similar effect, a major mainland US coast producer (Taylor Shellfish) has located its hatchery operations on the Hawaiian Islands, where warmer climate and surface waters coincide with the availability of deep and cool oceanic water from just offshore. This gives the capacity to regulate water temperatures with much reduced energy consumption. These Chinese and US solutions are advantageous both commercially and in terms of carbon emissions.

R. philippinarum aquaculture clearly lies at the sustainable end of the spectrum. Nevertheless, in comparison with East Asia and despite its advantages, LTA makes a relatively small contribution to food supply in the maritime countries around the Atlantic (Tett et al., 2025). Having noted a 20% decline in aquaculture production of indigenous mussels (*Mytilus edulis*) in the European Union, a number of proposals have been formulated which could expedite LTA production. These include more conducive legal environments and technical support, improved availability of skills, and a more proactive capital investment and policy environment. In addition, they identify the need for public economic policy that encourages LTA through the creation of a "level playing field." In the United Kingdom at least, the policy environment relating to the inshore areas where LTA could be developed is crowded with competing priorities and dominated by more powerful interests than typically found within the generally smaller-scale inshore firms (Humphreys, 2022).

The science of the Manila clam reveals it to be a species capable of and preadapted for life in the altered environments of the Anthropocene. In well-managed estuarine habitats, *R. philippinarum* offers much-needed sustainable food and socio-economic benefit, while also providing for a thriving and enriched benthic ecology, not least in the context of warming temperate climates. Already, the irreversible impacts of human activity make the continued expansion of its cosmopolitan distribution inevitable. In these contexts, will we welcome the global benefits and opportunities that the preeminently cosmopolitan *Ruditapes philippinarum* offers?

References

Adam, D. (2024). Ditching "Anthropocene": Why ecologists say the term still matters. *Nature*. https://doi.org/10.1038/d41586-024-00756-9.

Cahill, A. E., Aiello-Lammens, M. E., Fisher-Reid, M. C., Hua, X., Karanewsky, C. J., Ryu, H. Y., Sbeglia, G. C., Spagnolo, F., Waldron, J. B., Warsi, O., & Wiens, J. J. (2013). How does climate change cause extinction? *Proceedings of the Royal Society B: Biological Sciences, 280*(1750), 20121890. https://doi.org/10.1098/rspb.2012.1890.

Chure, G., Banks, R. A., Flamholz, A. I., Sarai, N. S., Kamb, M., Lopez-Gomez, I., Bar-On, Y., Milo, R., & Phillips, R. (2022). A quantitative database of human impacts on planet Earth. *Patterns, 3*(9), 100552.

Crutzen, P. J. (2022). Geology of mankind. *Nature, 415*(23), 315–320.

Eno, N. C., Clark, R. A., & Sanderson, W. G. 1997. Non-native marine species in British waters: A review and directory. JNCC.

FAO. (2024). *The state of world fisheries and aquaculture 2024. Blue transformation in action*. FAO https://doi.org/10.4060/cd0683en.

Froehlich, H. E., Montgomery, J. C., Williams, D. R., O'Hara, C., Kuempel, C. D., & Halpern, B. S. (2023). Biological life-history and farming scenarios of marine aquaculture to help reduce wild marine fishing pressure. *Fish and Fisheries, 24*(6), 1034–1047. https://doi.org/10.1111/faf.12783.

Golden, C. D., Koehn, J. Z., Shepon, A., Passarelli, S., Free, C. M., Viana, D. F., Matthey, H., Eurich, J. G., Gephart, J. A., Fluet-Chouinard, E., Nyboer, E. A., Lynch, A. J., Kjellevold, M., Bromage, S., Charlebois, P., Barange, M., Vannuccini, S., Cao, L., Kleisner, K. M., … Thilsted, S. H. (2021). Aquatic foods to nourish nations. *Nature, 598*(7880), 315–320. https://doi.org/10.1038/s41586-021-03917-1.

Hirvonen, K., Bai, Y., Headey, D., & Masters, W. A. (2020). Affordability of the EAT–Lancet reference diet: A global analysis. *The Lancet Global Health, 8*(1), e59. https://doi.org/10.1016/s2214-109x(19)30447-4.

Hughes, A. D., Charalambides, G., Franco, S. C., Robinson, G., & Tett, P. (2022). Blue nitrogen: A nature-based solution in the blue economy as a tool to manage terrestrial nutrient neutrality. *Sustainability, 14*, 10182. https://doi.org/10.3390/su141610182.

Humphreys, J., Syvret, M., Horsfall, S., Williams, C., Woolmer, A., & Adamson, E. A. S. (2021). Why we should learn to love Pacific oysters. In *The marine biologist* (pp. 10–11).

Humphreys, J., Choi, K.-S., & Watanabe, S. (2025). *Ruditapes philippinarum*: The $7 billion Mollusc. *Estuarine, Coastal and Shelf Science, 325*(2025), 109481.

Humphreys, J. (2022). Marine conservation: Smoke and mirrors in the coastal zone. In J. Humphreys, & S. Little (Eds.), *Challenges in estuarine and coastal science*. Pelagic Publishing.

Kolbert, E. (2009). *The sixth extinction*. The New Yorker https://www.newyorker.com/magazine/2009/05/25/the-sixth-extinction.

Kowarik, I., & Starfinger, U. (2009). Neobiota: A European approach. In P. Pyšek, & J. Pergl (Eds.). In *Biological invasions: Towards a synthesis: 8* (pp. 21–28). Neobiota.

Malhi, Y. (2017). The concept of the anthropocene. *Annual Review of Environment and Resources, 42*(1), 77–104. https://doi.org/10.1146/annurev-environ-102016-060854.

Ray, G. C., & McCormick-Ray, J. (2014). *Marine conservation: Science, policy, and management*. Wiley Blackwell.

Riquet, F., Sanchez, F., Antajan, E., Bru, N., & Caill-Milly, N. (2026). The international Manila and European clams conference: Euro-Mediterranean overview and future directions to support sustainable clam fisheries. *Aquatic Living Resources, 39*(4), 5.

Soga, M., & Gaston, K. J. (2018). Shifting baseline syndrome: Causes, consequences, and implications. *Frontiers in Ecology and the Environment, 16*(4), 222–230. https://doi.org/10.1002/fee.1794.

Stoll, J. S., Leslie, H. M., Britsch, M. L., & Cleaver, C. M. (2019). Evaluating aquaculture as a diversification strategy for Maine's commercial fishing sector in the face of change. *Marine Policy, 107*, 103583. https://doi.org/10.1016/j.marpol.2019.103583.

Streftaris, N., Zenetos, A., & Papathanassiou, E. (2005). Globalisation in marine ecosystems: The story of non-indigenous marine species across European seas. *Oceanography and Marine Biology., 43*, 419–453.

Tett, P., Charalambides, G., Franco, S. C., Hughes, A. D., Mikkelsen, E., Nolde Nielsen, K. å, Arthur Bastos Routledge, E., Nielsen, P., & James, P. (2025). Leaving the niche: Recommendations for mainstreaming low trophic aquaculture in countries around the Atlantic basin. *Marine Policy, 171*, 106475. https://doi.org/10.1016/j.marpol.2024.106475.

Yonge, C. M. (1960). *Oysters (New Naturalist Special Volume)*. London: Collins.

Appendix: Notes on identification

In addition to information in this Appendix, Chapters 5 and 6 include figures of use in identification as follows: Fig. 5.1 labeled shell exterior, dorsal aspect; Fig. 5.2 labeled diagram of valve interior; Fig. 5.5 Lateral aspect of valve exteriors of three specimens of different sizes; Fig. 6.3 Exterior lateral aspect with siphons extended; Fig. 6.4 Manila clams for sale on the quay in Qingdao, China. The figures below provide further shell and siphon images. Table A.1 gives a technical description of the shell features.

Other genera in the Venerid subfamily Tapetinae include *Tapes; Polititapes; Paphia; Protapes; Gomphina; Gomphinella; Katelysia; Marcia; Venerupis; Ruditapes; and Irus.* In practice, many of these are easily distinguished from *Ruditapes* spp. Table A.2 lists those species that may be confused with the Manila clam. Identification errors are usually confined to other *Ruditapes* species. Misidentification is a particular risk where native sympatric species of similar appearance are found, such as *R. aspera in East Asia, R. decussatus* in Europe, and *Leukoma staminea* in North America. However, *L. staminea* belongs to a different subfamily (Venerinae), and distinguishing it from *R. philippinarum* is much less difficult than with the two congeneric species.

Technical species descriptions can sometimes be based on a limited number of specimens from a single site. As such, they can fail to accommodate the range of intraspecific variation in such a widely distributed and cultivated species. Moreover, reliance on general shape descriptions, shell color, and even morphometric ratios can be problematic for *R. philippinarum,* which is known to exhibit significant phenotypic plasticity in response to various environmental variables, even between distinct populations along the same coast (Caill-Milly et al., 2014).

This also creates unavoidable difficulties for key construction, forcing reliance on imprecise and relative morphologic interpretations. For example, the final dichotomy of Poutiers (1998) key for western Pacific bivalves is as follows:

a. Shell relatively short for its height, with a moderately strong sculpture; pallial sinus deep, nearly attaining mid-length of valves ... *R. philippinarum.*

b. Shell relatively long for its height, with a moderately fine sculpture; pallial sinus shallow, far from attaining mid-length of valves . . . . *R. variegatus.*[1]

This dichotomy is valid, but the use of relative terms such as "moderately strong sculpture" presupposes a level of comparative species knowledge sometimes unavailable.

Beyond conchological diagnostics, soft tissue anatomy can provide definitive information for species identification, especially with live and fresh specimens. In Europe, where the native venerid *R. decussatus* is easily confused with *R. philippinarum*, live specimens can be distinguished by the extent to which the inhalant and

[1] Now an unaccepted synonym of *Venerupis aspera*

Table A.1 Conchological features of *R. philippinarum*.

External shell features

- Strong and solid equivalent and inequilateral shell, of strong and broadly oval appearance variously described as subtriangular or elongate-sub-quadrate.
- Umbones anterior to midline (with no "ears" projecting either side)
- Smooth (rather than crenulate) shell margins
- Ligament external and entirely posterior to the umbones
- Distinct lunule of lanceolate shape
- Escutcheon asymmetrical
- External shell sculpture consists of both concentric and radial grooves/ridges/ribs. The radial ribs at the posterior part of valves have relatively larger intervals such that the concentric and radiating ridges interact especially posteriorly to form a decussate or "latticed" texture.

Internal shell features

- Hinge heterodont with three cardinal teeth in each valve (No lateral teeth).
- Each valve has two adductor scars of approximately equal size
- Pallial line with clear posterior sinus which does not extend beyond the midline of the shell
- Lower edge of the pallial sinus is distinct from, rather than confluent with, the pallial line for the whole of its length.

Table A.2 Selected similar venerids.

- *Ruditapes philippinarum* (A. Adams & Reeve, 1850). Subfamily Tapetinae.Synonyms include: *Venerupis philippinarum; Tapes japonica*; *T. philippinarum*; *T. semidecussata* (and see Chapter 1). Distribution. NE and NW Pacific; NE Atlantic; Mediterranean; NW Atlantic.
- *Ruditapes decussatus* (Linnaeus, 1758). Subfamily Tapetinae. Synonyms include: *Tapes decussatus; Venerupis decussata* Distribution: N. Atlantic; Mediterranean.
- *Venerupis aspera* (Quoy & Gaimard, 1835). Subfamily Tapetinae Synonyms include: *Ruditapes variegatus* Distribution: Southern Japan to NW Australia
- *Ruditapes bruguieri* (Hanley, 1845). Subfamily Tapetinae Synonyms include: *Tapes bruguieri*; *Venerupis bruguieri*. Distribution: Indian Ocean, Arabia; Sri Lanka; India; Gulf of Oman; Bay of Bengal.
- *Venerupis corrugata* (Gmelin, 1791). Subfamily Tapetinae. Synonyms include: *Venerupis pullastra*; *V. saxatilis*; *V. senegalensis*. Distribution: NE Atlantic; Baltic; Mediterranean; Southern Africa. AMBI Ecological Group: Very sensitive to disturbance
- *Leukoma staminea* (Conrad, 1837). Subfamily Venerinae Synonyms include: *Protothaca staminea*; *Tapes staminea*; *Paphia staminea*. Distribution: NE Pacific

exhalant siphons are separate or fused along much of their length (Fig. A.1). In the laboratory, subjecting live specimens to heat stress in a shallow bowl under a high-power bench lamp usually causes siphon extension. (Figs. 6.3 and 6.4 in Chapter 6 also show extended siphons of *R. philippinarum*).

In East Asia, a similar difference in the degree of siphon fusion distinguishes *R. philippinarum* from *R. aspera.* These species can also be distinguished on the basis of siphon tentacle morphology and the extent of the pallial sinus (Figs. A.2 A and B).

FIGURE A.1

Shell and siphon differences of *R. philippinarum* (above) and *R. decussatus* (below). Although the siphons of *R. philippinarum* are far from fully extended, the extent of their fusion compared to *R. decussatus* is visible. Typical differences in overall shell shape are also apparent. The *dark* color of *R. decussatus* is a result of staining from sediments rather than a diagnostic feature.

Photo credit: the author.

Table A.3, which can be used in conjunction with these figures, provides useful diagnostic features for distinguishing *R. philippinarum* from the two congeneric species with which it is most often confused.

Growing use of molecular identifiers facilitates reliable identification. Barcoding of the gene for the protein cytochrome oxidase I is increasingly assisting reliable identification, but neither morphological nor genetically derived identifications are helped by the legacy of approaches to molluscan specimens in which only hard shell material is retained.

FIGURE A.2 Distinguishing morphological siphon and shell features of *R. philippinarum* and *R. aspera*.

(A) Morphological features of *R. philippinarum* (left) and *R. aspera* (right), showing differences in shells (including the shape and extent of the pallial sinus) and inhalant siphon tentacle structure. (B) Diagrammatic representation of distinguishing features of *R. aspera* (1, 3 and 5) and *R. philippinarum* (2.4 and 6) on the basis of inhalant siphonal tentacle morphology (simplified). Differences in sperm morphology are also shown.

(A) From Tezuka. N. (2025) Increased impact of predators and coastal warming on Ruditapes philippinarum *populations in Japan.* Estuarine, Coastal and Shelf Science, 317, *109201; (B) from Sagara, J. (1952). On the taxonomic difference between* Venerupis semidecussata *and* V. variegata. Bulletin of the Japanese Society of Scientific Fisheries, 18 *(3), 41–42.*

Table A.3 Diagnostic features for distinguishing *Ruditapes philippinarum* from *R. aspera* and *R. decussatus*. Individual specimens may occasionally range beyond stated metrics; therefore, identification using shells alone should be based on an assessment of a number of features, ideally across a series of specimens with other species available for comparison. Whenever possible, the siphons of whole specimens should be examined.

Conchology		***Ruditapes philippinarum***	***Ruditapes aspera***	***Ruditapes decussatus***	**Sources**
	Shell length				
		Up to 5.5 cm	Up to 3.6 cm	Up to about 8 cm	Lam (1980)
	Sculpture				
		Moderately strong sculpture	Moderately fine sculpture	Moderately fine sculpture	Poutiers (1998)
		Radial ribs raised, 90–100+	Radial ribs more flattened, 50–70	Radial and concentric ribs of equal strength	Tang et al. (2022) Alonso Suarez and Raven (2020)
	Pallial sinus				
		Approx. half shell length (46% [±3%] of shell length) (Fig. A.1)	Less than one-third of the shell length (Fig. A.1)	Usually more than half of the shell length	Lam (1980) Alonso Suarez and Raven (2020)
		Rounded; longer than high	About as long as high	Pointed toward the umbo or rounded	Poutiers (1998) Alonso Suarez and Raven (2020)
	Lunule				
		About twice as long as wide Often dark Sometimes asymmetrical	About 3–4 times longer than wide	Usually clear colored	Poutiers (1998) Alonso Suarez and Raven (2020)
	Escutcheon				
		Often asymmetrical. More distinct in the left valve	Narrow. Well-marked on both valves.		Poutiers (1998) Albayrak et al. (2001)
Anatomy					
	Siphons				
		Fused for most of the length	Separate for most of their length	Separate for most of their length	Tang et al. (2022) Alonso Suarez and Raven (2020)

FIGURE A.3 Evidential images of a newly discovered naturalized Manila clam population in Sweden.

Evidential images corroborating the arrival and naturalization of *Ruditapes philippinarum* in the Swedish Skagerrak. (A) Preserved whole specimen. Lateral shell aspect showing also siphon fusion. (B) Distinctive year groups indicating an established reproducing population.

From Greeve et al. (2025).

Since 2010 environmental DNA (eDNA) has become an increasingly common basis for the identification of macrofauna in ecosystems, especially non native species. The "biphasic" life history of bivalves such as *R. philippinarum* (Chapter 10) results in the presence of larval DNA in the water column. While it can be difficult to reliably identify bivalve larvae to species level with microscopy, molecular genetic PCR-based techniques can now be used to detect the arrival of nonnatives even before they have been otherwise reported. eDNA techniques can be used either for specific species searches or in more general assessments of biodiversity through the simultaneous identification of many taxa from the same environmental sample using metabarcoding techniques (Chapter 4).

For preliminary and informal reports of first arrival and naturalization, or any reports of ostensible distribution in tropical waters, persuasive photographic evidence should include images of both shell and siphons, along with an indication of multiple year groups as indicated by Fig. A.3 from the Swedish Skagerrak (Chapter 3). Genetic verification can then follow.

References

Albayrak, S., Aslan, H., & Balkis, H. (2001). A contribution to the Aegean Sea fauna: *Ruditapes philippinarum* (Adams and Reeve, 1850) (Bivalvia: Veneridae). *Israel Journal of Zoology, 47*, 299–300.

Alonso Suarez, A., & Raven, J. G. M. (2020). First records of naturalised Ruditapes philippinarum (Adams & Reeve, 1850) (Bivalvia: Veneridae) in Asturias (NW Spain), keys for identification, and evidence of hybridisation with native *Ruditapes decussatus* (Linnaeus, 1758). *Elona, Revista de Malacologia Iberica, 2*, 67–83.

Caill-Milly, N., Bru, N., Barranger, M., Gallon, L., & D'amico, F. (2014). Morphological trends of four Manila clam populations (*Venerupis philippinarum*) on the French Atlantic Coast: Identified spatial patterns and their relationship to environmental variability. *Journal of Shellfish Research, 33*(2), 355–372. https://doi.org/10.2983/035.033.0205.

Greeve, Y., Lindegarth, M., Panova, M., & Strand, A. (2025). First report of the Manila clam *Ruditapes philippinarum* on the Swedish Skagerrak coast. *Estuarine Coastal and Shelf Science, 325*, 109487.

Lam, V. W. W. (1980). Shell form and diagnostic differences in the structure of the siphons and ciliary currents of the ctenidia in coastal species of the *Tapetinae* (Bivalvia: Veneracea) in Hong Kong. In B. Morton (Ed.), *The Malacofauna of Hong Kong and Southern China: Proceedings of the first international workshop on the malacofauna of Hong Kong and the South China Sea* (pp. 11–32). Hong Kong University Press.

Poutiers, J. M. (1998). Bivalves. In K. E. Carpenter, & V. H. Niem (Eds.), *The living marine resources of the Western Central Pacific. Volume 1. Seaweeds, corals, bivalves, and gastropods. FAO species identification guide for fishery purposes* (pp. 123–362). FAO.

Sagara, J. (1952). On the taxonomic difference between Venerupis semidecussata and V. variegata. *Bulletin of the Japanese Society of Scientific Fisheries, 18*(3), 41–42.

Tang, Y., Huo, Z., Liu, Y., Wang, Y., Zuo, L., Fang, L., Zhao, W., Tan, Y., & Yan, X. (2022). Full mitochondrial genomes reveal species differences between the Venerid clams *Ruditapes philippinarum* and *R. variegatus*. *Genes, 13*, 2157. https://doi.org/10.3390/genes13112157.

Tezuka, N. (2025). Increased impact of predators and coastal warming on *Ruditapes philippinarum* populations in Japan. *Estuarine Coastal and Shelf Science, 317*, 109201.

Index

Page numbers followed by "*f*" and "*t*" indicate, figures and tables respectively.

9780443367045